Dermato-Oncology Study Guide

Vincent Liu
Editor

Dermato-Oncology Study Guide

Essential Text and Review

 Springer

Editor
Vincent Liu
Department of Dermatology
University of Iowa Health Care
Iowa, IA
USA

ISBN 978-3-030-53436-3   ISBN 978-3-030-53437-0   (eBook)
https://doi.org/10.1007/978-3-030-53437-0

This Springer imprint is published by the registered company Springer Nature Switzerland AG
The registered company address is: Gewerbestrasse 11, 6330 Cham, Switzerland

Acknowledgments

Just as the subject of this book focuses upon the busy intersection between skin and systemic malignancy, the book itself is a product of the intersection of efforts by numerous contributors. First, I am indebted to all the distinguished authors, an eclectic panel of experts from across the country, each of whom has graciously dedicated valuable time and unparalleled expertise to this work. As this book was directly inspired by the forum covering these topics conducted at the annual American Academy of Dermatology meetings through the years, I owe particular thanks to the forum's original founder, Seemal Desai, MD. For translating our copious research and writing into a useful, attractive final product, Springer deserves many thanks, especially to Grant Weston, for helping with the conception and maturation of the project, and to Maha SathishBabu and Sathyapriya Venkatesan for invaluable technical assistance. This book does not come to fruition, however, without Chris Huber, who kindly committed herself to countless meetings, phone calls, e-mails, revisions, edits, reformats, and reminders, all performed with attention to detail, reliability, and diplomacy. Special gratitude goes to Paula, Linnea, and Lydia, for inspiring me and for tolerating and supporting my prolonged diversion with this passion. Ultimately, this work is about and for our patients, who privilege us by allowing us to care for them.

Iowa City, IA, USA   Vincent Liu, MD

Contents

Introduction

Truth is revealed at intersections. The intersection between skin and systemic malignancy is a rich, varied, and dynamic territory, familiarity with which is critical for the clinician (Fig. 1). Such is the focus of this book, designed to explore the world of dermatooncology in a practical manner which should equip the reader with useful tools for navigation.

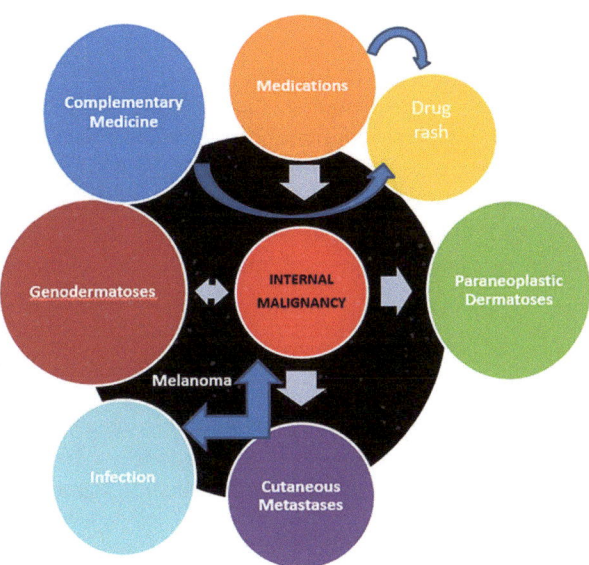

Fig. 1 Intersection between skin and systemic malignancy

One broad path from the intersection carries signs of internal malignancy to the skin. As "window to the soul," the skin offers unique views into internal health. Specifically, systemic malignancy can manifest in the skin in a variety of ways. Cutaneous metastases represent the most direct of these manifestations, identification of which carries significant implications for diagnosis, prognosis, and management (Chap. 1). Alternatively, genodermatoses with malignant potential, encompassing syndromes whose genetic defects result in parallel development of both malignancy and cutaneous signs, can be initially detected by skin examination (Chap. 3). One additional means by which internal cancer can present on the skin is captured by paraneoplastic dermatoses, conditions whereby mucocutaneous findings reflect systemic malignancy via indirect mechanisms (Chap. 2).

Attempts to control these traffic patterns can create their own consequences. Efforts to treat dermatologic conditions and systemic malignancy can lead to complications, knowledge of which is crucial for those caring for these afflicted patients. The spectrum of medical, biologic, and radiation anticancer therapies can elicit a spectrum of adverse mucocutaneous reactions, all of which demand recognition and management by the clinician (Chaps. 8–11). Furthermore, medications used to treat dermatologic disease can also increase risk for subsequent development of malignancy, a factor which must be weighed in treatment deliberations (Chap. 6). Complementing medical therapy, surgery remains an indispensable modality in diagnosis and therapy in cutaneous oncology, associated with its own unique risks requiring careful consideration (Chap. 5).

Upon closer inspection, the complexity of the intersection between skin and systemic malignancy is exposed, and a rotary is discovered, allowing for multidirectional and cyclical traffic. Infections epitomize this bidirectional influence as certain oncogenic infections can give rise to tumors both internal and of the skin, and conversely, systemic malignancy can induce immunosuppression predisposing to cutaneous infection (Chap. 7). Melanoma, the prototypical skin cancer with systemic impact, not only has a significant risk of metastasis, but its treatment, including revolutionary targeted and immunotherapeutic agents, can yield significant skin effects (Chap. 4). Finally, complementary and alternative medicine, ever growing in popularity, is often used to treat skin conditions but may carry carcinogenic risk, or is sometimes used in an effort to treat malignancy but with risk of skin reactions (Chap. 12).

This book is provided as a roadmap to this exquisitely fascinating intersection between skin and systemic malignancy. As dermatologists, oncologists, radiation oncologists, surgical oncologists, primary care physicians, nurses, and medical students all journey through this busy intersection, this material is presented with them in mind. Ultimately, it is hoped that this knowledge will translate into enhanced effectiveness by the clinical provider, thereby translating into optimized patient care.

Iowa City, IA, USA   Vincent Liu, MD

Cutaneous Metastases

1

Martin Dittmer and Vincent Liu

Abbreviations

AA	Alopecia areata
AFP	Alpha-fetoprotein
AFX	Atypical fibroxanthoma
BCC	Basal cell carcinoma
BMP	Basic metabolic panel
CBC	Complete blood count
CEA	Carcinoembryonic antigen
CK	Cytokeratin
CM	Cutaneous metastasis
CMBC	Cutaneous metastatic breast cancer
CMChC	Cutaneous metastatic choriocarcinoma
CMCRC	Cutaneous metastatic colorectal cancer
CMHCC	Cutaneous metastatic hepatocellular carcinoma
CMLC	Cutaneous metastatic lung cancer
CMM	Cutaneous metastatic melanoma
CMRC	Cutaneous metastatic renal cancer
CMTC	Cutaneous metastatic thyroid carcinoma
CT	Computed tomography
CTC	Circulating tumor cell
CXR	Chest x-ray

M. Dittmer (✉)
SSM Health Saint Louis University Hospital—South, St. Louis, MO, USA

V. Liu
Department of Dermatology, University of Iowa Health Care, Iowa, IA, USA
e-mail: Vincent-liu@uiowa.edu

© Springer Nature Switzerland AG 2021
V. Liu (ed.), *Dermato-Oncology Study Guide*,
https://doi.org/10.1007/978-3-030-53437-0_1

DDx	Differential diagnosis
DLE	Discoid lupus erythematosus
EAC	Erythema annulare centrifugum
EIC	Epidermal inclusion cyst
EMPD or EMP	Extramammary Paget disease
ER	Estrogen receptor
FFPE	Formalin fixed paraffin embedded
GCDFP-15	Gross cystic disease fluid protein 15
GI	Gastrointestinal
GU	Genitourinary
H&E	Hematoxylin and eosin
H&N	Head and neck
HCC	Hepatocellular carcinoma
HMWCK	High molecular weight cytokeratin
IHC	Immunohistochemistry
IMF	Inframammary fold
LDH	Lactate dehydrogenase
LMWCK	Low molecular weight cytokeratin
MITF	Melanogenesis associated transcription factor
MRI	Magnetic resonance imaging
NMSC	Non-melanoma skin cancer
pcSCC	Primary cutaneous squamous cell carcinoma
PET	Positron emission tomography
PR	Progesterone receptor
PSA	Prostate specific antigen
RCC	Renal cell carcinoma
SCC	Squamous cell carcinoma
SMJN	Sister Mary Joseph Nodule
TTF1	Thyroid transcription factor 1
UA	Urine analysis
β-hCG	Beta-human chorionic gonadotropin

Learning Objectives

1. To recognize the clinical features of cutaneous metastases, with specific reference to how their cutaneous manifestation is influenced by primary tumor, gender, and anatomic location.
2. To recognize the critical role of pathologic tools (histomorphology, immunohistochemistry, cytogenetics, molecular analysis) in diagnosis and prognosis of cutaneous metastases.
3. To understand how identification of skin metastases may impact prognosis & management.

Introduction

- Significance: Clinical recognition of cutaneous metastases offers the opportunity for diagnosis of the underlying malignancy, treatment of local symptoms, more accurate prognosis and staging, and optimal management of the neoplasm
- Frequency:
- Cutaneous metastases occur in <0.1–9% of cancer patients [1–8]
- Incidence likely to increase with increasing cancer survivorship
- Timing: Most often present following diagnosis of the primary tumor in late-stage disease
  - Mean latency from primary diagnosis to skin metastasis is 2–3 years [9–12]
  - Present prior to or concurrent with the primary tumor in 13–43% of cases [9, 10, 13]
- Presentation:
- Most present as a firm nodule, but may mimic a wide variety of skin lesions from neoplastic to inflammatory dermatoses, and are not clinically suspected in upwards of half of cases (higher for special sites such as eyelid) [11]
- Select CMs may overlap primary skin diseases both clinically and by histomorphology
- Approach: Thus, accurate characterization requires a high index of suspicion followed by a careful weighing of clinical and pathologic findings

History

- Evidence of primary and secondary skin tumors exists in mummified remains dating back millennia [14]
- 1829—Term **'metastasis'** is coined by Jean Claude Recamier [15]
- 1856—French surgeon, Alfred Velpeau, describes advanced CMBC, likening the appearance to the breastplate of the cuirassier (known today as carcinoma *en cuirasse*) [16]
- 1874—Sir James Paget describes eczematous nipple changes in breast cancer patients [17]
- 1889—Stephen Paget, son of Sir James Paget, coins 'seed and soil' hypothesis to explain organotropic metastasis [15]
- 1893—Hutchinson describes 'cancer erythema' in a patient of Kaposi [17]

- 1928—Dr. William James Mayo publishes description of umbilical metastasis, first observed by Sister Mary Joseph Dempsey [18]
- 1931—Rasch renames 'cancer erythema' to 'carcinoma erysipelatoides'
- 1933—Parkes Weber details telangiectatic metastasis of breast cancer [17]
- 1977—Isaiah Fidler reinvigorates the seed and soil hypothesis and study of tumor microenvironment [15]
- 1993—Lookingbill *et al.* publish the largest review of cutaneous metastasis of the time [1]

Pathophysiology

- Metastasis in general is an inefficient, but nonrandom process
  - The metastatic success rate is less than 0.01% for circulating tumor cells (CTCs) [19]
- A number of adaptations are important for most tumor cells prior to metastasis
  - Recruitment of proteolytic factors, E-cadherin loss, altered integrin and intermediate filament expression, activation of podoplanin, and cytoskeletal rearrangement [20]
  - In carcinomas, these adaptations promote locomotion, reshaping cellular morphology toward that seen in embryogenesis through a reversible process known as the **epithelial mesenchymal transition** (EMT)
  - Once resistant to *anoikis* (apoptosis with loss of contact to extracellular matrix), metastasizing cells spread to distant sites via blood, lymphatics, subendothelial or perineural spaces, and body cavities [20]
- Some primary tumors initiate specific metastases with increased frequency (organotropism)
- Two prevalent theories on organotropism:
  1. *Mechanical hypothesis*: championed by Virchow and James Ewing; states that lymph-vascular anatomy directs tumor cells until they arrest at their respective destinations
     - *Supporting observations*:
           Colorectal liver metastasis (via portal venous system)
           Vertebral metastasis from breast and prostate cancer (via Batson's plexus)
           Lung metastasis of lower rectal cancer (via inferior rectal veins) [21]
           Adrenal metastasis of lung cancer (via retrograde lymphatic migration)
  2. *Seed and soil hypothesis*: coined by Stephen Paget; postulates the interaction of tumor cells (seed) and host site (soil) determines metastatic efficiency (Fig. 1.1)
     - Tumor cells prepare recipient sites for their arrival even prior to metastasis by formation of the so-called **'pre-metastatic niche'**
           Tumor-secreted factors and exosomes target select organs and prompt vascular leakiness, recruitment of bone-marrow derived cells, and stromal remodeling [22]
     - Evidence from proteomic, antibody, and *in vivo* phage display studies suggests the presence of organ-specific, peptide 'zip codes' on endothelial cells of capillary beds which select for certain CTCs [23, 24]
     - The peptide motif (CVALCREACGEGC) is thought to assist CTC extravasation into skin at endothelial cells [23]

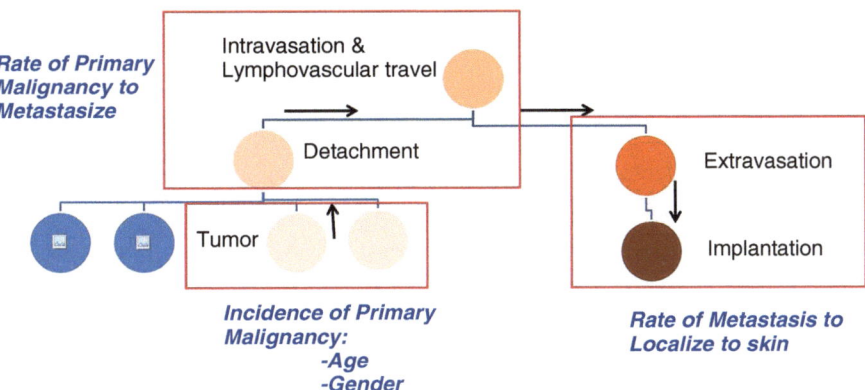

Fig. 1.1 Seed and soil hypothesis of cutaneous metastases

- Chemokine signaling may also be important in directing cells from solid tumors
    CXCR4: dermal-homing for Sezary and carcinomatous cells
    CCR10: ligand is a keratinocyte chemokine; important in epidermotropic metastasis, including CMM
    Likely of secondary importance
- Later steps of metastasis (cell survival post-extravasation, growth initiation, and growth maintenance) pose the highest hurdles for relocating tumor cells [25]
- **Immune surveillance** plays a pivotal role in thwarting these later steps
- Localization of metastatic deposit dependent upon local immunologic milieu: It is known that high Treg density in nodal basins increases nodal metastasis in several carcinomas
- In a study of skin metastasis, Treg cell percentage was highest in the head and neck (following hair follicle density) which was also the most common body site for metastasis by surface area [26]
- *Supporting observations*:
    Melanoma brain metastasis (via specific endothelial cell adhesion)
    Melanoma cells injected into syngeneic mice become established in specific organs even when surgically ectopic
    Peritoneovenous shunts in ovarian cancer patients do not greatly increase hematogenous metastasis

Clinical Themes

Epidemiology

- Reporting highly variable, but melanoma, breast, colorectal, and lung CMs are most frequent (Table 1.1)

Table 1.1 Most common metastases to the skin by primary tumors

2017 tumor incidence in US (excluding NMSC)		Frequent cutaneous metastases[a]	Skin preference[b]
Male	Female		
1) Prostate	1) Breast	**Breast**	N/A
2) Lung and Bronchus	2) Lung and Bronchus	**Melanoma**	N/A
3) Colon and Rectum	3) Colon and Rectum	**Lung**	++
4) Urinary bladder	4) Uterus	**Colorectal**	+/−
5) Cutaneous melanoma	5) Thyroid	Oral cavity	N/A
6) Renal	6) Cutaneous melanoma	Gyn tract	+/−
7) NHL	7) NHL	Upper GI	+/−
8) Leukemia	8) Leukemia	Renal	++
9) Oral cavity	9) Pancreas	MUP	N/A
10) Liver	10) Renal	Lower GU	N/A
		Liver	−−
		Prostate (uncommon)	−−

[a]Bolded are most common; Order does not exactly reflect CM incidence owing to diverse inclusion criteria, decade, and country of study
[b]++ strong positive, +/− average, and −− strong negative preference

Relative Incidence of Cutaneous Metastasis by Primary Tumor

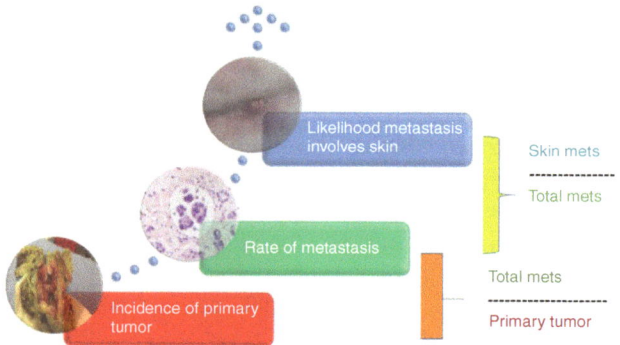

Fig. 1.2 Relative incidence of cutaneous metastasis by primary tumor

- CM epidemiology reflects cancer epidemiology and biology
  - Epidemiology: generally, the most common cancers supply the most CMs
  - Biology: internal malignancies exhibit variable proclivity for sending metastases to skin (Fig. 1.2)
  - Example: Prostatic adenocarcinoma has high incidence, but lower rates of high stage disease and poor skin preference, so CM is rare [10]
    78% of prostate cancers are localized at diagnosis compared to 16% for lung cancer [27]

Prostate cancer is more common, more aggressive, and higher stage at diagnosis in African American men, but it is unclear whether prostatic CMs, too, are more common in this population

- Despite variable reporting, general knowledge of CM incidence helps guide history-taking and workup

Age

- Most CMs present in adults, between the fifth and seventh decade of life
- Stratification by age and tumor type: Fig. 1.3 demonstrates CMs by decade in men and women from two series [28, 29]
- Compare to overall cancer incidence by age (Fig. 1.4 from 2005 SEER data *(approximate time of CM dx in most cases from [28–30])*)
- In young adults, testicular cancer should be considered in men and breast cancer and melanoma in women [31, 32]
- Pediatric CMs, like pediatric malignancies, are uncommon
  - Rhabdomyosarcoma and neuroblastoma are relatively common followed by choriocarcinoma, osteosarcoma, chondrosarcoma, and Ewing's sarcoma [33]

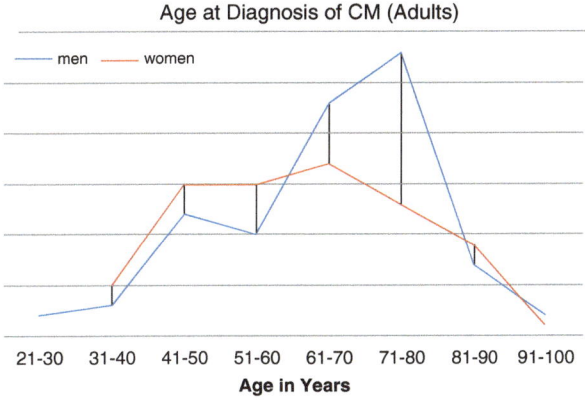

Fig. 1.3 Cutaneous metastasis diagnosis distribution by age

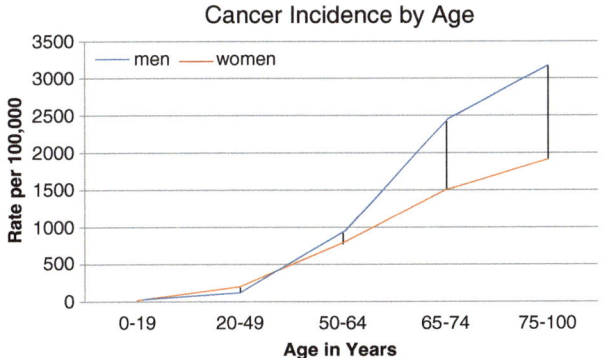

Fig. 1.4 Cancer incidence by age

Sex

- Overall cancer incidence and mortality is greater in men than women [27]
- Several non-sex specific cancers are also more frequent in men, including oral cavity/pharyngeal SCC, gastric cancer, HCC, bladder cancer, and RCC
- These gaps between the sexes are often reflected in literature on CM, with CMRC, CMHCC, and bladder urothelial carcinoma showing male preference [10, 34]
- Conversely, breast cancer is 100 times more common in women than men and is by far the most common cause of CM in this group with few reported cases of CMBC among males [35]

Clinical Morphology

- CMs have mimicked many primary skin diseases, from neoplasms to inflammatory and infectious dermatoses (Table 1.2)

Primary/Secondary Morphology

Primary or CM?

CMs are more likely to be multiple than primary skin tumors and tend to grow more rapidly than their adnexal tumor mimickers.

Table 1.2 Spectrum of primary skin diseases simulated by cutaneous metastases

Keratinocytic	Fibrohistiocytic	Inflammatory
Epidermal cyst	Fibroma	Contact dermatitis
Pilar cyst	Keloid	Lymphedema
Papilloma	Xanthoma	Eczema
Basal cell carcinoma	Xanthelesma	Calcified scar
Squamous cell carcinoma	**Vascular**	Hidradenitis suppurativa
Cutaneous horn	Pyogenic granuloma	Discoid lupus
Callus	Lymphangioma	Morphea
Keratoacanthoma	Hemangioma	Lichen planopilaris
Neuroendocrine	Kaposi sarcoma	Pseudopelade
Merkel cell carcinoma	Angiosarcoma	Lichen planus-like keratosis
Neural	**Infectious**	Alopecia areata
Granular cell tumor	Herpes zoster	Chalazion
	Cellulitis	Prurigo
	Condyloma accuminatum	Erythema annulare centrifugum
	Paronychia	Intertrigo
	Kerion	Rhinophyma/Rosacea
	"Blueberry muffin baby"	Vasculitis
Melanocytic	**Adnexal**	Erythema nodosum
Melanoma	EMP	**Other**
Nevus	Cylindroma	Generalized pruritus

- Most CMs are firm, even "stony" nodules [36]
- Grow rapidly then plateau at several centimeters in diameter
- Others barely palpable or rarely gigantic (up to 20 cm) [37]
- May be single or multiple, wide-spread or agminated
- Surface change is rare, but ulceration may occur [38]
- A significant portion (15% in one series) are clinically labeled "rash" [11] and biopsy is warranted when:
  - Unexplained erythema lasts >2 weeks in patients with cancer history [38]
  - Erythematous lesions form discrete nodules
  - Reasonable initial treatment (e.g. antibiotics) fails

Color
- Commonly skin-colored or varying hues of red to blue, brown, or black
- Violaceous/vascular-appearing: CMRC, CMHCC, CMChC, CMTC, and metastatic angiosarcoma
- Hyperpigmentation is especially common to CMM and CMBC
- Multiple CM nodules may vary in color from one another within the same patient [36]

Distribution [10]
- Most CMs appear on ventral skin in proximity to the tumor of origin
- The trunk has the highest absolute tumor burden
- Correcting for percent surface area, the head and neck are most frequently involved [26]
- *The diaphragm rule*: CMs from organs below the diaphragm tend to metastasize to infradiaphragmatic skin while those above the diaphragm tend to produce supradiaphragmatic metastases
- Notable exceptions: RCC which favors supradiaphragmatic spread and upper GI tract, liver, and rectal cancers which may show no clear preference in this regard
- In colon and paired organs, contralateral metastases are as common as ipsilateral metastases
- Figure 1.5 summarizes frequent sources for visceral CM by body site
- Special sites and frequent sources
  - Scalp: breast, lung, renal, and MUP
  - Nasal tip: renal and lung, so-called **'clown nose'**
  - Eyelid: breast (up to 50%), lung, GI, renal, MUP [39]
  - Orbit: breast, prostate, lung
    Accompanied by reduced ocular motility (54%), proptosis (50%), and palpable mass (43%) [40]
    Enophthalmos may be more common with CMBC [41]
  - Umbilicus: stomach, pancreas, colon, renal, ovary, breast, MUP [42]
  - Penis: bladder, prostate, colorectal
    Skin nodules may be accompanied by priapism [43]
  - Scrotum: colorectal, prostate, lung
  - Vulva: cervical SCC, EMPD-like metastatic adenocarcinoma

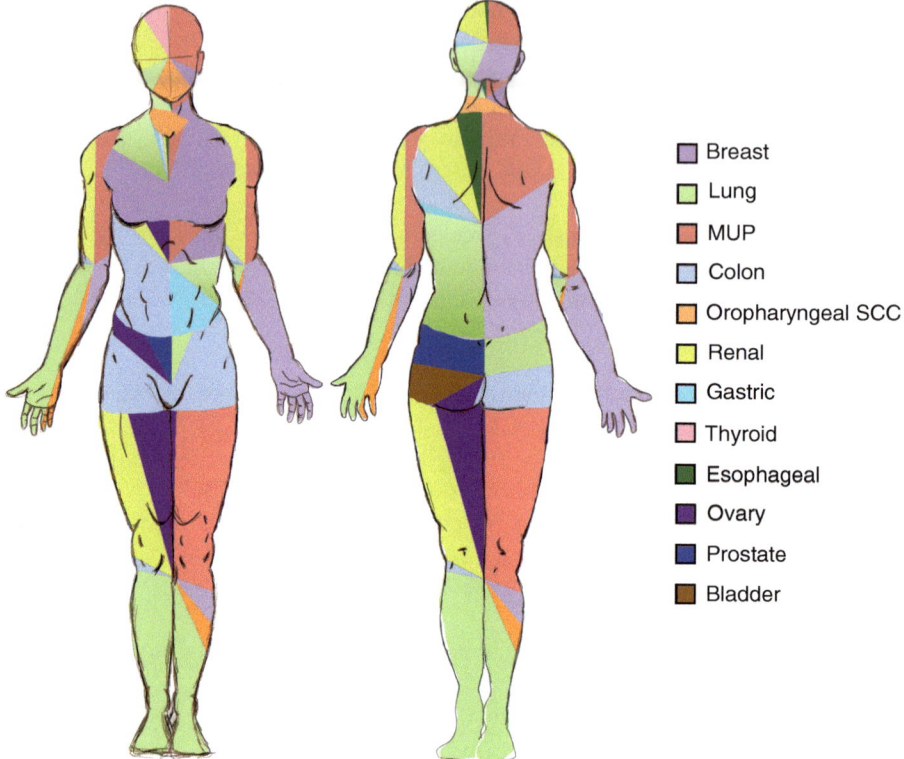

Fig. 1.5 Anatomic map of cutaneous localization of metastases by internal tumor type

- Acrometastases (lung, kidney, breast, GI) [44, 45]
    Hand: lung, breast, renal
    Foot: renal, colorectal, lung
- Previous scar: many carcinomas (and lymphomas) appear to home to scar tissue
    Locus minoris resistentiae/chemotactic factors or 'seeding phenomenon'
    New lesions developing in old scars should be considered for biopsy

Associated Symptoms/Signs
- Pain—uncommon, but may be severe, spontaneous, and difficult to manage
    - Tumor entrapment or inflammation of sensory nerves
    - However, also described with carcinoid tumors even in absence of marked neural involvement
    - Some cases only relieved by local excision
- Pruritus/burning—Described with Paget's disease and perianal/genital metastases mimicking Paget's disease (sometimes called secondary EMPD)
- Bleeding—RCC, HCC, choriocarcinoma, and metastatic angiosarcoma

- Odor—Necrotic, exudative, or infected lesions may be odorous and impact patients' quality of life
- Other—Large, fibrotic thoracic CM plaques have interfered with deep inspiration and sleep
- Adenopathy—Regional lymph nodes should be palpated during general examination
- Careful examination of the neck (including thyroid), ears, nose, and throat, as well as the breasts, testes, rectum, and prostate is prudent
- Constitutional symptoms/signs—May or may not accompany; when present should alert the clinician to possibility of progressing, recurrent, or occult malignancy

Noninvasive Testing

- Dermoscopy [46]
  - Dermoscopic features classic for both melanocytic and non-melanocytic origin have been described in CMs
  - **Vascular pattern** most common
      Frequent vascular patterns were linear irregular, polymorphous, and arborizing
  - CMs may convincingly pose as melanocytic lesions by dermoscopy, with pigment streaks, globules, and even a blue-white veil
      Attributed to melanocyte colonization of CMs, which may be facilitated by tumor cell epidermotropism
      Relatively common with CMBC; other pigmented CMs reported include prostatic, anorectal, and medullary thyroid carcinomas
  - Larger series are needed for future application of dermoscopy to CMs
- In vivo reflectance confocal microscopy
  - May have use if initial biopsy is declined by patient [47]

Work-Up/Investigation

Pathologic Examination

Technique
- Punch biopsy typically obtains diagnostic tissue
- Deeper lesions may require telescoping or excisional biopsies
- Fine needle aspiration is also successfully employed

Histomorphology

Primary or CM?
Marked vascular invasion, dermal placement without epidermal involvement, and disorganized collagen dissection are soft clues for CM [48]

- **Melanoma** and **adenocarcinoma** are the most common histopathologies underlying CMs [48]
  - Dermal and subcutaneous involvement most common
  - Epidermotropism is rare, but often causes ulceration, with occasional pigmentation, and may be difficult to differentiate from primary cutaneous tumors
- Histology alone, in general, poorly discriminates primary and secondary skin neoplasms
  - In one series, 27 of 41 CMs (66%) were correctly diagnosed as such by four of four pathologists, blinded to clinical information and IHC staining; organ of origin was not correctly predicted in most of these cases [11]
- Some tumors are more amenable to histopathological recognition
  - e.g. HCC, mesothelioma, seminoma, choriocarcinoma, papillary and follicular thyroid carcinoma, ameloblastoma, and Leydig cell tumors, which have unique microscopic appearances [28]
- Most have a broader differential or are indistinguishable from primary tumors by histopathology
  - Some CMs and primary adnexal tumors
  - Many primary and secondary cutaneous sarcomas
  - Small cell lung cancer and Merkel cell carcinoma
- Histological comparison to any previous (potential primary) tumor specimen is helpful whenever possible
- Table 1.3. Histomorphologic clues to tumor origin. *No one feature is specific*, but together may direct further workup
- Up to 5% of metastatic tumors lose histomorphologic similarity to the primary tumor [49] and many of these are not identified with further available testing (MUP)

Ancillary Testing

Immunohistochemistry
- Immunohistochemistry uses antibodies to report on expression of specific cellular antigens, primarily for the classification of tissue origin, but also to aid in diagnosis of invasion, prognostication, determination of therapeutic response, and identification of micrometastasis, among other uses
- Critically important testing method which is complementary to histopathological and clinical investigation
  - Clinical evaluation and morphology determine IHC panels and the results may also lead to further clinical studies
- In metastasis of unknown origin, panels of stains are typically ordered sequentially to screen for lines of differentiation (Table 1.4) and identify cell-specific antigens
- In diagnosis of tumor origin, **cytokeratins** (CKs) are the most fundamental of epithelial markers useful in diagnosis of carcinoma; they are organized into high molecular weight (HMW) and low molecular weight (LMW) based on electrophoresis migration and tissue expression patterns

Table 1.3 Histomorphologic clues to tumor origin of skin metastases

Table 3. Histomorphologic clues to tumor origin

Histologic pattern/features						
Squamous cell carcinoma	lung	cervix	esophagus	oropharynx		some primary skin tumors
Adenocarcinoma						
Papillary	thyroid (colloid, foamy macrophages)	ovary	lung	stomach	colorectal	urinary bladder
Micropapillary	ovary	urinary bladder	salivary gland	lung		breast
Epidermotropism	melanoma	breast	prostate	colon	ovary	larynx	primary cutaneous tumors

Cellular features						
Epithelioid	melanoma	carcinoma			lymphoma	sarcoma
Clear cell	RCC	lung	liver	ovary	endometrium	cervix, vagina	clear cell sarcoma	some primary cutaneous
Spindle cell	sarcoma	melanoma				carcinomas with sarcomatous differentiation
Muchous/myxoid	bowel	appendix	pancreas	breast	lung	many sarcomas
Signet ring cell	stomach	breast	gallbladder	melanoma	pancreas	colorectal	lung	urinary bladder
Apocrine	breast		renal			primary cutaneous apocrine carcinoma
Small Round Blue Cell Tumor						
• Children	lymphoma	neuroblastoma	ES/PNET	solid alveolar rhabdomyosarcoma	wilms tumor	medulloblastoma	retinoblastoma
• Adults	lymphoma	small cell neuroendocrine tumor	desmoplastic small round cell tumor	synovial sarcoma	merkel cell carcinoma	mesenchymal chondrosarcoma
			small cell neuroendocrine tumors			

Nuclear features						
Molding	papillary thyroid	HCC	meningioma	melanoma	adrenocortical carcinoma	pheochromocytoma
Intranuclear inclusions						
Salt and pepper chromatin	neuroendocrine tumors					
Large eosinophilic nucleoli	prostate		melanoma			

Stromal features/collections/deposits						
Intratumoral PMN with karyorrhexis		colon and rectum				
Intratumoral melanin	melanoma	breast, other carcinomas	clear cell sarcoma	pigmented DFSP	melanotic schwannoma	melanotic medulloblastoma	pigmented Paget and EMP disease
Psammoma bodies			meningioma	mesothelioma	papillary (thyroid, serous ovarian, lung, RCC)	duodenal somatistatinoma
Intratumoral bile	HCC		cholangiocarcinoma			

PMN-polymorphonuclear (cells), DFSP-dermatofibrosarcoma protuberans, EMP-extramammary Paget disease, PNET-primitive neuroectodermal tumor, ES-Ewing sarcoma, HCC-hepatocellular carcinoma, RCC-renal cell carcinoma

PMN polymorphonuclear (cells), *DFSP* dermatofibrosarcoma protuberans, *EMP* extramammary Paget disease, *PNET* primitive neuroectodermal tumor, *ES* Ewing sarcoma, *HCC* hepatocellular carcinoma, *RCC* renal cell carcinoma

Table 1.4
Immunohistochemical
screening panel to elucidate
line of differentiation of skin
metastases

Melanocytic/Neural crest	S100 (neural crest), NSE
Epithelial	Pankeratin[a]
Hematolymphoid	LCA
Mesenchymal	vimentin (least specific)
Germ cell	hCG, AFP
Neuroendocrine	NSE, chromogranin

NSE neuron specific enolase, *LCA* leukocyte common antigen, *AFP* alpha fetoprotein, *hCG* human chorionic gonoadotropin
[a]General term for an IHC cocktail staining a broad number of CKs

- HMWCKs are expressed in squamous epithelia and tumors while LMWCKs are expressed in all other epithelia and tumors
- As with all IHC markers, exceptions to the rule, especially in undifferentiated tumors which may 'lose' characteristic markers or 'gain' uncharacteristic markers, necessitate combination of these stains into "cocktails" to increase diagnostic yield
  - One discerning combination often utilized with metastatic adenocarcinomas is that of CK7 and CK20; resulting staining patterns direct investigation toward additional, more specific markers (Fig. 1.6)
- Still, state-of-the-art IHC fails to suggest a primary site in 60% of MUP cases; the success rate may be as low as 25% in some populations [50]

Molecular Testing [49, 51]
- A significant proportion of metastases lose phenotypic similarity to their primary tumor by light microscopy
- When IHC, too, fails to suggest a primary site, molecular testing may prove helpful
- The gene expression profile of a MUP can be compared to several known tumors using **microarrays** or **RT-PCR**
- Previously impractical, current technology allows for this testing on FFPE tissue, the processing method for most tumor biopsies
- Several such assays are now commercially available (Table 1.5)
  - Some claim improved accuracy over IHC
- Initial validation appears promising, however, evidence that these tests improve clinical outcomes is lacking
- Randomized prospective trials are necessary
- Several societies do not recommend their routine use (NCCN, national institute for health and care excellence, European Society of Medical Oncology) [52]
- Additional barriers to routine use:
  - Limited tumor library, quality of the specimen (ideal tissue has maximal tumor volume and minimal necrosis), cost (many of these tests are on the order of $3000)

Laboratory Testing [52–54]
- No current standard exists
- Dictated by positive findings from history and physical, but in general includes at least CBC and BMP
- UA for hematuria or stool occult blood often performed

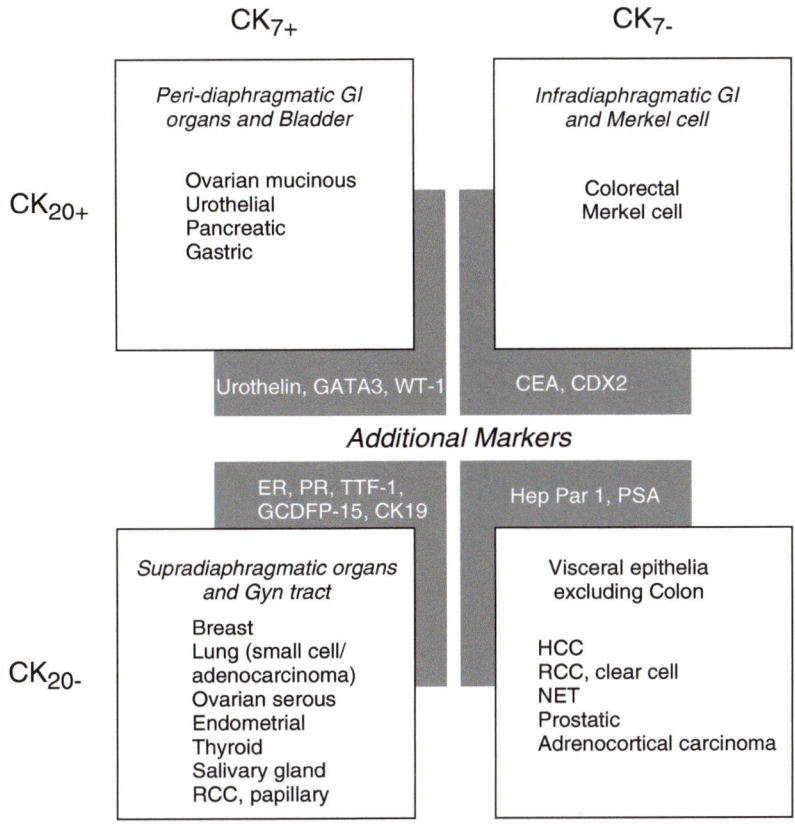

CK_{7+} CK_{7-}

CK_{20+}

Peri-diaphragmatic GI organs and Bladder

Ovarian mucinous
Urothelial
Pancreatic
Gastric

Infradiaphragmatic GI and Merkel cell

Colorectal
Merkel cell

Urothelin, GATA3, WT-1   CEA, CDX2

Additional Markers

ER, PR, TTF-1, GCDFP-15, CK19   Hep Par 1, PSA

CK_{20-}

Supradiaphragmatic organs and Gyn tract

Breast
Lung (small cell/adenocarcinoma)
Ovarian serous
Endometrial
Thyroid
Salivary gland
RCC, papillary

Visceral epithelia excluding Colon

HCC
RCC, clear cell
NET
Prostatic
Adrenocortical carcinoma

HCC-hepatocellular carcinoma, RCC-renal cell carcinoma, NET-neuroendocrine tumor,
ER-estrogen receptor, PR-progesterone receptor, TTF-1-thyroid transcription factor 1,
CEA-carcinoembronic antigen, PSA-prostate specific antigen, WT-1-Wilms tumor protein

Fig. 1.6 Immunohistochemical identification of metastatic epithelial tumors: example algorithm

Table 1.5 Gene expression testing for work-up of metastases of unknown primary

Test	Company	Technology platform	Specimen	Classification accuracy (%)	Tumor types tested	Regulatory status
RosettaGX cancer origin	Rosetta Genomics	miRNA profile	FFPE	85	49	Available under CLIA
Tissue of origin	Cancer Genetics Inc.	Oligonucleotide microarray	FFPE block or unstained slides	84	15	Cleared by FDA
Cancer type ID	bioTheranostics	RT-qPCR	FFPE	87	54	Available under CLIA

- Serum tumor markers, in general, are not sufficiently sensitive or specific to be of diagnostic value, but ought to be considered in the appropriate clinical
  - e.g. PSA for men with CM and osteoblastic bone metastases
  - β-hCG and AFP in poorly differentiated mediastinal/retroperitoneal midline tumors (potential germline origin)
  - CA 15–3 in women with CM and axillary involvement
- LDH increases in cancer due to *Warburg effect* (preferential fermentation metabolic pathway by cancer cells resulting in increase in lactate production) and is used as nonspecific tumor marker
  - Classically elevated in dysgerminoma
  - Not specific for cancer or cancer type
  - Prognostic value in several metastatic solid tumors and hematologic malignancies [55]
- Serum antibody-based platforms show promise in detection of CTCs in patients with several types of metastatic cancer
  - May one day aid in ddx of primary vs secondary skin tumors [56]

Imaging/Procedures [54]

- CXR (at least) indicated in all patients lacking a primary site
- Abdominal CT may identify primary tumor in one-third of patients lacking other localizing findings [53]
- Chest and pelvis CT increases sensitivity of detection
- Mammogram in female patients
- Testicular ultrasound if indicated by history and physical
- Endoscopy is not indicated for all patients without other localizing signs, symptoms, or laboratory abnormalities
  - Colonoscopy useful in patients with IHC suggesting colonic source (CK7-CK20+CDX2+)
  - Bronchoscopy when lung primary is possible (CK7+TTF1+)
- MRI- data limited for broad application, but useful in detection of occult primary breast cancers when other methods fail
- FDG PET/CT—Detects primary tumor site in 37% of MUP
  - Especially lung and head and neck cancers [57]
- Utility of bone scan in identifying primary site is unknown
- Upper GI radiographs, barium enema, and IV pyelogram not useful unless supported by other findings (positive UA or heme occult test)
- Acrometastasis should also include X-ray and/or MRI (gold standard) of the extremity [45]
- OctreoScan or ^{68}Ga-DOTA-NOC receptor PET/CT are sensitive methods for detecting sources of neuroendocrine tumor metastasis

Management and Prognosis

CMs of Known Primary Tumors

- Treatment of the primary tumor guides management of the CM

- Some CMs readily respond to systemic therapy (e.g. CMBC improving with HER2-targeted chemotherapy, etc.)
- However, fibrotic lesions (such as carcinoma *en cuirasse*) limit delivery of IV chemotherapeutic drugs and resist systemic treatment [58]
- *Abscopal response* wherein primary tumor irradiation results in CM regression [59]
- Local therapy sometimes performed for palliation or cosmesis
  - Options include cryotherapy, topical miltefosine, imiquimod, radiation and brachytherapy, intralesional chemotherapy, isolated limb perfusion, electroporation by electrochemotherapy (especially melanoma and sarcoma), local excision, and Mohs micrographic surgery (to preserve function/cosmesis)
- Local surgical treatment of isolated CMs may improve prognosis in patients with lung and gastric cancer [42], and some authors recommend metastasectomy in all cases where tumor burden might be decreased and function improved
  - 1 cm margins are often taken, although there is no evidence supporting this practice [59]
- For cutaneous adenocarcinomas in which a primary tumor is not identified, but suspected, initial management does warrant complete, conservative excision
- Diagnosis of CM typically carries a poor prognosis
  - The average survival time of patients with CMs is 7.5 months [12]
- Patients with CMs from prostate, breast, and previously treated colorectal cancers may have more favorable course than other CMs [9, 32]
- In children, a subset of neuroblastomas (stage IVS) spontaneously regress or differentiate to benign ganglioneuromas even in the setting of CM
  - Infants with metastasis limited to skin and liver (without major bone marrow involvement)

Metastasis of Unknown Primary [60]

- Prognosis of MUP is generally guarded, with 1 year survival ranging from 15 to 35%
- MUP patients are typically treated with empiric, platinum or taxane-based combination chemotherapy
- New efforts to subclassify or 'stage' MUPs have distinguished favorable and unfavorable groups
  - Grouped into broad prognostic categories by histology, involved sites, and immunohistochemical profile
- In some MUP, eventual discovery of the primary site is associated with improved prognosis [54, 61]
  - Hence, increasing efforts to molecularly profile tumors of unknown primary site
- In the case of melanoma, however, lack of a primary site appears to carry a more favorable prognosis [62]
  - Presumably as a result of better immune control and primary tumor regression

Specific Cutaneous Metastases

Melanoma

Epidemiology
- In adults over 50, the incidence of melanoma has increased at 2.6% per year since 1996 without a proportionate rise in mortality, reflecting more aggressive surveillance and potentially therapeutic progress [63, 64]
- Incidence of CMM also appears to have increased, from 10% of all cutaneous metastases in 1972 to 18.2% in 2004 [12, 34]
- Cutaneous metastases occur in nearly half of metastatic melanoma patients and herald disease 5% of the time [65, 66]
- Ocular melanomas send subcutaneous metastases in 17% of cases [17], but are not considered further here

Location [65]
- Scalp, face, and extremities host most secondary sites when in-transit and satellite metastases are excluded
- Primary site is not identified in up to 8% of CMM

Clinical Patterns
- Frequently solitary dermal or subcutaneous nodules
  - 36–68% are clinically pigmented; amelanotic metastases do occur [1, 66] (Fig. 1.7a)
- DDx includes nevi including blue nevi, primary melanoma, dermatofibroma, angiokeratoma, other pigmented and some vascular skin lesions
- Very rarely, cytolytic metastases produce diffuse gray to blue or black hyperpigmentation (**diffuse melanosis cutis**) and may turn urine black when oxidized (melanuria) [36, 67]

Histopathology
- Highly variable morphology from epithelioid to clear cell to spindled to small cell morphology, however, tumor cells often contain prominent nucleoli (Fig. 1.7b)
- Melanin pigment is helpful when present
- CMM is suspected when dermal or subcutaneous tumor nodules lack a related junctional intra-epidermal component or other circumstantial evidence (in situ melanoma, preexisting nevi, regression, etc.)
- To be differentiated from 'primary dermal melanoma,' which are less aggressive
- When epidermotropism is exhibited, may be confused with primary site melanoma
- Histopathology has mimicked dermatofibromas, sarcomas, AFX, and blue nevi, among many others
- Amelanotic lesions with acinar or single cell invasion simulate lobular breast cancer

Fig. 1.7 (**a**) Black dermal-based papules and plaques of cutaneous metastatic melanoma. (**b**) Cutaneous metastatic melanoma histology demonstrating large epithelioid to spindled cells with conspicuous nucleoli

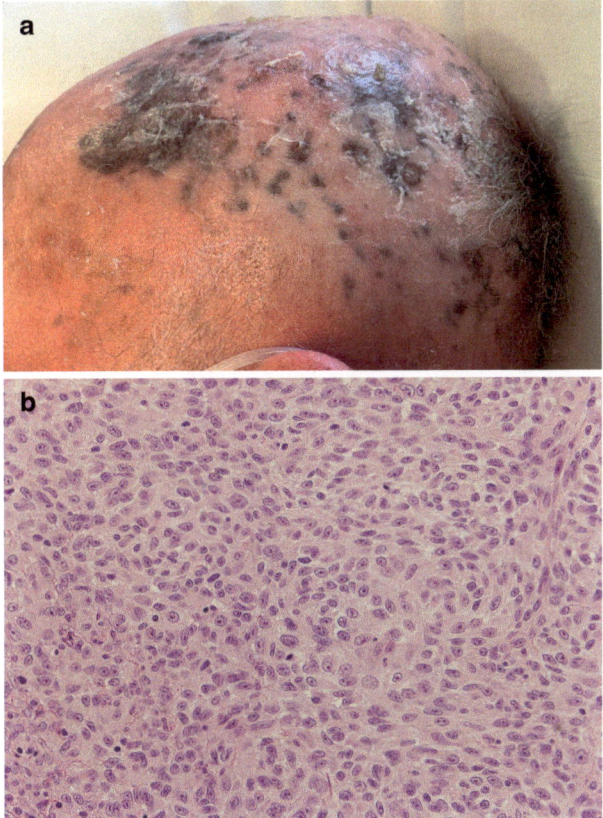

- Immunohistochemical staining positive for melanocytic markers (many of which are also expressed by neural tissues): S100, Melan-A, SOX10, HMB45, and MITF (Melan-A, MITF, and HMB45 usually negative in desmoplastic melanoma while SOX10 and S100 are positive)

Breast Cancer

Epidemiology
- Excluding skin, most common primary cancer site in women
- Most common cause of skin metastasis in women
    - >½ to 2/3 of all cutaneous metastases in women [1, 28]
- Rates of breast cancer have equalized among white and black women [31]
- Improved cancer survivorship has been associated with higher CNS metastasis [68] and potentially greater rates of cutaneous metastasis
- Often the result of direct extension rather than bona fide metastasis

Location
- Thorax and scalp most commonly affected

Clinical patterns
The following clinical patterns imply a histopathology, but clinical and pathologic phenotypes may overlap in a single patient, synchronously or metachronously

- Nodular
  - Dermal and subcutaneous papulonodules (Fig. 1.8a).
  - 80% of CMBC [69]
  - Painless, rubbery, less often with ulceration, bullae, or tenderness
  - DDx is broad, and is the least specific of the cutaneous metastatic breast cancer presentations
  - Ductal and lobular histomorphologies most common (Fig. 1.8b).

Fig. 1.8 (**a**) Pink nonscaly papulonodules of cutaneous metastatic breast carcinoma. (**b**) Cutaneous metastatic breast carcinoma pathologic image showing nests of basophilic cells without prominent palisading

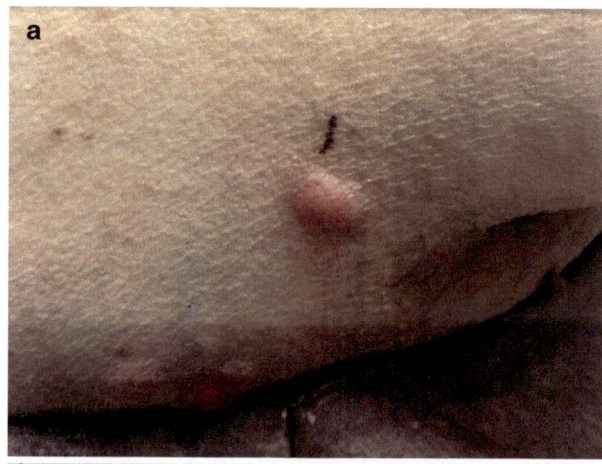

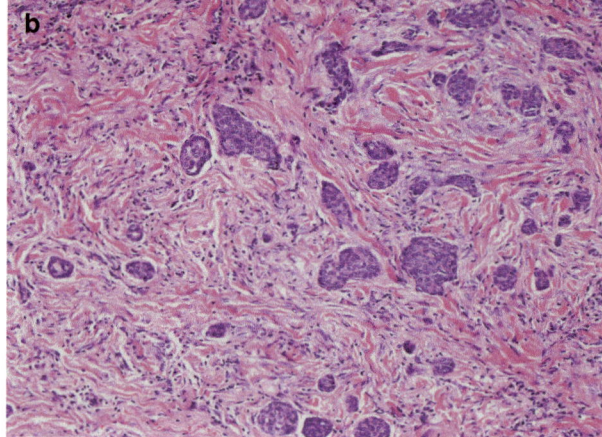

- Tumor islands in fibrotic stroma; occasionally infiltrative cords; rare cases of intra-tumoral melanin mimicking melanoma clinically and by histopathologic examination

Inflammatory metastasis
Refers to lesions with inflammation clinically and does not indicate presence of an infiltrate on microscopy.

- Carcinoma erysipelatoides (inflammatory)
  - Dermal lymphatic and vascular congestion forms erythematous, erysipelas-like patches and plaques
  - 3–11% of cutaneous metastatic breast cancer [1, 69]
  - DDx includes cellulitis, EAC, and benign lymphedema, among others
  - Benign lymphedema tends to be painless and slowly progressive
  - Thus, early biopsy should be considered in patient with breast cancer history who rapidly develops severe pain, vascular congestion, and necrosis at the site in absence of fever and leukocytosis
  - Regional edema in treated breast cancer patients signifies recurrence in a quarter of patients [70]
  - Phenotype documented with many other visceral and, rarely, primary skin tumors (porocarcinoma, primary cutaneous apocrine carcinoma, and lymphoma)
  - Histopathologically, tumor cells invade and expand dermal lymphatic spaces, often with a relative paucity of acute inflammation
- Carcinoma telangiectaticum
  - Yellow to purple papulovesicles or plaques also known as carcinoma telangiectoides
  - 11.2% of CMBC [69]
  - DDx includes lymphangioma circumscriptum, vasculitis, and angiosarcoma
  - Other visceral tumors may infrequently present as telangiectatic skin lesions
  - Histopathology demonstrates blood vessel dilatation in the papillary dermis with intra-vascular tumor rafts
- Carcinoma *en cuirasse*
  - Sclerotic, indurated chest wall plaques which resemble the armor breastplates of the *cuirassier*
  - Patients often asymptomatic, but may report pain, discharge, bleeding, and, if end-stage, difficulty with inspiration
  - 3% of CMBC [69]
  - DDx includes morphea, keloids, and breast fibromatosis (desmoid tumor of the breast)
  - Documented with many other malignancies (lung, prostate, pcSCC)
  - Although the chest wall is the most common site, clinically and histopathologically similar lesions have occurred at distant sites (e.g. forearm) [71]

- Characteristic single-filing of lobular tumor cells through a desmoplastic stroma on histopathology
- Single-file cells may be mistaken for fibroblasts at scanning magnification
- Alopecia neoplastica (AN)
  - Smooth alopecic plaques resulting from scalp metastasis
  - Resemble alopecia areata in their sharp demarcation, but varying degrees of inflammation and fibrosis more frequently suggest a scarring alopecia
  - 2% of all CMBC [69]
  - Breast cancer is the underlying tumor in 84% of AN cases, but other carcinomas have produced the phenotype [72]
  - DDx may range from AA to DLE, LPP, pseudopalade, morpheaform BCC, keloid, EMPD [73]
- Inframammary crease metastasis
  - CMBC involving the fibrofatty tissue in the inframmamary fold (IMF)
  - Exophytic callus-like growth often with fissuring and skin retraction
  - The IMF contains glandular breast tissue in a minority of women, thus breast cancer of the IMF may be primary or metastatic [74]
  - BMI, age, and breast mass do not predict presence of glandular tissue in the IMF
  - <2% of breast cancers originate here [75]
  - DDx includes callus, epidermal inclusion cyst, intertrigo, and BCC
  - Histopathology demonstrates tumor cell nests which may exhibit collagen retraction simulating BCC at scanning magnification
    Larger biopsies are recommended in this area to avoid this pitfall by demonstration of cellular atypia and lack of peripheral palisading
- Eyelid metastasis
  - Three types: nodular, diffuse, and ulcerative
  - Nodular: identical to nodular metastases at other sites
  - Diffuse: pattern resembles morphea with red-brown to lilac indurated plaques
  - Ulcerative: epidermal involvement in diffuse or nodular type
  - Bilateral in some cases; Mask-like four-eyelid involvement is exceptional
  - Breast cancer accounts for 75% of orbital and 50% of eyelid metastases [39, 76]
  - CMs from all sites comprise 0.3% of eyelid biopsies [39]
  - Clinical DDx: chalazion, morphea, contact dermatitis, blepharoconjunctivitis, angioedema, orbital cellulitis, sarcoidosis, fat hernia, sebaceous neoplasms, and lymphoproliferative disorders
  - Esophageal, lung, gastric, and renal primary tumors are other relatively common causes, and the origin remains unknown in as many as 11% of cases [39]
  - Pain, diplopia, and other vision changes suggests involvement of deeper orbital structures
  - Enophthalmos occurs more commonly with breast cancer (10% of cases) relative to other sources of orbital metastasis [76]
  - Histiocytoid variant of lobular breast carcinoma and primary cutaneous signet-ring cell/histiocytoid carcinoma of the eyelid may be indistinguishable clinically and histopathologically

- Paget disease [77]
  - Tumor cell extension (not 'true' metastasis) to the nipple areola complex via in situ intraductal or invasive migration
  - Unilateral erythematous, eroded, ulcerated, or scaly plaques may weep and bleed
  - Pruritus, burning, and pain are common
  - Partial response to topical steroids may further mislead the clinician toward an eczematous process
  - <5% of patients with breast carcinoma
  - Half of cases accompanied by palpable mass
  - Most harbor DCIS, then invasive ductal carcinoma, lobular cancers less frequent
  - Underlying mass may be distant from the nipple
  - Remaining cases have no palpable mass or are subclinical and discovered incidentally
  - May present in ectopic breast tissue
  - DDx may include atopic dermatitis, contact dermatitis, nipple adenoma, psoriasis, BCC, ruptured galactocele, cavernous hemangioma, florid papillomatosis of the nipple, and radiation changes
  - Clinically pigmented Paget disease has also mimicked melanoma
  - A breast or areolar rash notably sparing the nipple is unlikely to be Paget disease [36]
  - Healed nipple rashes do not preclude the presence of disease, thus strongly suspect histories warrant biopsy even with healed skin on presenting examination [78]
  - Histopathology shows epidermis partially replaced by enlarged Paget cells demonstrating macronucleoli and abundant pale cytoplasm
- Male breast cancer [35, 58]
  - <1% of all breast cancers
  - Specific CMBC phenotypes are not well described in men
  - Predisposing factors include BRCA1/2 mutations, Klinefelter syndrome, history of radiation to the chest, hyperestrogenism of various causes, etc.
  - Majority are invasive ductal carcinomas

Lung Cancer

Epidemiology [31]
- Second most common cancer in men and women, excluding primary skin cancer
- Highest cancer-related mortality in both genders
  - Incidence and mortality declining
- Smoking rates converging between men and women
- Rates of CMLC appear to be on a relative incline, especially in women
  - Lung cancer was second most common cause of CM in men behind melanoma and sixth in women in Lookingbill's 1993 report [1]

- A Taiwanese review from 1986 to 2006 which excluded melanoma showed lung cancer was the most common cause of CM in men and second in women [79]
- A US review from 1990 to 2005 also excluding melanoma showed similar findings [11]
• Overall, CMLC is still uncommon, representing 1.7–12% of all lung cancers [80, 81]
• Although reports are conflicted, lung cancer is thought to metastasize to skin early in disease relative to other cancers which show late-stage CM
• CMLC is the first sign of disease in 14% of cases [82]

Location [81, 83]
• CMLC commonly affects the thorax, but is also discovered at distant sites (head and extremities)
• Tumors in upper lung lobes supply a disproportionately high number of CMs
  - Attributed to hematogenous accessibility of these well-vascularized regions

Clinical Patterns
• Nodular metastasis
  - Multiple painless skin colored few centimeter mobile nodules
  - Some solitary and ulcerated
• Breast cancer-like metastasis
  - Erysipelatoid, telangiectatic, and *en cuirasse* lung cancer are all described
• Acrometastasis [44, 45]
  - Metastasis to bone and soft tissue of hands and feet
  - In one review, most common sources in order of frequency were lung, kidney, breast, and GI
      Lung cancer shows preference for upper extremity acrometastasis
      Acrometastases of the feet are frequently sourced to genitourinary cancers
  - Present as an erythematous enlarging digit or mass
      Frequently accompanied by a deep-seated pain, which is resistant to over-the-counter analgesic medications
  - 10% present before the primary tumor
  - Lung cancers cause 41% of subungual metastases [84]
  - DDx includes primary skin tumors as well as rheumatoid arthritis, osteomyelitis, pyogenic granuloma, paronychia, digital mucous cyst, gout, dactylitis, and primary bony neoplasms
• Zosteriform metastasis [85, 86]
  - Dermatomally distributed cutaneous metastasis reminiscent of herpes zoster infection
  - Also described with breast, kidney, prostate, bladder, melanoma, lymphoma, and pcSCC
  - CM may also preferentially seed previous zoster scars due to local immunosuppression (**locus minoris resistentiae**)

- Scalp metastasis
  - Classically described with renal metastasis, however, lung was responsible for the most cases in a retrospective study of primary and secondary scalp neoplasms [87]
  - In women, CMBC is most common
  - MUP is also commonly responsible by other reports
  - Warmth and vascularity of scalp implicated

Histopathology
- Adenocarcinoma produces more CMLC than do SCC and small cell carcinomas
- Large cell carcinoma is variably reported in the literature, but produces CMLC infrequently [88]
- When undifferentiated on H&E, several IHC markers are employed:
  - Lung adenocarcinoma is typically CK7+20– and TTF-1+
  - Lung SCCs are usually CK5/6+, CK7–, TTF-1–, p63+, and p40+
- Nuclear molding and 'salt and pepper' chromatin of small cell lung cancer in skin may raise suspicion for Merkel cell carcinoma, which often shares expression of neuroendocrine markers (CD56, chromogranin, synaptophysin)
  - CK20 staining in Merkel cell carcinoma is characteristically paranuclear and dot-like, but this classic staining pattern is not always observed
  - Merkel cell polyomavirus (MCPyV) and TTF-1 immunohistochemical stains may help differentiate such cases
- Other lung cancers very seldom produce CMs (mesothelioma, mucoepidermoid carcinoma, pulmonary sarcomas) [88]

Renal Cancer

Epidemiology [31]
- Most are RCCs derived from proximal tubular cells
- Incidence now stabilizing after decades of elevated rates owing to prevalence of cross sectional imaging
- Mortality on the decline since 2002
- M:F ratio is 2:1
- CMRCs comprise 3–6.8% of all CMs [89]
- CMRCs are also more common in men [34]

Location
- Head and neck most common sites
- Second most common cause of acrometastasis
- Nephrectomy scars may also be involved

Clinical Patterns
- Nodular metastasis
  - Most common
  - Asymptomatic, well-defined, skin-colored to red-violaceous (Fig. 1.9a)

Fig. 1.9 (**a**) Cutaneous metastatic renal cell carcinoma presenting as a red vascular-appearing papule on the chin. (**b**) Cutaneous metastatic renal cell carcinoma pathologic image showing vascularized, closely packed, atypical, variably clear, epithelioid tumor cells

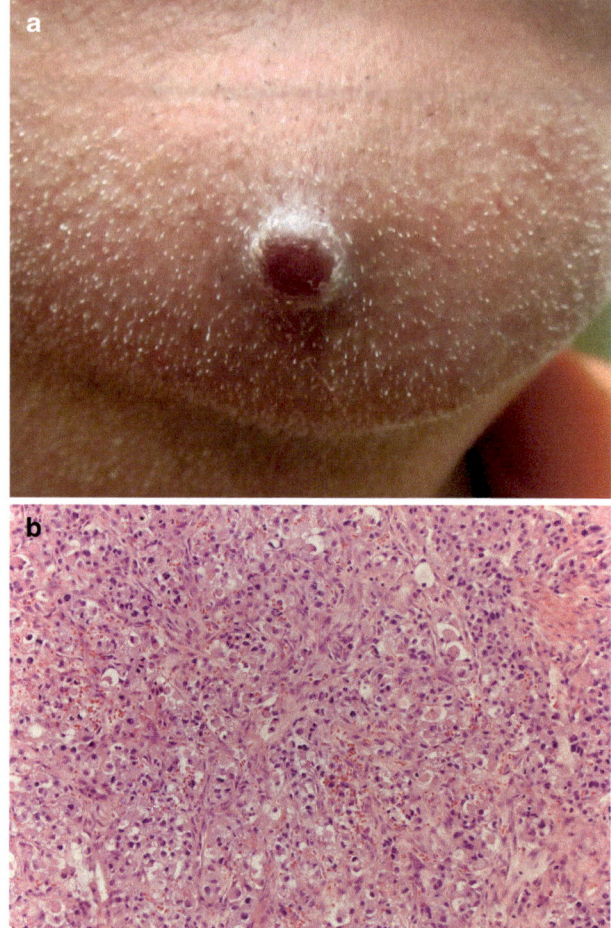

- Nasal tip metastasis
  - Classic for lung cancer, but RCC underlies most cases [82]
  - DDx includes rhinophyma, furuncle, hemangioma, granuloma faciale, and lupus pernio, among others
- Vascular-appearing
  - Vascularity of RCC may be clinically apparent in CMRCs
  - Visible vessels and hemosiderin create violaceous to black nodules
  - Pulsatile lesions prone to bleeding have occurred
  - In nares, has presented with epistaxis
  - DDx includes pyogenic granuloma, Kaposi sarcoma, hemangioendothelio-mas, many other vascular tumors, leukemia cutis, vasculitis, and certain drug eruptions

Histopathology
- Over 2/3 are clear cell type [32] showing cells with abundant, clear cytoplasm arranged in glandular configuration with highly vascular stroma (Fig. 1.9b).
- Other histologies such as papillary carcinoma and Wilms tumor less frequent
- Pathologic DDx includes primary skin and metastatic tumors with clear cell morphology
- In one review, 8 of 15 occult RCCs with CM were initially diagnosed as primary skin neoplasms on histopathology (sweat gland tumors, adenoma sebaceum, xanthoma, and hemangioma) [89]
- The presence of psammoma bodies may indicate a papillary RCC or one of several other characteristic cancers (meningioma, mesothelioma, papillary thyroid carcinoma, uterine papillary serous carcinoma)
- CK7−CK20− (CK7+ with papillary subtype), positive for CD10, PAX-8
- P40 may be helpful in differentiating from some adnexal tumors [90]

Gastric Cancer

Epidemiology [27]
- Fifeetenth most common cancer in U.S. men and women, comprising 1.5% of all new cancer cases in the U.S.
- 2:1 male predominance
- Gastric adenocarcinomas represent 5–7% of CMs; Higher in Japan where tumor incidence is higher [91]
- In Western countries, CMs occur in <1% of upper GI carcinomas [13]
- Other upper GI cancers (esophagus and duodenum) less common

Location
- Abdomen (including SMJN)
- Surgical sites

Clinical Patterns
- Nonspecific nodules
- Zosteriform
- Breast cancer-like (inflammatory, AN, or *en cuirasse*)
- Exophytic growths or ulcers
- DDx includes pyogenic granuloma, epidermal cyst, hemangiomas, neurofibromas, and herpes zoster [91]

Histopathology
- Often poorly-differentiated adenocarcinomas with some percentage of signet-ring cells
- The mucin is colloidal iron positive and hyaluronidase-resistant

Colorectal Cancer

Epidemiology
- Colorectal cancer is the third most common cancer in both men and women [27]
- 0.81–4.4% of colorectal cancers produce CMs [79, 92]
- Metastases from rectal cancers may show greater skin preference than colonic tumor metastases [32]

Location
- Perineum and abdomen (including SMJN)
- Scalp metastasis may be more common with rectal tumors [10]

Clinical Patterns [32, 38, 93]
- Dome-shaped nodules, plaques with or without ulceration, and subcutaneous tumors have mimicked neurofibromas, lipomas, granulomas, and inflammatory alopecias
- Rarely, zosteriform
- Cystic lesions
- EMPD-like metastases-
  - EMPD is an in situ manifestation of apocrine cancer demonstrating well-defined erythematous, scaly patches and plaques clinicopathologically identical to mammary Paget disease, but in apocrine gland containing skin away from breast tissue
    Disproportionately affects postmenopausal Caucasian women
    Majority are *not* associated with an underlying invasive component
  - Many authors also include the similar pagetoid metastases from GI, GU, and breast cancers to anogenital skin (sometimes called secondary EMPD)
    These EMPD-like metastases are uncommon and carry a poor prognosis relative to 'bona fide' primary EMPD without invasion [93]
  - Like Paget disease of the breast, rare cases may exhibit clinically apparent pigment and are to be differentiated from primary melanocytic lesions of the anogenital region

Histopathology
- Cigar-shaped nuclei with well-differentiated, mucin-producing columnar epithelium
- Primary and metastatic tumors often show infiltrating neutrophils and intraluminal debris (dirty necrosis) which may be a helpful diagnostic clue [48]
- Histology closely overlaps with primary mucinous carcinoma of skin which is usually CK20− (in contrast to CRC)
- EMPD-like metastases of colorectal origin are often CK7−CK20+ unlike primary EMPD which is typically CK7+CK20−; additionally, primary EMPD is more often CDX2− and GCDFP-15+; GATA-3 expression also suggests primary EMPD unless the metastasis is urothelial in origin [94, 95]

SCC of the Upper Aerodigestive Tract

Epidemiology
- SCC of the oral cavity and pharynx is the ninth most common NMSC in men
- CMs occur in 1–2% of H&N mucosal SCCs [96]
- Skin involvement is more frequent owing to direct extension and iatrogenic implantation
- Oral cavity is the most common site [1]

Location
- Face and neck
- Neck dissection suture lines, tracheostomy site

Clinical Patterns
- Nodules (often multiple) and ulcers
- Palpable neck disease is common and should be sought on physical examination [96]

Histopathology
- Often metastasize to skin with actinic damage, creating confusion with pcSCC
- pcSCCs more commonly express p63 than metastatic SCC [97]

Urinary Bladder Cancer (Lower GU)

Epidemiology
- Fourth most common malignancy in men, excluding NMSC [31]
- Bladder CMs are more common in men [34], reflecting bladder cancer epidemiology [27]
- Comprise 0.45–1.7% of CMs [1, 26]

Location
- Inguinal and pubic area
- Lower extremities

Clinical Patterns [8, 32]
- Multiple or single, red papulonodules
- Cystic lesions or subcutaneous nodules which are fixed or freely mobile
- Unusual cases mimic dermatitis, cellulitis, or keratoacanthoma

Histopathology
- Urothelial carcinoma most common
- May show squamous differentiation
- Most cases stain for CK7, CK20, Uroplakin III, and GATA3 (also notable expression in breast carcinomas, some NMSCs) [98]

Ovarian Cancer

Epidemiology
- 1.3% of all new cancer cases in the U.S [27].
- Of these patients, up to 5% develop CMs [99]
- CM from other more common gynecological tumors (uterine cancers) are less frequent likely owing to earlier stage at diagnosis [27]

Location
- Abdomen
  - Umbilical metastasis (SMJN) is frequently due to gynecologic and gastrointestinal malignancy; DDx for such lesions includes omphalomesenteric duct polyps, lipomas, endometriosis, fibromas, abscesses, and EICs [100]
- Scars secondary to drains, ports, catheters, laparotomy sites

Clinical Patterns [32, 99]
- Multiple small nodules, sclerodermoid plaques, and herpetiform lesions
- Typically occur late in disease course

Histopathology
- Epithelial cancers most common
- Papillary serous cystadenocarcinoma may be more common cause of CM than mucinous or endometrioid types
- Psammoma bodies provide helpful diagnostic clue for ovarian serous tumor, but are not specific
- Endometrioid metastasis must be distinguished from cutaneous endometriosis, which is typified by endometrial glands and stroma with chronic hemorrhage and usually no cellular atypia
- Serous type is typically CK7+CK20−, CA-125 positive, and PAX8 positive

Prostate Cancer

Epidemiology
- Most common malignancy in males excluding NMSC
- Most cases localized at diagnosis
- Rare cause of CM, with overall rate of 0.09% [32, 101]
- Unclear whether specific ethnic and socioeconomic populations have greater burden of CMs as has been observed for prostate cancer incidence and all-site metastasis

Location
- Perineum
- Abdomen, especially suprapubic skin

- Anterior thighs
- Rarely, chest, scalp, SMJN

Clinical Patterns
- Multiple skin-colored to pink nodules
- Zosteriform metastasis attributed to propensity for perineural invasion

Histopathology
- Adenocarcinoma most common; nucleoli prominent
- Small cell carcinoma and other histologies very rare
- Prostatic adenocarcinoma stains for PSA, prostatic acid phosphatase, and AMACR

Liver Cancer

Epidemiology
- Cancers of the liver and intrahepatic bile ducts are the 13th most common cancer type among men and women, comprising 2.4% of all new cancer cases in the U.S.
- CMs are seen with 0.3% of liver cancers in high-prevalence populations (e.g. Asians, Pacific Islanders, American Indians) and up to 2.7% of cirrhotic HCCs [79, 102]
- Liver cancer metastasis, thus, has fairly low skin preference [10, 79]
- 3:1 male predominance [27], CMHCC also appears to show male predominance [10]

Location [102]
- Head and neck
- Chest and shoulders
- Supradiaphragmatic metastasis may be as common as infradiaphragmatic metastasis, but data is very limited [10]

Clinical Patterns [32, 102]
- Erythematous expanding, firm, nonulcerated nodules which may be umbilicated [103]
- Painless and may bleed easily
- Vascular-appearing simulating pyogenic granuloma, hemangioma, hemangiosarcoma, etc.

Histopathology
- Polygonal neoplastic cells with prominent nucleoli and variably granular to vacuolated cytoplasm arranged in trabeculae, which are invested by fine capillaries

- Nuclear pseudoinclusions, hyaline bodies, and intratumoral bile pigment are diagnostic clues [104]
- CK7-CK20-; positive for AFP, Hep Par 1, and Arginase 1

Thyroid Cancer

Epidemiology
- Most common endocrine malignancy [105]
- Cutaneous metastasis is rare with prevalence reported at <1% for the papillary subtype [106]

Location [32, 105, 106]
- Scalp, face, FNA site, thyroidectomy scar, and distant locations

Clinical Patterns
- Isolated or multiple skin-colored nodules and plaques sometimes associated with pruritus or pain [107]
- Rarely, vascular-appearing, even pulsatile [32]
- DDx often includes primary adnexal tumors [106]

Histopathology
- Papillary carcinomas comprise most cases of thyroid cancer, but there is disagreement in the literature as to the most common subtype underlying cutaneous metastasis
- Papillary and follicular subtypes as a group more commonly produce cutaneous metastases than anaplastic and medullary subtypes [106, 108]
- Psammoma bodies in papillary subtype
- CK7+CK20−; TTF-1 and thyroglobulin positive; calcitonin for medullary subtype

Metastasis of Unknown Primary Site

Epidemiology
- MUPs are diagnosed when a primary site is not elucidated by histopathology and available immunohistochemistry or body imaging
- 2–10% of all cancer diagnoses [60]
- Of these, 7.4% occur in skin [1]

Fig. 1.10 Skin metastasis of unknown primary site presenting as an eroded papule with surrounding intertrigo-like changes in the inframammary fold

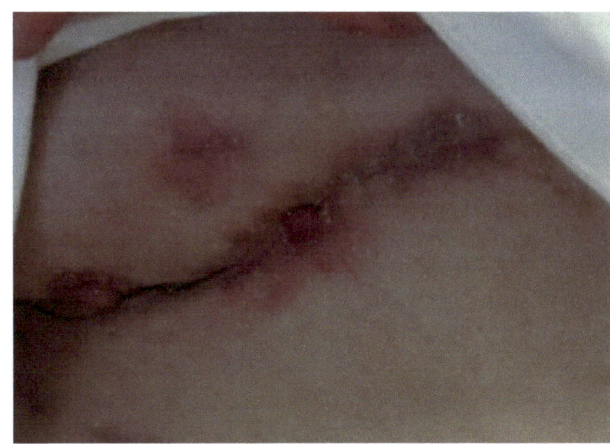

Location
- Head and neck
- Lower extremities
- Abdomen

Clinical Patterns
- Nonspecific nodules, ulcers, indurated plaques (Fig. 1.10)

Histopathology
- Most are well-differentiated or moderately well-differentiated adenocarcinomas [53]

Conclusions

- Identification of cutaneous metastases is important for accurate diagnosis, precise staging, and optimal management of underlying unsuspected new or recurrent malignancy
- The incidence of cutaneous metastasis is 0.1–9% of cancer patients
- While rare, incidence is likely to increase as end-stage cancer survivorship improves
- Personal history of skin *and* visceral cancer should be elicited in all patients with a new, rapidly growing skin nodule
- Advances in molecular diagnostics may improve future identification of unknown primary sites underlying CMs, allowing for directed therapy

Review Questions and Answers

1. A 65-year-old male patient presents with a growing painful nodule and swelling of the right hand, refractory to NSAIDs. Plain films of the hand show soft tissue involvement of the second distal phalanx with lytic bone lesions and periosteal reaction. A CT scan of the chest, abdomen, and pelvis is most likely to show which of the following?
   (a) Mass in the RUL of the lung
   (b) Mass in the LLL of the lung
   (c) Mass in the right renal cortex
   (d) Mass in the distal colon

   Answer: (a) Lung cancer is the most common cause of acrometastasis to the hand, likely owing to arterial access, which bypasses pulmonary and hepatic capillary beds. (Long *et al*) Similarly, the high vascularity of the upper lung lobes makes acrometastasis more likely with tumors originating here.

2. Which of the following scenarios is least likely to represent Paget disease of the breast?
   (a) 50-year-old woman with erythematous, scaly, weeping plaque of the nipple, areola, and breast, showing clinically apparent pigment
   (b) 50-year-old woman with erythematous, scaly, weeping plaque of the nipple, areola, and breast, without palpable breast mass
   (c) 50-year-old woman with erythematous, scaly, weeping plaque of the nipple, areola, and breast, partially responding to topical steroids
   (d) 50-year-old woman with erythematous, weeping plaque of the areola and breast, sparing the nipple

   Answer: (d) Paget disease of the breast may mimic melanoma with clinically apparent pigment, is associated with a palpable breast mass in only half of cases, and may partially respond to topical steroids. However, a breast rash sparing the nipple is unlikely to represent Paget disease.

3. Which of the following warrants biopsy in consideration of cutaneous metastasis?
   (a) New nodule growing in nephrectomy scar
   (b) Two week cellulitis in a patient with history of gastric cancer
   (c) Scarring alopecia in a patient with history of breast cancer
   (d) All of the above

   Answer: (d) All of the above raise suspicion for cutaneous metastasis and warrant consideration of biopsy.

4. Which of the following most accurately describes CM distribution?
   (a) The H&N foster the highest absolute number of CMs, and have the highest Treg percentage

(b) The abdomen fosters the highest absolute number of CMs, and has the highest Treg percentage
(c) The abdomen fosters the highest number of CMs per body surface area, and has the highest Treg percentage
(d) The H&N foster the highest number of CMs per body surface area, and have the highest Treg percentage

Answer: (d) In a study of 1984 cases, the trunk hosted the most cutaneous metastases in absolute terms, however, the head and neck had the highest metastasis burden, then trunk, then upper and lower extremities, when corrected for surface area. The H&N also contained the highest density of Treg cells, suggesting metastatic patterns in the skin are heavily influenced by local immunity and tolerance (Schulman *et al*).

5. An elderly male patient comes to the clinic with a firm nodule on the left lower back. He has a 4-month history of night sweats and unintentional weight loss. Remaining ROS is negative. The differential diagnosis includes cutaneous metastasis. If this suspicion is confirmed, which of the following primary tumors is most likely?
   (a) Colorectal cancer
   (b) Prostatic adenocarcinoma
   (c) Prostatic small cell carcinoma
   (d) Testicular germ cell tumor
   (e) Lung adenocarcinoma

Answer: (e) Little information is given regarding the patient's past medical history, so knowledge of cutaneous metastasis epidemiology is required to identify the most likely primary site. In an elderly man with a lower back nodule and without other localizing symptoms, lung adenocarcinoma is most likely. Lung cancer is the second most common non-melanoma skin cancer in men, is one of the most common causes of cutaneous metastasis in both sexes, and has a positive skin preference (especially for skin of the back). Colorectal cancer is a distinct possibility, but produces fewer CMs and more characteristically affects the abdomen and pelvis.

References

1. Lookingbill DP, Spangler N, Helm KF. Cutaneous metastases in patients with metastatic carcinoma: a retrospective study of 4020 patients. J Am Acad Dermatol. 1993;29(2 Pt 1):228–36.
2. Gates O. Cutaneous metastases of malignant disease. 1937.
3. McWhorter JE, Cloud AW. Malignant tumors and their metastases: a summary of the necropsies on eight hundred sixty-five cases performed at the Bellevue hospital of New York. Ann Surg. 1930;92(3):434–43.
4. Abrams HL, Spiro R, Goldstein N. Metastases in carcinoma; analysis of 1000 autopsied cases. Cancer. 1950;3(1):74–85.

5. Enticknap JB. An analysis of 1,000 cases of cancer with special reference to metastasis. Guys Hosp Rep. 1952;101(4):273–9.
6. Reingold IM. Cutaneous metastases from internal carcinoma. Cancer. 1966;19(2):162–8.
7. Spencer PS, Helm TN. Skin metastases in cancer patients. Cutis. 1987;39(2):119–21.
8. Mc DJ, Heckel NJ, Kretschmer HL. Cutaneous metastases secondary to carcinoma of urinary bladder; report of two cases and review of the literature. Arch Dermatol Syphilol. 1950;61(2):276–84, illust.
9. Hu SC, Chen GS, Lu YW, Wu CS, Lan CC. Cutaneous metastases from different internal malignancies: a clinical and prognostic appraisal. J Eur Acad Dermatol Venereol. 2008;22(6):735–40.
10. Kovacs KA, Kenessey I, Timar J. Skin metastasis of internal cancers: a single institution experience. Pathol Oncol Res. 2013;19(3):515–20.
11. Sariya D, Ruth K, Adams-McDonnell R, Cusack C, Xu X, Elenitsas R, et al. Clinicopathologic correlation of cutaneous metastases: experience from a cancer center. Arch Dermatol. 2007;143(5):613–20.
12. Saeed S, Keehn CA, Morgan MB. Cutaneous metastasis: a clinical, pathological, and immunohistochemical appraisal. J Cutan Pathol. 2004;31(6):419–30.
13. Lookingbill DP, Spangler N, Sexton FM. Skin involvement as the presenting sign of internal carcinoma. A retrospective study of 7316 cancer patients. J Am Acad Dermatol. 1990;22(1):19–26.
14. Lowenstein EJ. Paleodermatoses: lessons learned from mummies. J Am Acad Dermatol. 2004;50(6):919–36.
15. Talmadge JE, Fidler IJ. AACR centennial series: the biology of cancer metastasis: historical perspective. Cancer Res. 2010;70(14):5649–69.
16. Velpeau A. A treatise on the diseases of the breast and mammary region. London: Sydenham Society; 1856. p. 340–1.
17. Schwartz RA. Cutaneous metastatic disease. J Am Acad Dermatol. 1995;33(2 Pt 1):161–82; quiz 83–6.
18. Abu-Hilal M, Newman JS. Sister Mary Joseph and her nodule: historical and clinical perspective. Am J Med Sci. 2009;337(4):271–3.
19. Kovacs KA, Hegedus B, Kenessey I, Timar J. Tumor type-specific and skin region-selective metastasis of human cancers: another example of the "seed and soil" hypothesis. Cancer Metastasis Rev. 2013;32(3–4):493–9.
20. Liu W, Vivian CJ, Brinker AE, Hampton KR, Lianidou E, Welch DR. Microenvironmental influences on metastasis suppressor expression and function during a metastatic cell's journey. Cancer Microenviron. 2014;7(3):117–31.
21. Pan HD, Zhao G, An Q, Xiao G. Pulmonary metastasis in rectal cancer: a retrospective study of clinicopathological characteristics of 404 patients in Chinese cohort. BMJ Open. 2018;8(2):e019614.
22. Hoshino A, Costa-Silva B, Shen TL, Rodrigues G, Hashimoto A, Tesic Mark M, et al. Tumour exosome integrins determine organotropic metastasis. Nature. 2015;527(7578):329–35.
23. Rajotte D, Arap W, Hagedorn M, Koivunen E, Pasqualini R, Ruoslahti E. Molecular heterogeneity of the vascular endothelium revealed by in vivo phage display. J Clin Invest. 1998;102(2):430–7.
24. Ruoslahti E. Vascular zip codes in angiogenesis and metastasis. Biochem Soc Trans. 2004;32(Pt3):397–402.
25. Chambers AF, Naumov GN, Vantyghem SA, Tuck AB. Molecular biology of breast cancer metastasis. Clinical implications of experimental studies on metastatic inefficiency. Breast Cancer Res. 2000;2(6):400–7.
26. Schulman JM, Pauli ML, Neuhaus IM, Sanchez Rodriguez R, Taravati K, Shin US, et al. The distribution of cutaneous metastases correlates with local immunologic milieu. J Am Acad Dermatol. 2016;74(3):470–6.
27. SEER Cancer stat facts [internet]. National Cancer Institute: Surveillance, Epidemiology, and End Results Program 2008–2014. https://seer.cancer.gov/statfacts/.

28. Fernandez-Flores A. Cutaneous metastases: a study of 78 biopsies from 69 patients. Am J Dermatopathol. 2010;32(3):222–39.
29. Nashan D, Muller ML, Braun-Falco M, Reichenberger S, Szeimies RM, Bruckner-Tuderman L. Cutaneous metastases of visceral tumours: a review. J Cancer Res Clin Oncol. 2009;135(1):1–14.
30. SEER*Explorer: an interactive website for SEER cancer statistics [Internet]. Surveillance Research Program, National Cancer Institute. Cited 2017 Apr 14. https://seer.cancer.gov/explorer/.
31. American Cancer Society. Cancer facts and figures 2017. Atlanta: American Cancer Society; 2017.
32. Alcaraz I, Cerroni L, Rutten A, Kutzner H, Requena L. Cutaneous metastases from internal malignancies: a clinicopathologic and immunohistochemical review. Am J Dermatopathol. 2012;34(4):347–93.
33. Wesche WA, Khare VK, Chesney TM, Jenkins JJ. Non-hematopoietic cutaneous metastases in children and adolescents: thirty years experience at St. Jude Children's Research Hospital. J Cutan Pathol. 2000;27(10):485–92.
34. Brownstein MH, Helwig EB. Metastatic tumors of the skin. Cancer. 1972;29(5):1298–307.
35. Karakuzu A, Koc M, Ozdemir S. Multiple cutaneous metastases from male breast carcinoma. J Am Acad Dermatol. 2006;55(6):1101–2.
36. Braverman I. Skin signs of systemic disease. Philadelphia: W. B. Saunders Company; 1970.
37. Lee JH, Lee PK, Ahn ST, Oh DY, Rhie JW, Han KT. Unusally huge metastatic cutaneous renal cell carcinoma to the right buttock: case report and review of the literature. Dermatol Surg. 2006;32(1):159–60.
38. Rendi MH, Dhar AD. Cutaneous metastasis of rectal adenocarcinoma. Dermatol Nurs. 2003;15(2):131–2.
39. Martorell-Calatayud A, Requena C, Diaz-Recuero JL, Haro R, Sarasa JL, Sanmartin O, et al. Mask-like metastasis: report of 2 cases of 4 eyelid metastases and review of the literature. Am J Dermatopathol. 2010;32(1):9–14.
40. Shields JA, Shields CL, Brotman HK, Carvalho C, Perez N, Eagle RC Jr. Cancer metastatic to the orbit: the 2000 Robert M. Curts lecture. Ophthal Plast Reconstr Surg. 2001;17(5):346–54.
41. Rao RC, Elner VM, Demirci H. A red and swollen eyelid. Breast carcinoma metastasis to left lacrimal gland. JAMA Oncol. 2015;1(4):537–8.
42. Fernandez-Anton Martinez MC, Parra-Blanco V, Aviles Izquierdo JA, Suarez Fernandez RM. Cutaneous metastases of internal tumors. Actas Dermo-Sifiliograficas. 2013;104(10):841–53.
43. Mearini L, Colella R, Zucchi A, Nunzi E, Porrozzi C, Porena M. A review of penile metastasis. Oncol Rev. 2012;6(1):e10.
44. Long LS, Brickner L, Helfend L, Wong T, Kubota D. Lung cancer presenting as acrometastasis to the finger: a case report. Case Rep Med. 2010;2010:234289.
45. Stomeo D, Tulli A, Ziranu A, Perisano C, Maccauro VS. Acrometastasis: a literature review. Eur Rev Med Pharmacol Sci. 2015;19(15):2906–15.
46. Chernoff KA, Marghoob AA, Lacouture ME, Deng L, Busam KJ, Myskowski PL. Dermoscopic findings in cutaneous metastases. JAMA Dermatol. 2014;150(4):429–33.
47. Richtig E, Gerger A, El-Shabrawi-Caelen L, Szkandera J, Hofmann-Wellenhof R. Reflectance confocal microscopy in early diagnosis of cutaneous metastases of breast cancer. Gynecol Oncol. 2009;115(3):510–1.
48. Leonard N. Cutaneous metastases: where do they come from and what can they mimic? Curr Diagn Pathol. 2007;13:320–30.
49. Bender RA, Erlander MG. Molecular classification of unknown primary cancer. Semin Oncol. 2009;36(1):38–43.
50. Ma XJ, Patel R, Wang X, Salunga R, Murage J, Desai R, et al. Molecular classification of human cancers using a 92-gene real-time quantitative polymerase chain reaction assay. Arch Pathol Lab Med. 2006;130(4):465–73.
51. Dabbs DJ. Diagnostic immunohistochemistry: theranostic and genomic applications. 5th ed. Philadelphia: Elsevier; 2018. 47 p.

52. Fizazi K, Greco FA, Pavlidis N, Daugaard G, Oien K, Pentheroudakis G. Cancers of unknown primary site: ESMO clinical practice guidelines for diagnosis, treatment and follow-up. Ann Oncol. 2015;26(Suppl 5):v133–8.
53. Carroll MC, Fleming M, Chitambar CR, Neuburg M. Diagnosis, workup, and prognosis of cutaneous metastases of unknown primary origin. Dermatol Surg. 2002;28(6):533–5.
54. Pavlidis N, Khaled H, Gaafar R. A mini review on cancer of unknown primary site: a clinical puzzle for the oncologists. J Adv Res. 2015;6(3):375–82.
55. Wulaningsih WHL, Garmo H, et al. Serum lactate dehydrogenase and survival following cancer diagnosis. Br J Cancer. 2013;113(9):1389–96.
56. Gold BCM, Furtado LV, et al. Do circulating tumor cells, exosomes, and circulating tumor nucleic acids have clinical utility? A report of the association for molecular pathology. J Mol Diagn. 2015;17(3):209–24.
57. Pavlidis N, Pentheroudakis G. Cancer of unknown primary site. Lancet. 2012;379(9824):1428–35.
58. Lauren CT, Antonov NK, McGee JS, de Vinck DC, Hibshoosh H, Grossman ME. Carcinoma en cuirasse caused by pleomorphic lobular carcinoma of the breast in a man. JAAD Case Rep. 2016;2(4):317–9.
59. Wong CY, Helm MA, Kalb RE, Helm TN, Zeitouni NC. The presentation, pathology, and current management strategies of cutaneous metastasis. N Am J Med Sci. 2013;5(9):499–504.
60. Greco FA, Oien K, Erlander M, Osborne R, Varadhachary G, Bridgewater J, et al. Cancer of unknown primary: progress in the search for improved and rapid diagnosis leading toward superior patient outcomes. Ann Oncol. 2012;23(2):298–304.
61. Davis KS, Byrd JK, Mehta V, Chiosea SI, Kim S, Ferris RL, et al. Occult primary head and neck squamous cell carcinoma: utility of discovering primary lesions. Otolaryngol Head Neck Surg. 2014;151(2):272–8.
62. Bae JM, Choi YY, Kim DS, Lee JH, Jang HS, Lee JH, et al. Metastatic melanomas of unknown primary show better prognosis than those of known primary: a systematic review and meta-analysis of observational studies. J Am Acad Dermatol. 2015;72(1):59–70.
63. American Cancer Society. Cancer facts and figures 2016. Atlanta: American Cancer Society; 2016.
64. Apalla Z, Nashan D, Weller RB, Castellsague X. Skin cancer: epidemiology, disease burden, pathophysiology, diagnosis, and therapeutic approaches. Dermatol Ther. 2017;7(Suppl 1):5–19.
65. Plaza JA, Torres-Cabala C, Evans H, Diwan HA, Suster S, Prieto VG. Cutaneous metastases of malignant melanoma: a clinicopathologic study of 192 cases with emphasis on the morphologic spectrum. Am J Dermatopathol. 2010;32(2):129–36.
66. Rubegni P, Lamberti A, Mandato F, Perotti R, Fimiani M. Dermoscopic patterns of cutaneous melanoma metastases. Int J Dermatol. 2014;53(4):404–12.
67. Sebaratnam DF, Venugopal SS, Frew JW, McMillan JR, Finkelstein ER, Martin LK, et al. Diffuse melanosis cutis: a systematic review of the literature. J Am Acad Dermatol. 2013;68(3):482–8.
68. Bendell JC, Domchek SM, Burstein HJ, Harris L, Younger J, Kuter I, et al. Central nervous system metastases in women who receive trastuzumab-based therapy for metastatic breast carcinoma. Cancer. 2003;97(12):2972–7.
69. Mordenti C. Cutaneous metastatic breast carcinoma: a study of 164 patients. Acta Dermatovenerologica. 2000;9(4):143–8.
70. Damstra RJ, Jagtman EA, Steijlen PM. Cancer-related secondary lymphoedema due to cutaneous lymphangitis carcinomatosa: clinical presentations and review of literature. Eur J Cancer Care. 2010;19(5):669–75.
71. Farahat A, Mohamed S, Vijay A, Magdy N, Elaffandi A. Invasive duct carcinoma of the forearm: a rare case of distant, isolated 'carcinoma en cuirasse'. J Surg Case Rep. 2015;2015(6):rjv062.

72. Conner KB, Cohen PR. Cutaneous metastasis of breast carcinoma presenting as alopecia neoplastica. South Med J. 2009;102(4):385–9.
73. Haas N, Hauptmann S. Alopecia neoplastica due to metastatic breast carcinoma vs. extramammary Paget's disease: mimicry in epidermotropic carcinoma. J Eur Acad Dermatol Venereol. 2004;18(6):708–10.
74. Gui GP, Behranwala KA, Abdullah N, Seet J, Osin P, Nerurkar A, et al. The inframammary fold: contents, clinical significance and implications for immediate breast reconstruction. Br J Plast Surg. 2004;57(2):146–9.
75. Waisman M. Carcinoma of the inframammary crease. Arch Dermatol. 1978;114(10):1520–1.
76. Jakobiec FA, Stagner AM, Homer N, Yoon MK. Periocular breast carcinoma metastases: predominant origin from the lobular variant. Ophthalmic Plast Reconstr Surg. 2016;33(5):361–6.
77. Sandoval-Leon AC, Drews-Elger K, Gomez-Fernandez CR, Yepes MM, Lippman ME. Paget's disease of the nipple. Breast Cancer Res Treat. 2013;141(1):1–12.
78. Masters RK, Robertson JF, Blamey RW. Healed Paget's disease of the nipple. Lancet. 1993;341(8839):253.
79. Hu SC, Chen GS, Wu CS, Chai CY, Chen WT, Lan CC. Rates of cutaneous metastases from different internal malignancies: experience from a Taiwanese medical center. J Am Acad Dermatol. 2009;60(3):379–87.
80. Marcoval J, Penin RM, Llatjos R, Martinez-Ballarin I. Cutaneous metastasis from lung cancer: retrospective analysis of 30 patients. Australas J Dermatol. 2012;53(4):288–90.
81. Pajaziti L. Skin metastases from lung cancer: a case report. 2015.
82. Chun SM, Kim YC, Lee JB, Kim SJ, Lee SC, Won YH, et al. Nasal tip cutaneous metastases secondary to lung carcinoma: three case reports and a review of the literature. Acta Derm Venereol. 2013;93(5):569–72.
83. Molina Garrido MJ, Guillen Ponce C, Soto Martinez JL, Martinez YSC, Carrato MA. Cutaneous metastases of lung cancer. Clin Transl Oncol. 2006;8(5):330–3.
84. Cohen PR. Metastatic tumors to the nail unit: subungual metastases. Dermatol Surg. 2001;27(3):280–93.
85. Kato N, Aoyagi S, Sugawara H, Mayuzumi M. Zosteriform and epidermotropic metastatic primary cutaneous squamous cell carcinoma. Am J Dermatopathol. 2001;23(3):216–20.
86. Niiyama S, Satoh K, Kaneko S, Aiba S, Takahashi M, Mukai H. Zosteriform skin involvement of nodal T-cell lymphoma: a review of the published work of cutaneous malignancies mimicking herpes zoster. J Dermatol. 2007;34(1):68–73.
87. Chiu CS, Lin CY, Kuo TT, Kuan YZ, Chen MJ, Ho HC, et al. Malignant cutaneous tumors of the scalp: a study of demographic characteristics and histologic distributions of 398 Taiwanese patients. J Am Acad Dermatol. 2007;56(3):448–52.
88. Mollet TW, Garcia CA, Koester G. Skin metastases from lung cancer. Dermatol Online J. 2009;15(5):1.
89. Connor DH, Taylor HB, Helwig EB. Cutaneous metastasis of renal cell carcinoma. Arch Pathol. 1963;76:339–46.
90. Lee JJ, Mochel MC, Piris A, Boussahmain C, Mahalingam M, Hoang MP. p40 exhibits better specificity than p63 in distinguishing primary skin adnexal carcinomas from cutaneous metastases. Hum Pathol. 2014;45(5):1078–83.
91. Avgerinou G, Flessas I, Hatziolou E, Zografos G, Nitsios I, Zagouri F, et al. Cutaneous metastasis of signet-ring gastric adenocarcinoma to the breast with unusual clinicopathological features. Anticancer Res. 2011;31(6):2373–8.
92. Krathen RA, Orengo IF, Rosen T. Cutaneous metastasis: a meta-analysis of data. South Med J. 2003;96(2):164–7.
93. Karam A, Dorigo O. Increased risk and pattern of secondary malignancies in patients with invasive extramammary Paget disease. Br J Dermatol. 2014;170(3):661–71.
94. Shu B, Shen XX, Chen P, Fang XZ, Guo YL, Kong YY. Primary invasive extramammary Paget disease on penoscrotum: a clinicopathological analysis of 41 cases. Hum Pathol. 2016;47(1):70–7.

95. Zhao MZL, Sun L, et al. GATA3 is a sensitive marker for primary genital extramammary paget disease: an immunohistochemical study of 72 cases with comparison to gross cystic disease fluid protein 15. Diagn Pathol. 2017;12(1):51.
96. Pitman KT, Johnson JT. Skin metastases from head and neck squamous cell carcinoma: incidence and impact. Head Neck. 1999;21(6):560–5.
97. Kanitakis J, Chouvet B. Expression of p63 in cutaneous metastases. Am J Clin Pathol. 2007;128(5):753–8.
98. Amin MB, Trpkov K, Lopez-Beltran A, Grignon D. Best practices recommendations in the application of immunohistochemistry in the bladder lesions: report from the International Society of Urologic Pathology consensus conference. Am J Surg Pathol. 2014;38(8):e20–34.
99. Cormio G, Capotorto M, Di Vagno G, Cazzolla A, Carriero C, Selvaggi L. Skin metastases in ovarian carcinoma: a report of nine cases and a review of the literature. Gynecol Oncol. 2003;90(3):682–5.
100. Yan LSS, Bitterman P, et al. Umbilical lesions: clinicopathologic features of 99 tumors. Int J Surg Pathol. 2018;26(5):417–22.
101. Brown GT, Patel V, Lee CC. Cutaneous metastasis of prostate cancer: a case report and review of the literature with bioinformatics analysis of multiple healthcare delivery networks. J Cutan Pathol. 2014;41(6):524–8.
102. Reuben S, Owen D, Lee P, Weiss A. Hepatocellular carcinoma with cutaneous metastases. Can J Gastroenterol. 2009;23(1):23–5.
103. Nggada HA, Ajayi NA. Cutaneous metastasis from hepatocellular carcinoma: a rare presentation and review of the literature. Afr J Med Med Sci. 2006;35(2):181–2.
104. Isa NM, Bong JJ, Ghani FA, Rose IM, Husain S, Azrif M. Cutaneous metastasis of hepatocellular carcinoma diagnosed by fine needle aspiration cytology and Hep Par 1 immunopositivity. Diagn Cytopathol. 2012;40(11):1010–4.
105. Alwaheeb S, Ghazarian D, Boerner SL, Asa SL. Cutaneous manifestations of thyroid cancer: a report of four cases and review of the literature. J Clin Pathol. 2004;57(4):435–8.
106. Reusser NM, Holcomb M, Krishnan B, Rosen T, Orengo IF. Cutaneous metastasis of papillary thyroid carcinoma to the neck: a case report and review of the literature. Dermatol Online J. 2015;21(2):13030/qt78v2d22d.
107. Nashed C, Sakpal SV, Cherneykin S, Chamberlain RS. Medullary thyroid carcinoma metastatic to skin. J Cutan Pathol. 2010;37(12):1237–40.
108. Dahl PR, Brodland DG, Goellner JR, Hay ID. Thyroid carcinoma metastatic to the skin: a cutaneous manifestation of a widely disseminated malignancy. J Am Acad Dermatol. 1997;36(4):531–7.

Paraneoplastic Dermatoses and Related Conditions

2

Martin Dittmer and Vincent Liu

Abbreviations

AN	Acanthosis nigricans
ANM	Acanthosis nigricans maligna
BMZ	Basement membrane zone
DIF	Direct immunofluorescence
EAC	Erythema annulare centrifugum
EBRT	Electron beam radiation therapy
ECP	Extracorporeal photopheresis
EGFR	Epidermal growth factor receptor
EGR	Erythema gyratum repens
EM	Erythema multiforme
GVHD	Graft versus host disease
HLA	Hypertrichosis lanuginosa acquisita
IIF	Indirect immunofluorescence
IVIg	Intravenous immune globulin
LM	Lichen myxedematosus
MDS	Myelodysplastic syndrome
MEN1	Multiple endocrine neoplasia type 1

M. Dittmer (✉)
SSM Health Saint Louis University Hospital—South, St. Louis, MO, USA

V. Liu
Department of Dermatology, University of Iowa Health Care, Iowa, IA, USA
e-mail: Vincent-liu@uiowa.edu

MGUS	Monoclonal gammopathy of undetermined significance
MM	Multiple myeloma
MTX	Methotrexate
NET	Neuroendocrine tumor
NHL	Non-hodgkin lymphoma
NME	Necrolytic migratory erythema
NSF	Nephrogenic systemic fibrosis
NXG	Necrobiotic xanthogranuloma
PCT	Porphyria cutanea tarda
PG	Pyoderma gangrenosum
pHTN	Pulmonary hypertension
PND	Paraneoplastic dermatosis
PRP	Pityriasis rubra pilaris
SCC	Squamous cell carcinoma
SLE	Systemic lupus erythematosus
SPEP	Serum protein electrophoresis
TEN	Toxic epidermal necrolysis
UPEP	Urine protein electrophoresis

Learning Objectives

1. To become familiar with the key clinical and pathologic features of the most common paraneoplastic dermatoses.
2. To identify the relationship between internal malignancies and cutaneous manifestations in paraneoplastic dermatoses.
3. To recognize essential features of uncommon paraneoplastic dermatoses as well as atypical presentations of commonly benign dermatoses which may offer the key clue to an occult malignancy.
4. To stratify eruptions as classically obligate or facultative paraneoplastic dermatoses in order to direct appropriate investigation into possible underlying tumor.
5. To describe predictors of prognosis and current management options including modern therapeutic modalities for paraneoplastic dermatoses and their underlying tumors.

Introduction

- Definition: PNDs are cancer related skin lesions which occur at a distance from direct tumor contact (in the absence of cutaneous metastasis, infection due to immunosuppression, mass effect, or therapy-related effect) [1]
- Significance: Timely clinical recognition of paraneoplastic dermatoses can afford opportunities toward diagnosis of the underlying malignancy, expedited treatment in early stages, and monitoring for disease recurrence
- Frequency: Collectively, paraneoplastic syndromes occur in 7–15% of cancer patients [2, 3]
  - Skin is the second most commonly affected system, following endocrine [4]
  - Still, paraneoplastic dermatoses are uncommon

- Timing: By definition, PNDs occur after their associated cancers are established (occult or not); however, their diagnosis may precede or follow diagnosis of the malignancy [5]
  - A critical tumor mass or mediator-elaborating subclone may be requisite before skin changes become apparent [5]
  - Skin findings often garner clinical attention before malignancy is suspected, for example:
      In ANM, 58% present prior to cancer diagnosis [2]
      Bazex syndrome precedes a cancer diagnosis by an average of 2–6 months, but latency periods of up to 6 years are reported [6]
  - The tumor may be localized or metastatic at the time of PND diagnosis
- Presentation: Paraneoplastic syndromes in skin have been categorized as papulosquamous disorders, reactive erythemas, neutrophilic dermatoses, dermal proliferative disorders, deposition disorders, interface dermatitides, among other dermatoses [7]
  - Very rarely, two distinct PNDs present in a single patient (acquired ichthyosis and DM or pityriasis rotunda or EGR and Sweet's syndrome and relapsing polychondritis), suggesting shared pathomechanisms

- Approach: As most PNDs have benign counterparts, clinical decisions to pursue an underlying malignancy rest on a practical knowledge of the types and strengths of paraneoplastic associations in the skin

History [4, 5, 8]

- 1868—Hebra describes hyperpigmentation related to malignancy
- 1890—French physician M. Auche details first paraneoplastic neurological syndrome
- 1893—Darier first describes ANM
- 1965—Bazex publishes on paraneoplastic hyperkeratosis of the extremities
- 1976—Curth proposes her postulates which suggest a causal relationship between a neoplasm and skin disorder (excluding malignancy-associated genodermatoses)
  - The cancer and dermatosis start at the same time
     Reformed by McLean—"the dermatosis may or may not precede the diagnosis of the tumor"
  - They follow a parallel course
     Remit and recur with the tumor
  - The dermatosis associates with specific tumors*
  - The dermatosis is usually not common*
  - The two conditions are associated in a high percentage*
- When these criteria are fulfilled, a causal association is highly probable, but "[statistical] proof may be lacking" according to McLean et al
- Several are debated as nonessential by some authors and their absence does not rule out true PND*
- Well-designed case-control or population-based cohort studies are needed to solidify the standing of a PND
  - Some skin disorders historically associated with internal malignancy (e.g. Bowen disease) have fallen short of statistical significance under such scrutiny [9]

Pathophysiology

- Mechanisms underpinning PNDs are incompletely described, although many theories exist, ranging from immune mimicry to mediator production and consumption (Fig. 2.1)
- Paraneoplastic dermatoses may not share the same mechanism as their common benign counterparts
  - Malignancy-associated acquired ichthyosis vs ichthyosis vulgaris: epidermal lipid synthesis differs
  - HLA has not shown consistent hormone elevation
  - Some immune targets in PNP are not seen in PV
- As mentioned previously, two distinct PNDs can present in the same patient or one PND may have features of another (10% of EGR patients have palmoplantar keratoderma similar to Bazex syndrome), pointing to potential mechanistic overlap [10]

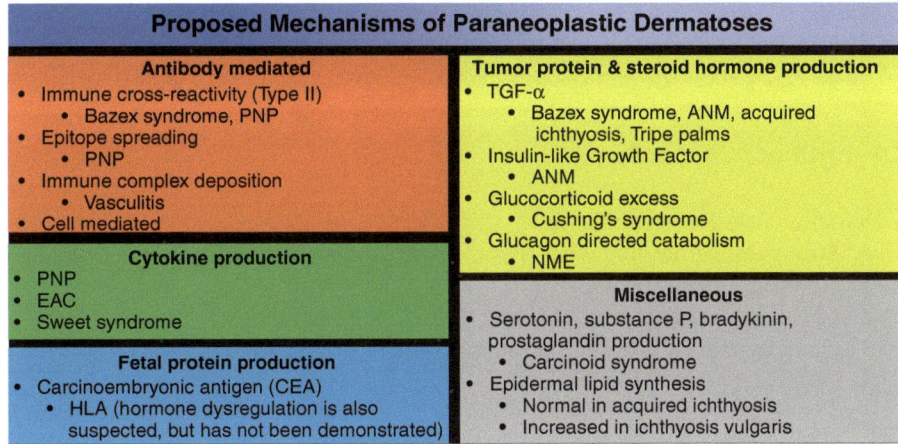

Fig. 2.1 Proposed mechanisms of paraneoplastic dermatoses

Clinical Themes

Epidemiology

- Incidence of most PNDs is not well described in the literature
- For most of the entities discussed, case series and small clinicopathologic or literature reviews provide the best available estimates of PND incidence within a population demonstrating the cutaneous phenotype [11–13]
  - Malignancy-associated dermatoses are probably overrepresented in the literature
  - True incidence also affected by paucity of epidemiological studies of corresponding benign dermatosis
- Paraneoplastic syndromes, in general, expected to become more common as cancer patients live longer [3]
- Ratio of PND to benign counterpart highly variable
  - Very high percentage of benign dermatosis makes the PND association difficult to establish as more than chance occurrence (e.g. seborrheic keratoses of the Leser-Trelat sign)
  - Very high percentage of malignant dermatosis such that non-PND cases were met with skepticism (EGR-like eruptions)

Age

- As cancer is primarily a disease of middle and old age, so too are PNDs
- Excluding genodermatoses, reports in children are quite rare
  - Show unique primary tumors/proliferations (Castleman's syndrome in PNP [14], Wilms tumor in malignant AN [15])

Gender

- Some PNDs show sex predilection, discussed further under individual dermatoses

Strength of Association

- Variably classified qualitatively
  1. Strong, moderate, and weak association (excessively arbitrary, anecdotal, and not clinically actionable)
  2. Classical vs nonclassical (similar to 3, used in neurology literature)
  3. Obligate vs facultative (proposed by others and preferred in this text) (Fig. 2.2) (Table 2.1)

     Obligate PND is defined herein as those dermatoses which associate with a neoplasm in >1/2 of cases
- Several paraneoplastic neurological syndromes are known to produce characteristic onconeural antibodies which are used to support their diagnosis
  - Further study of the mechanisms of PNDs will likely yield similar assays in the future, functioning as an additional sort of Curth postulate (as is already in use for PNP)
- In the case of classic paraneoplastic neurological syndromes, if a tumor is found that is not that typically appearing as part of the syndrome, further search for a second more characteristic tumor is recommended by the Paraneoplastic Neurological Syndrome Euronetwork [16]

Temporal Association

- PND may present prior to, concurrent with, or following diagnosis of cancer
- Abrupt onset of extensive cutaneous disease or an atypical distribution should raise suspicion for a PND
- Timing of PND in some cases depends on the type of underlying tumor
  - For example, paraneoplastic PG tends to present after MDS but before MGUS [17]

Diagnostic Considerations/Workup

- Review of systems including constitutional symptoms and investigation for cancer risk factors (smoking, smokeless tobacco, alcohol, radiation, family history, etc.) mandatory in all patients
- Comprehensive physical examination including examination of the oral cavity, thyroid, liver, spleen, and lymph nodes
- Benign causes of the dermatosis in question in general should be excluded first
  - In HLA, porphyria, thyroid disease, and medications such as penicillamine, cyclosporine, and phenytoin should be considered before a malignancy workup [18]

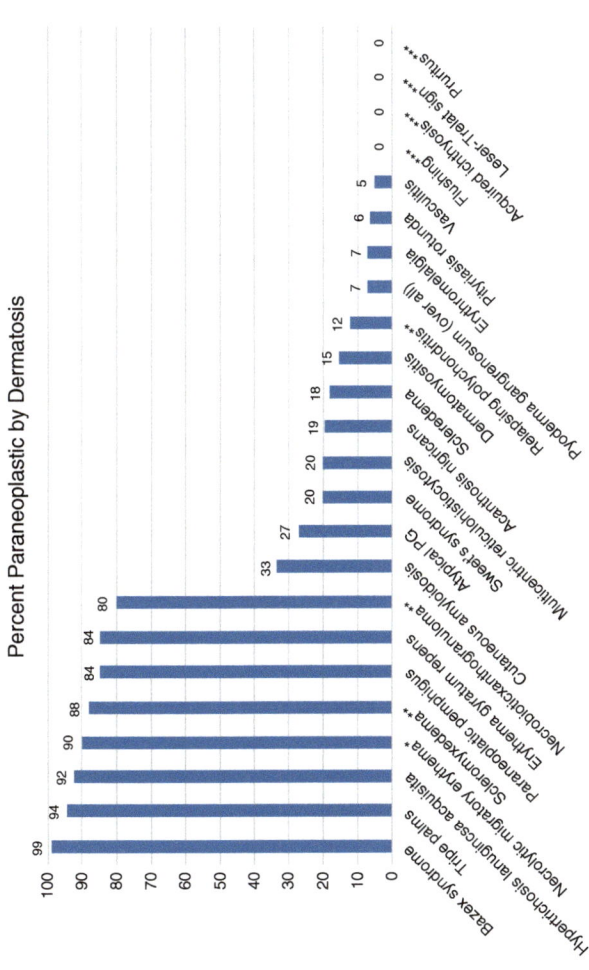

Fig. 2.2 Paraneoplastic association frequency per dermatosis. *At least 24 pseudoglucagonoma cases with ~300 PND cases, **includes MDS or MGUS, ***not well established

Table 2.1 Obligate paraneoplastic dermatoses with associated neoplasms

Bazex syndrome	SCC of aerodigestive tract, esophagus, lung; cervical nodal metastasis of unknown primary site
Tripe palms	Lung cancer (especially if no concurrent AN), GI malignancies
HLA	Colorectal, lung, breast cancer in women; lung followed by colorectal cancer in men
NME	Pancreatic alpha-cell tumor
Scleromyxedema	IgG lambda gammopathy, MM, Waldenstrom macrogloblinemia, Hodgkin lymphoma, NHL
PNP	NHL, CLL, Castleman's disease
EGR	Lung cancer most common, followed by esophageal and breast cancer
NXG	MM, CLL, Hodgkin lymphoma, NHL

- CBC with differential and serum chemistries reasonable if PND suspected
  - Long-term screening with exam, CBC, and other bloodwork suggested for some PNDs citing potential for seemingly delayed-onset malignancy (e.g. Sweet syndrome, NXG, scleromyxedema, Bazex syndrome)
- For scleromyxedema, PG, scleredema, PNP, Sweet syndrome, amyloidosis, relapsing polychondritis, recalcitrant or unusual vasculitis, pityriasis rotunda, consider SPEP with immunofixation, UPEP, quantitative immunoglobulins
- For obligate PNDs with very characteristic tumors (e.g. PNP), CT of the chest, abdomen, and pelvis are recommended
- Further studies such as endoscopy, somatostatin receptor scintigraphy, etc. directed by symptoms and presenting skin findings

Specific Paraneoplastic Dermatoses

Obligate (Strong Association) PNDs

- Thorough cancer workup mandatory in all patients (Table 2.1)

Tripe Palms (aka Acquired Pachydermatoglyphia and Acanthosis Palmaris)

Epidemiology
- Malignancy identified in 94% of cases in a series of 90 patients [11]

Clinical Features [14, 19]
- Wrinkled palmar skin resembling tripe (muscular wall of bovine foregut)
- Soles of feet sometimes involved
- Patients can have concurrent AN (77% of cases) and, less frequently, the sign of Leser-Trelat [19]
  - Circulating cytokines such as EGF-α and TGF-α implicated in both ANM and Tripe palms [4]
  - Some consider it a palmar form of AN
- *Associated tumor:* GI adenocarcinoma, **lung cancer (especially if no AN present)** [14, 20]

- Precedes cancer diagnosis in nearly half of cases [19]
- Ddx includes acromegaly, hypertrophic pulmonary osteoarthropathy, thyroid acropathy, and palmoplantar keratodermas (tylosis, Howel-Evans syndrome, etc.) [21]

Histopathology
- Acanthosis, papillomatosis, sometimes, mucin deposition

Management and Prognosis
- 1/3 remit with treatment of the underlying tumor

Bazex Syndrome (aka Acrokeratosis Paraneoplastica)

Epidemiology
- Caucasian men over 40 years of age, women rarely affected [6, 22]
- >110 reported cases in (mostly French) literature [6]
- Nearly all associated with malignancy
  - An acrokeratosis paraneoplastica-like eruption has been reported in SLE without malignancy [23]

Clinical Features [8, 24]
- Symmetric psoriasiform, hyperkeratotic, erythematous lesions which progress through three stages
  - first stage: asymptomatic and ill-defined acral macules with minimal, adherent scale and occasional crust of fingers, toes, and nose; paronychia
  - second stage: fingertip and toe lesions spread over palmoplantar surfaces forming keratoderma with **sparing of the central palm and foot arch** [1]; nasal lesions may spread to form violaceous, scaly plaques on cheeks; entire external ears may be involved; variable nail dystrophy (horizontal/longitudinal ridging, discoloration, subungual debris, thickening) [18]; underlying tumor usually becomes symptomatic
  - third stage: extension of psoriasiform plaques to trunk, extensor surfaces, dorsal hands/feet, scalp; some lesions with vesiculation/bullae
- In a large series, 79% had ear involvement and 63% had lesions on the nose, whereas knee and elbow sites were affected in <25% of cases and occurred in later stages of disease [22, 24]
- Dermatosis precedes the tumor diagnosis by 2–11 months on average, depending on the report [6, 22]
- Pruritus is not common
- *Associated tumor:* SCC of aerodigestive tract most common; lung and esophageal SCC; cervical nodal metastases of unknown primary site may occur

Pearl: Involvement of nasal tip and ear helices favors Bazex syndrome over psoriasis; both show Auspitz sign, but borders are classically ill-defined in the former [8].

Histopathology [6, 8, 24]

- Pathologic appearance likely evolves with the disease and may be nonspecific in early stages
- Ortho- and parakeratosis
- Psoriasiform epidermal hyperplasia
- Dyskeratosis, vacuolar degeneration, pigment incontinence
- Bullae, when present, are usually subepidermal and PCT may be suggested by the pathologist
- Rare basal layer immunoglobulin deposition on DIF suggests possible immune cross-reactivity between squamous tumor antigens and components of the basal layer [25]
- Tumor-elaborated TGF-α may underlie hyperkeratotic skin lesions, but this remains speculative

Management and Prognosis [24]

- Usually remits and recurs with the tumor
- Dystrophic nails may persist despite treatment
- Retinoids and PUVA have been employed for cutaneous disease

Paraneoplastic Pemphigus

Epidemiology

- Men between ages 45 and 70 [26]
- In a literature review of 163 cases, 84% were associated with hematopoietic neoplasms and another 16% of patients had solid tumors [27]

Clinical Features [27]

- Clinical manifestations are protean: Lichen planus-like, GVHD-like, pemphigus vulgaris-like, bullous pemphigoid-like, TEN-like, EM-like [28]
  - Multiple forms can occur in synchronous or metachronous fashion
- Consistent clinical features include:
  - **Painful stomatitis**
      Oral ulcer harbinger in 45% of cases [27]
  - Lesions of nasopharynx, tonsils, anogenital mucosa common
  - Conjunctival lesions resembling cicatricial pemphigoid in 2/3 [27]
  - Nearly all develop skin findings (especially of trunk, proximal extremities) (Fig. 2.3a).
- Diagnostic criteria were proposed by Anhalt *et al* in 1990 and later revised by Camisa and Helm (Table 2.2) [24, 29].
- Notably, up to 1/3 of cases do not demonstrate malignancy at the time of diagnosis [26]
- *Associated tumor:* Hematopoietic neoplasms including NHL, CLL, Castleman's disease [28]
- 1/2 present after tumor diagnosis [30]

Fig. 2.3 Paraneoplastic pemphigus. (**a**) Clinical image showing erosive stomatitis. (**b**) Histopathology showing suprabasilar acantholysis

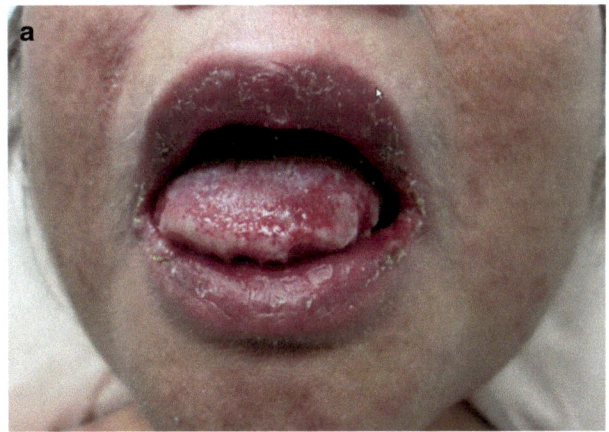

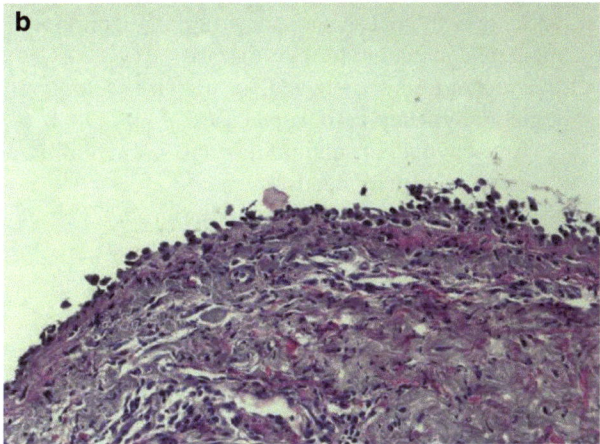

Table 2.2 Revised criteria for PNP diagnosis (adapted from Camisa and Helm)

Major	Polymorphous mucocutaneous eruption
	Concurrent neoplasm
	Specific immunoprecipitation pattern on sera
Minor	Acantholysis
	Intercellular and BMZ staining with IgG and C3 on DIF
	Staining of rat urothelium on IIF (desmoplakin antibodies)
3 major or 2 major and 2 minor criteria are required for diagnosis	

Histopathology [26]

- Lichenoid interface dermatitis with dyskeratosis ± blister cavity
- Suprabasilar acantholysis (Fig. 2.3b)
- Several autoantibodies have been described targeting plakins (envoplakin, periplakin, desmoplakin, BP230, and plectin), desmogleins, desmocollins, BP180, and α-2 macroglobulin-like protein (anti-A2ML1)
  - Accounting for variable IIF and DIF studies

- Interestingly, isolated low-positive titers of anti-A2ML1 also occur in TEN [31]
- Multiple immunoglobulins (IgG, IgA, IgM, and C3) with granular or linear BMZ and/or intercellular staining on DIF
- No single mechanistic theory perfectly unites all the clinicopathologic findings
  - May relate to molecular mimicry by tumor antigens, cytokine dysregulation with desmolgein autoantibodies followed by intracellular plakin targeting, and/or an initial lichenoid reaction with epitope spreading [32]
- An eruption with histopathology and DIF consistent with PNP has been reported following pembrolizumab (the patient had SCC of the tongue and notably lacked the characteristic stomatitis) [26]
  - Desmoglein 3 autoantibodies were not tested

Management and Prognosis [24, 28]
- Patients with NHL have poor prognosis, commonly develop pulmonary involvement, and succumb to their disease
- In a recent multivariate analysis for risk factors of death and survival in hematologic malignancy-associated cases:
  - Anti-envoplakin antibiodies and bronchiolitis obliterans showed significant association with death
  - TEN-like pattern, BP-like pattern (only 3 cases), and bronchiolitis obliterans were associated with decreased survival

Erythema Gyratum Repens

Epidemiology [24, 33]
- Caucasian
- 2:1 male to female predominance
- Average onset in sixth decade of life
- In a literature review of 49 cases, 84% were associated with malignancy

Clinical Features [24, 33]
- Described in 1953 by Gammel as resembling "knotty cypress wood grain"
- Arcuate, serpiginous, and whorled concentric bands of erythema
- Usually macular, rarely palpable
- Trailing scale
- Rapid growth up to 1 cm per day is characteristic
  - *repens* comes from the Latin to crawl or creep
- Pruritus is common, often severe
- Diagnosis of rash precedes tumor by 9 months on average
- Cases of EGR-like eruptions without malignancy have been increasingly reported
  - Many cases appear in the wake of other cutaneous dermatoses (PRP, psoriasis, CREST syndrome, BP, etc.) representing morphologic transformation or an independent eruption [34] (Fig. 2.4)

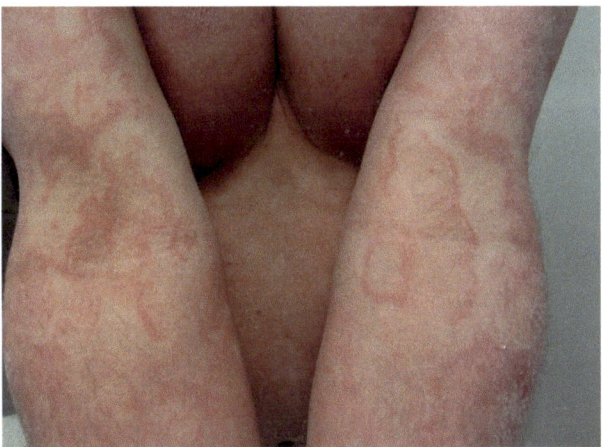

Fig. 2.4 Erythema gyratum repens-like eruption presenting with serpiginous and polycyclic erythematous patches on the extremities and trunk. Note the trailing scale [34]

- A portion of these may have represented other figurate erythemas
- *Associated tumor:* Lung cancer, followed by esophageal and breast cancer

> ***Pearl:*** Within the ddx, superficial EAC can also be pruritic but is usually localized and slow-growing [24].

Histopathology [33]
- Hyperkeratosis, parakeratosis
- Variable spongiosis
- Superficial perivascular lymphohistiocytic infiltrate with occasional eosinophils

Management and Prognosis
- EGR usually remits with anti-tumor therapy [33]
- Treatment with oral retinoids, MTX, infliximab, secukinumab, and ixekizumab have been implicated in transformation to EGR-like eruptions [34]

Necrolytic Migratory Erythema

Epidemiology [7]
- Generally occurs in the fifth and sixth decades of life
- No gender predilection
- There are at least 24 cases of pseudoglucagonoma syndrome reported (18 of which were benign) with ~300 reports of NME in glucagonoma syndrome [35]

Clinical Features [4, 24, 35]
- Polymorphous rash

- Pink polycyclic maculopapules progress to vesicles and erosions
- Knees and intertriginous sites, including perineum; distal extremities and central face
- May be painful, pruritic
- Usually signals glucagonoma syndrome
  - Elevated serum glucagon
  - Histologic or radiographic findings of pancreatic neuroendocrine tumor
  - Associated weight loss, steatorrhea, anemia, new-onset diabetes mellitus [36]
- *Associated tumor:* NME is the presenting sign of glucagonoma (α-cell pancreatic tumor) syndrome in ~90% of cases [36]
- Rarely, occurs with cirrhosis, celiac disease, pancreatitis, and other malignancies without a pancreatic α-cell tumor (pseudoglucagonoma syndrome)
- Pathogenesis likely relates to depletion of several nutrients via glucagon's catabolic signaling including amino acid depletion due to increased gluconeogenesis
  - Angular cheilitis and glossitis often accompany NME
- Increased epidermal arachidonic acid also implicated
- DDx includes pseudoglucagonoma syndrome, acrodermatitis enteropathica, chronic mucocutaneous candidiasis (secondary infection with *C. albicans* is common in NME), psoriasis, pemphigus foliaceus [37]

Histopathology [4, 35]
- Marked parakeratosis
- Hypogranulosis
- Vacuolar interface dermatitis: Keratinocyte vacuolization with superficial necrosis

Management and Prognosis [4, 35, 36, 38]
- NME resolves with successful tumor resection; complete resolution within 48 hours has been reported
- Somatostatin analogues and radionuclide treatment when surgery is not possible
- Some but not all patients clear with amino acid infusion
- Discovery often delayed and more than 1/2 have metastatic disease at diagnosis
- Mean survival 3–7 years
- **Thromboembolic phenomenon** in 24% of patients
- Minority of glucagonomas (3%) occur in setting of MEN1
  - Other NETs should be sought
  - Detailed family history obtained

Scleromyxedema

Epidemiology [7, 13]
- No gender predilection
- Occurs in third through fifth decade of life
- In review of 26 patients, 88% had an abnormal paraprotein; 5 died from hematologic malignancy

Clinical Features [13, 14, 39]

> **Pearl:** Unlike scleroderma, calcinosis and telangiectasias are not present [39].

- Few millimeter firm, waxy papules symmetrically distributed, often in **linear array**, coalescing into shiny **widespread** sclerodermoid plaques
- Requires differentiation from scleroderma, scleredema, NSF, thyroid myxedema, and localized lichen myxedematosus (Table 2.3. Diagnostic criteria of scleromyxedema vs localized LM)
- May affect any skin site, characteristically...
  - Deep furrowing of glabella produces leonine facies
  - Annular plaques over PIP joints create the "doughnut sign"
- *Associated tumor:* monoclonal gammopathy in 88%
  - Usually **IgG lambda** type, most of undetermined significance
  - Minority declare multiple myeloma, Waldenstrom macroglobulinemia, Hodgkin lymphoma, NHL
    These may be therapy-related
- Localized lichen myxedematosus or papular mucinosis describe mucinoses involving only a few sites with otherwise similar morphology of linear waxy papules
  - Localized forms follow a benign course without blood dyscrasia

Histopathology [14, 39–41]
- Increased fibroblasts, sclerosis, dermal mucin ('microscopic triad')
- Dermal mucin may be minimal and overlooked without dedicated histochemical staining
  - Colloidal iron or Alcian blue
- Factors other than the paraprotein alone likely cause the skin findings
  - Paraprotein levels do no not correlate with disease severity
  - Purified paraprotein sera does not cause fibroblast proliferation *in vitro*
- Fibromucinous changes extend deeper (subcutaneous tissue) in NSF compared to scleromyxedema
  - Otherwise, these two show very similar histomorphology and require clinical distinction

Table 2.3 Scleromyxedema diagnostic criteria [39]

Localized LM	Scleromyxedema	Atypical LM
Limited distribution of papulonodules	Generalized papular, sclerodermoid rash	Scleromyxedema without gammopathy
Mucin ± increased fibroblasts	Triad of mucin, increased fibroblasts, fibrosis	Localized LM with gammopathy
No gammopathy	Monoclonal gammopathy	Localized LM with features of multiple subtypes
No thyroid disease	No thyroid disease	

- CD34+ fibrocytes present in scleromyxedema and early NSF are uncommon in scleroderma
- Localized LM subtypes show increased mucin ± fibroblast proliferation

Management and Prognosis [13, 14, 39]
- Monitor with serum protein electrophoresis, immunofixation electrophoresis, Ig levels every 6 months
- Characteristically difficult to treat
  - Retinoids, corticosteroids, plasmapheresis, ECP, melphalan, MTX, cyclophosphamide often tried
  - Chemotherapy associated with high morbidity, mortality
- Extracutaneous manifestations numerous and include:
  - Inflammatory myopathy with proximal muscle weakness, peripheral neuropathy, carpal tunnel syndrome, arthralgias, migratory arthritis, Raynaud phenomenon, pHTN with vascular mucin deposition, restrictive and obstructive lung disease, dysphagia, scleroderma-like renal disease
  - Rare fatalities and coma reported following dysarthria and flu-like prodrome, however, no mucin deposition has been observed in the brain at autopsy
- Localized LM not associated with paraproteinemia usually requires no treatment

Hypertrichosis Lanuginosa Acquisita

Epidemiology [14, 42, 43]
- Women disproportionately affected
- In a review of 24 cases, 92% had underlying malignancy and the remaining patients may have had incomplete evaluation
- Since this review, more than twice as many paraneoplastic cases have been reported

Clinical Features [14, 24, 43]
- Aka malignant down
- Fine, nonpigmented *lanugo hair* growth mainly on face
  - Acquired *terminal hair* growth in the hirsute patient requires endocrine evaluation, and search for adrenal or ovarian tumor if serum testosterone levels are very high (>200 ng/dL)
- Lanugo hair occasionally involves trunk, limbs
- May grow inches long in areas
- Often craniocaudal spread
- Painful glossitis, angular cheilitis, and swelling of tongue papillae may also occur
- *Associated tumor:* Colorectal carcinoma most common in women followed by lung, then breast cancer; in men, colorectal cancer is second to lung cancer
- Benign cases may be seen in AIDS, anorexia nervosa, prescription drug use (minoxidil, cyclosporin, phenytoin), PCT, thyrotoxicosis

Histopathology
- Lanugo hairs are nonpigmented and have no medulla (compared to vellus hairs with variable medullation and pigmented terminal hair) [43]

Management and Prognosis [14, 43]

- GI (colonoscopy) and chest evaluation including chest x-ray and mammography
- Electrolysis, shaving for cosmesis; definitive treatments are those directed at the tumor
- Prognosis is poor with <2 year survival for most reported patients

Necrobiotic Xanthogranuloma

Epidemiology

- In a case series and literature review totaling 46 cases, 80% showed paraprotein-emia, however, most did not have multiple myeloma [44]

Clinical Features [44]

- Indurated dermal yellow to red-orange nodules and plaques (Fig. 2.5a)

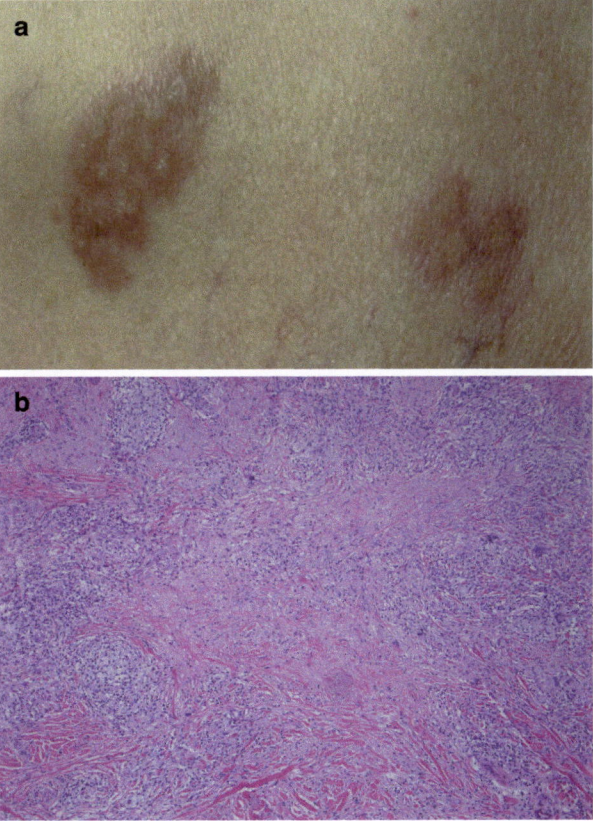

Fig. 2.5 Necrobiotic xanthogranuloma. (**a**) Clinical appearance showing red-yellow dermal plaques. (**b**) Histopathology of NXG with granulomatous inflammation, foreign-body and rare Touton giant cells

- Periorbital, flexures, trunk
- Anterior orbit may be affected with proptosis and restricted extraocular motility
- Rare extracutaneous findings including involvement of lung, myocardium, aerodigestive tract, skeletal muscle, ovary, and kidneys
- *Associated tumor:* Multiple myeloma, CLL, Hodgkin disease, and NHL

Histopathology
- Granulomatous reaction pattern with foreign-body and Touton type giant cells (Fig. 2.5b)
- Cholesterol clefts within necrobiosis

Management and Prognosis [44–47]
- Individual lesions frequently recur or grow following surgical debulking, removal, or biopsy
- Intralesional and systemic corticosteroids, alkylating agents, antimetabolites, and plasmapheresis met with variable success; IVIg with or without EBRT for refractory cases
- May have relatively indolent course even if MM develops
- NXG may precede malignancy by a decade or longer, and patients require life-long follow-up
- Screening echo also recommended due to reports of **giant cell myocarditis**

Facultative (Moderate, Weak, and Poorly Established Association) PNDs

- Malignancy detected in a significant minority of cases or strength of association is not well established
- Thorough history, exam, age-appropriate screening, symptom-directed evaluation indicated
- Further workup on cases-by-case basis
  - Rapid onset AN with prominent mucosal involvement
  - Screening for paraprotein in vasculitis, scleredema, PG, etc.
- Several other PNDs are especially rare to encounter or their association with internal malignancy is not broadly accepted (Tables 2.4 and 2.5)

Paraneoplastic Syndromes Caused by Skin Tumors

Melanoma-associated retinopathy

- Tumor-derived retinal antigens drive an autoimmune-mediated retinopathy [48]

Table 2.4 Clinicopathologic features of facultative and proposed paraneoplastic dermatoses

Disorder	Key clinical features	Key pathological features	Malignancy association	Comments	References
Cutaneous amyloidosis	Eyelid, facial "pinch" purpura, ecchymoses (due to vessel wall infiltration and Factor X absorption by amyloid fibrils), waxy papulonodules; carpal tunnel syndrome, macroglossia; nodular localized amyloidosis presents with waxy skin-colored to yellow nodules and plaques on the trunk/extremities	Of primary cutaneous forms, macular and lichen amyloidosis derive from the keratonocyte while nodular amyloidosis contains immunoglobulin light chains; Primary systemic amyloid is usually AL fig light chains) while secondary systemic amyloid is derived from acute phase reactants (AA amyloid); on histology, amyloid appears as amorphous eosinophilic, fissured deposits, which stain with Congo Red (apple-green birefringence if polarized), Thioflavin T, +/– PM	13% of patients with primary systemic amyloidosis have MM; as many as 33% with skin findings have MM; 7–50% of nodular amyloidosis cases progress to systemic disease based on 2 small series of 10 and 15 patients, respectively	In a review of 236 patients with biopsy-proven amyloidosis (of any site), 42 had skin findings, excluding macroglossia; 23% had purpura or skin lesions in MM group, 21% had such without MM	[7, 50, 51]
PG	Papulopustules or vesicles progress to ulcers with undermined edges; Lower extremities, perineum, peristomal; often multifocal	Early- folliculocentric neutrophilic infiltrate Later- extensive necrosis with mononuclear infiltrate No primary vasculitis	7–16% (including MGUS); AML thought MCC historically, MM, MDS relatively common, rarely adenocarcinomas; gammopathies usually IgA	In a clincopathologic review of 86patients with PG, MGUS was most common; Of 340 hematologic malignancy-associated cases recently reviewed, MDS>MGUS>AML	[17, 52, 53]
Atypical PG	Superficial, nonulcerated, bullous lesions; Upper extremities, especially dorsal hands; Closely resemble Sweet syndrome	Same features as typical PG, but confined to superficial dermis	In a series of 22 patients with atypical PG, 27% had hematologic disease or malignancy	Overlaps with bullous Sweet syndrome and Neutrophilic Dermatosis of the Dorsal Hands	[53, 54]

(continued)

Table 2.4 (continued)

Disorder	Key clinical features	Key pathological features	Malignancy association	Comments	References
Sweet syndrome	More severe than idiopathic Sweet syndrome with bullous lesions, ulcers in addition to papulonodules/plaques; often oral cavity, trunk, and lower≥upper extremities (unlike idiopathic cases); Paraneoplastic Sweets less likely to have preceding URI syndrome or neutrophilia; Respond to CS regardless of cause	Dense upper and mid dermal neutrophilic infiltrate Papillary dermal edema Endothelial swelling, neutrophilic debris, without true vasculitis	20% have underlying malignancy, usually hematologic (AML); of solid tumors, urogenital carcinomas common	Review of 79 malignancy associated cases which were felt to represent "nearly 1/5" of all Sweet syndrome; extracutaneous manifestations more likely when paraneoplastic (MSK, renal, ocular, liver, lung)	[12]
Multicentric reticulohistiocytosis	Skin colored to reddish brown-yellow papulonodules of face, hands, joints; pathognomonic nail fold papules ("coral beads sign") in 1/3; palpebral xanthelasma-like lesions; vermicular erythematous lesions of nostrils; symmetric arthritis, especially hands (arthritis mutilans)	Nodular infiltrate with multinucleated, oncocytic, and other histiocytes; CD68-positive, S100- and CD1a-negative	15–31% have underlying malignancy	Review of 96 cases showed 20% were malignancy-associated; breast cancer and other adenocarcinomas most common	[55]
ANM	Hyperpigmented velvety plaques more widespread than benign AN: knuckles, eyelids, other atypical sites; Mucosal involvement in 1/2 of cases including palate	Hyperkeratosis, acanthosis, papillomatosis, paucinflammatory	19% of cases associated with malignancy, usually gastric adenocarcinoma	Review of 90 cases	[2, 3, 11]

Disorder	Key clinical features	Key pathological features	Malignancy association	Comments	References
Scleredema	Classically 3 types: infectious (streptococcal) associated Type 1, gammopathy associated Type 2, and diabetes associated Type 3; Non-pitting induration +/− erythema of upper back/neck/shoulders, sparing distal extremities; range of motion sometimes limited in shoulders; Erythema may be especially common in scleredema diabeticorum	Thickened reticular dermis, abundant interstitial mucin, without fibroblast proliferation	18% of cases associated with underlying neoplastic/proliferative disorder; MM, MGUS, rarely solid tumors	Case series of 44 patients; spontaneous resolution less likely if paraneoplastic; tx underlying MM, UVA1/PUVA used with occasional success	[41]
DM	Same features as idiopathic DM: heliotrope rash, Gottron papules, Shawl sign, Holster sign +/− episodic proximal muscle weakness and/or tenderness, dysphagia; however, age>40, constitutional changes, absence of Raynaud phenomenon, rapid-onset myositis, higher ESR, and higher CK level indicate elevated risk (consider cross-sectional imaging in addition to age-appropriate screening if present)	Lichenoid interface dermatitis +/− mucin deposition In muscle, mixed B- and T-cell infiltrate + perifascicular muscle fiber atrophy	13–25%; no increased risk in juvenile DM; Associated with ovarian, breast, lung, gastric, and prostatic carcinomas	In a population-based cohort study including 392 DM patients, RR cancer was 1.8 (95% CI, 1.1–2.7) in males and 1.7 (95% CI, 1.0–2.5) in females; Less responsive to GC therapy alone compared to idiopathic cases; up to 1/3 have persistent weakness despite tumor remission	[3, 14, 56]

(continued)

Table 2.4 (continued)

Disorder	Key clinical features	Key pathological features	Malignancy association	Comments	References
Relapsing polychondritis	Recurrent inflammation of cartilage and similar connective tissue resulting in tenderness and mutliation (pseudocellulitis of auricles sparing lobes, saddle nose deformity, tracheal obstruction); blood vessels, ocular, cardiac, and inner ear tissues also affected; Skin manifestations vary from aphthosis to LR to vasculitis	Mixed infiltrate at the chondrodermal junction; degeneration of cartilage and fibrosis in later lesions; Granular IgG and C3 of perichondrial tissue on DIF; anti-type II collagen antibodies	12% a/w MDS or MM	200 case single-center series	[57]
Erythromelalgia	Burning pain of hands or feet (rarely face), erythema, and elevated skin temperature; symptoms are worse with heat, relieved with cooling; Usually unilateral; UE involvement suggests secondary cause	Unknown/NS	7% in one review had vassociated myeloproliferative neoplasms	Clinical review of 87 patients; Treat w/ASA, vasodilators	[58]
Pityriasis rotunda	Hyper- or hypopigmented ovoid scaly plaques with follicular prominence often seen in South African black patients and in Japan; number and size of the lesions vary considerably	Ortho- or parakeratotic hyperkeratosis, loss of granular later, elongated rete ridges, follicular plugging; negative fungal studies	6% a/w malignancy; HCC, MM, AML, CML	Review of 181 cases within the Japanese literature; Can accompany acquired ichthyosis and considered a variant by some authors	[59]

Disorder	Key clinical features	Key pathological features	Malignancy association	Comments	References
Vasculitis	Clinical features of vasculitis of different calibers, however, palpable purpura appears most common	Usually CSVV, but larger vessels may be affected with or without ANCA antibodies	5% a/w neoplasms; Hematologic malignancy>solid tumors, except IgA vasculitis which is a/w lung cancer	Literature review of 222 patients	[60, 61]
Flushing	Carinoid syndrome is the prototypical PND which produces flushing; Foregut tumors (lung, gastric, pacreas, billiary) a/w salmon-pink or red flush while midgut tumors (ileum, appendix) a/w cyanotic hue classically; chronic flushing yields telangiectasias/rosacea-like changes; wheezing and abdominal pain suggest carcinoid syndrome over idiopathic flushing	Unknown/NS	Incidence of paraneoplastic flush is not well established; Flush may be a component of carcinoid syndrome (95% of cases), pheochromocytoma, mastocytosis, medullary thyroid carcinoma, renal cell carcinoma, VIPoma, POEMS syndrome		[62]

(continued)

Table 2.4 (continued)

Disorder	Key clinical features	Key pathological features	Malignancy association	Comments	References
Acquired ichthyosis	Clinical appearance is that of ichthyosis vulgaris (rhomboidal scales on extensor extremities and/or trunk), but occurring after age 20	Orthokeratosis, diminished granular layer	Hodgkin lymphoma is most common paraneoplastic cause; numerous benign conditions (sarcoid, leprosy, AIDS, nutritional deficinecy, cholesterol lowering medications) also implicated	Unlike ichthyosis vulgaris which shows increased epidermal lipid synthesis, paraneoplastic cases have normal epidermal lipid production	[1, 7, 18]
Leser-Trelat sign	Acute onset or growth of numerous seborrheic keratoses (often over several days)	Identical to idiopathic seborrheic keratoses	A/w gastric adenocarcinoma, gynecologic tumors, lymphoproliferative disorders; It's position as a PND remains controversial	43% a/w pruritus, 20% a/w AN	[14, 20, 63]
Pruritus	Chronic pruritus (>6wks) occurs in up to 1/5 of unselected populations; Features of paraneoplastic itch are NS, but hematologic malignancies are more likely to have concurrent skin eruptions than hepatobiliary and GI cancers	Pathologic features of isolated paraneoplastic pruritus are NS	In patients with pruritus with normal-appearing skin, adjusted HR for hematologic malignancy is 2.02 (1.48–2.75) and 3.73 (1.55–8.97) for incident bile duct malignancy	In a recent cross-sectional analysis of nearly 17,000 patients, the OR of malignancy was 5.76 in pruritic patients compared to those without pruritus (95% CI; 5.53–6.00); hepatic, billiary tract, hematologic, and cutaneous malignancies MCC	[64, 65]

Disorder	Key clinical features	Key pathological features	Malignancy association	Comments	References
Cushing syndrome	Hyperpigmentation, edema, acne vulgaris, ecchymoses, muscle wasting, menstrual changes, hirsutism, hypokalemia, hyperglycemia; striae, moon facies, truncal obesity less common than in Cushing disease (pituitary overproduction of ACTH)	NS	Small cell carcinoma of the lung; thymic, pancreatic carcinoid tumors, medullary thyroid cancer, pheochromocytoma	(1) Confirm hypercortisolism w/ late night salivary cortisol, 24 h urinary free cortisol; (2) ACTH-dependent? If no, image adrenals; (3) ACTH source? High dose dexamethasone suppression suggests pituitary disease and pituitary MRI indicated; non-suppression suggests ectopic Cushing syndrome necessitating CT chest/abdomen ± pentetreotide scyntigraphy	[1]

MCC most common cause, *NS* nonspecific, *CS* systemic corticosteroids, *HR* hazard ratio, *CK* creatine kinase, *UE* upper extremity, *a/w* associated with, *LR* livedo reticularis, *HCC* hepatocellular carcinoma, *CSVV* cutaneous small vessel vasculitis

Table 2.5 Other proposed or rare PNDs

Xanthoma
Trousseau's sign, Mondor's disease
Bullous pemphigoid
Pancreatic panniculitis
Papuloerythroderma of ofuju
Erythema ab igne
Hypertrophic osteoarthropathy
Florid cutaneous papillomatosis
Palmoplantar keratoderma
Erythema annulare centrifugum
Pityriasis rubra pilaris
Follicular spicules of multiple myeloma

Paraneoplastic cerebellar ataxia, encephalomyelitis, and Lambert-Eaton syndrome secondary to Merkel cell carcinoma

- As in small cell lung cancer, propensity for neurological paraneoplasia is likely due to a common neuroendocrine origin [49]

Melanoma also been rarely reported with paraneoplastic dermatoses, including PNP [27]

Conclusions

- Paraneoplastic syndromes collectively affect up to 15% of cancer patients, and skin is the second most commonly targeted organ system
- Many dermatoses have been associated with internal malignancy; Curth's postulates aim to describe paraneoplastic syndromes associated by more than chance alone
- As the mechanisms tying skin to internal malignancy become elucidated, the responsible factors will likely be incorporated into diagnostic criteria
- Large case-control and population-based cohort studies are needed to determine the real incidence and risk of most PNDs
- While further work is needed, familiarity with the current understanding of obligate and facultative PNDs allows for appropriate, efficient malignancy evaluation when confronted with these unusual phenomena

Review Questions and Answers

1. A 41 year old man presents with palmar rugosity and weight loss. Oral mucosa and remaining skin exam is within normal limits. Which of the following studies is most likely to reveal the underlying cause?
   (a) Upper endoscopy
   (b) Chest X-ray
   (c) Colonoscopy
   (d) Laryngoscopy

   Answer: (b) The vignette describes Tripe palms without associated acanthosis nigricans. Tripe palms without associated ANM is classically seen with lung cancer. When the two appear concurrently, GI adenocarcinoma is most likely.

2. A 54 year old woman presents with classic heliotrope rash, Gottron papules, and history of difficulty rising from a chair. You suspect malignancy-associated dermatomyositis. Which of the following increase the risk of paraneoplastic DM?
   (a) Presence of Raynaud phenomenon
   (b) Absence of Raynaud phenomenon
   (c) Lower CK level
   (d) Holster sign

   Answer: (b) In a retrospective study of 33 DM cases by Sparsa et al., the following factors were associated with malignancy: presentation to the Internal Medicine department ($P = 0.02$), presence of constitutional symptoms ($P < 0.01$), rapid onset of DM ($P = 0.02$), absence of Raynaud phenomenon ($P < 0.01$), higher mean ESR ($P < 0.01$), and creatine kinase level ($P < 0.01$).

3. Which of the following features suggests the diagnosis of Bazex syndrome over psoriasis?
   (a) Positive Auspitz sign
   (b) Involvement of the central palms and arches of the feet
   (c) Nail changes
   (d) Involvement of the nose and helices

   Answer: (d) Bazex syndrome presents with psoriasiform plaques which can show a positive Auspitz sign as in psoriasis, however, the distribution is atypical for psoriasis involving the nose and helices of the ears and sparing the central palm and foot arch. Nail changes occur in both diseases. In Bazex syndrome, horizontal and longitudinal ridging, discoloration, subungual debris, and onychauxis may be noted.

4. Which of the following is most concerning for a paraneoplastic phenomenon?
   (a) A 60 year old woman with terminal hair growth on the chin and upper lip and serum testosterone level of 100 ng/dL

(b) A 60 year old woman with lanugo hair on the face and extremities with painful glossitis
(c) A 62 year old obese man with velvety hyperpigmented plaques on the dorsal neck present for 9 months
(d) An 87 year old man with hundreds of stuck-on waxy papules on the trunk

Answer: (b) An elderly woman with lanugo hair and painful glossitis is highly suggestive of HLA, which carries a high association with underlying malignancy in the absence of drug-induced and metabolic causes. The remaining answers describe (a) hirsutism with mild elevation of testosterone (levels >200 ng/dL warrant search for ovarian or adrenal tumor), (b) benign AN, and (c) numerous seborrheic keratoses with unknown onset, and no features worrisome for malignancy.

5. Which of the following statements is TRUE?
   (a) 'Pinch purpura' of amyloidosis is due to amyloid fibril consumption of clotting factor XIII
   (b) Scleromyxedema may affect several organ systems, but characteristically spares the kidneys
   (c) Extracutaneous manifestations of Sweet syndrome are more common when paraneoplastic
   (d) NXG is most commonly associated with solid tumors

Answer: (c) MSK, renal, ocular, liver, and lung involvement is thought more likely in paraneoplastic Sweet syndrome. Amyloid is capable of binding and inactivating factor X. NXG is most commonly associated with hematopoietic neoplasms, not solid tumors. Scleromyxedema is associated with many common to rare extracutaneous manifestations *including* scleroderma-like acute renal crisis.

References

1. Poole S, Fenske NA. Cutaneous markers of internal malignancy. II. Paraneoplastic dermatoses and environmental carcinogens. J Am Acad Dermatol. 1993;28(2 Pt 1):147–64.
2. Caccavale S, Brancaccio G, Agozzino M, Vitiello P, Alfano R, Argenziano G. Obligate and facultative paraneoplastic dermatoses: an overview. Dermatol Pract Concept. 2018;8(3):191–7.
3. Pelosof LC, Gerber DE. Paraneoplastic syndromes: an approach to diagnosis and treatment. Mayo Clin Proc. 2010;85(9):838–54.
4. Da Silva JA, de Carvalho Mesquita K, de Souza Machado Igreja AC, et al. Paraneoplastic cutaneous manifestations: concepts and updates. An Bras Dermatol. 2013;88(1):9–22. https://doi.org/10.1590/S0365-05962013000100001.
5. McLean DI. Cutaneous paraneoplastic syndromes. Arch Dermatol. 1986;122(7):765–7.
6. Sarkar B, Knecht R, Sarkar C, Weidauer H. Bazex syndrome (acrokeratosis paraneoplastica). Eur Arch Otorhinolaryngol. 1998;255(4):205–10.
7. Chung VQ, Moschella SL, Zembowicz A, Liu V. Clinical and pathologic findings of paraneoplastic dermatoses. J Am Acad Dermatol. 2006;54(5):745–62; quiz 63–6.
8. Richard M, Giroux JM. Acrokeratosis paraneoplastica (Bazex' syndrome). J Am Acad Dermatol. 1987;16(1 Pt 2):178–83.

 9. Jaeger AB, Gramkow A, Hjalgrim H, Melbye M, Frisch M. Bowen disease and risk of subsequent malignant neoplasms: a population-based cohort study of 1147 patients. Arch Dermatol. 1999;135(7):790–3.
10. Boyd AS, Neldner KH. Erythema gyratum repens without underlying disease. J Am Acad Dermatol. 1993;28(1):132.
11. Brown J, Winkelmann RK. Acanthosis nigricans: a study of 90 cases. Medicine (Baltimore). 1968;47(1):33–51.
12. Cohen PR, Talpaz M, Kurzrock R. Malignancy-associated Sweet's syndrome: review of the world literature. J Clin Oncol. 1988;6(12):1887–97.
13. Dinneen AM, Dicken CH. Scleromyxedema. J Am Acad Dermatol. 1995;33(1):37–43.
14. Thiers BH, Sahn RE, Callen JP. Cutaneous manifestations of internal malignancy. CA Cancer J Clin. 2009;59(2):73–98.
15. Schwartz RA. Acanthosis nigricans. J Am Acad Dermatol. 1994;31(1):1–19; quiz 20–2.
16. Graus F, Delattre JY, Antoine JC, Dalmau J, Giometto B, Grisold W, et al. Recommended diagnostic criteria for paraneoplastic neurological syndromes. J Neurol Neurosurg Psychiatry. 2004;75(8):1135–40.
17. Montagnon CM, Fracica EA, Patel AA, Camilleri MJ, Murad MH, Dingli D, et al. Pyoderma gangrenosum in hematologic malignancies: systematic review. J Am Acad Dermatol. 2020;82(6):1346–59.
18. Boyce S, Harper J. Paraneoplastic dermatoses. Dermatol Clin. 2002;20(3):523–32.
19. Cohen PR, Grossman ME, Almeida L, Kurzrock R. Tripe palms and malignancy. J Clin Oncol. 1989;7(5):669–78.
20. Sabir S, James WD, Schuchter LM. Cutaneous manifestations of cancer. Curr Opin Oncol. 1999;11(2):139–44.
21. Cohen PR, Kurzrock R. Malignancy-associated tripe palms. J Am Acad Dermatol. 1992;27(2 Pt 1):271–2.
22. Bolognia JL, Brewer YP, Cooper DL. Bazex syndrome (acrokeratosis paraneoplastica). An analytic review. Medicine (Baltimore). 1991;70(4):269–80.
23. Huilaja L, Soronen M, Karjalainen A, Tasanen K. Acrokeratosis paraneoplastica-like findings as a manifestation of systemic lupus erythematosus. Acta Derm Venereol. 2019;99(3):333–4.
24. Stone SP, Buescher LS. Life-threatening paraneoplastic cutaneous syndromes. Clin Dermatol. 2005;23(3):301–6.
25. Pecora AL, Landsman L, Imgrund SP, Lambert WC. Acrokeratosis paraneoplastica (Bazex' syndrome). Report of a case and review of the literature. Arch Dermatol. 1983;119(10):820–6.
26. Chen WS, Tetzlaff MT, Diwan H, Jahan-Tigh R, Diab A, Nelson K, et al. Suprabasal acantholytic dermatologic toxicities associated checkpoint inhibitor therapy: a spectrum of immune reactions from paraneoplastic pemphigus-like to Grover-like lesions. J Cutan Pathol. 2018;45(10):764–73.
27. Kaplan I, Hodak E, Ackerman L, Mimouni D, Anhalt GJ, Calderon S. Neoplasms associated with paraneoplastic pemphigus: a review with emphasis on non-hematologic malignancy and oral mucosal manifestations. Oral Oncol. 2004;40(6):553–62.
28. Ouedraogo E, Gottlieb J, de Masson A, Lepelletier C, Jachiet M, de Chou CS, et al. Risk factors for death and survival in paraneoplastic pemphigus associated with hematologic malignancies in adults. J Am Acad Dermatol. 2019;80(6):1544–9.
29. Camisa C, Helm TN. Paraneoplastic pemphigus is a distinct neoplasia-induced autoimmune disease. Arch Dermatol. 1993;129(7):883–6.
30. Sklavounou A, Laskaris G. Paraneoplastic pemphigus: a review. Oral Oncol. 1998;34(6):437–40.
31. Poot AM, Diercks GF, Kramer D, Schepens I, Klunder G, Hashimoto T, et al. Laboratory diagnosis of paraneoplastic pemphigus. Br J Dermatol. 2013;169(5):1016–24.
32. Chan LS. Epitope spreading in paraneoplastic pemphigus: autoimmune induction in antibody-mediated blistering skin diseases. Arch Dermatol. 2000;136(5):663–4.

33. Boyd AS, Neldner KH, Menter A. Erythema gyratum repens: a paraneoplastic eruption. J Am Acad Dermatol. 1992;26(5 Pt 1):757–62.
34. Richey PM, Fairley JA, Stone MS. Transformation from pityriasis rubra pilaris to erythema gyratum repens-like eruption without associated malignancy: a report of 2 cases. JAAD Case Reports. 2018;4(9):944–6.
35. Mullans EA, Cohen PR. Iatrogenic necrolytic migratory erythema: a case report and review of nonglucagonoma-associated necrolytic migratory erythema. J Am Acad Dermatol. 1998;38(5 Pt 2):866–73.
36. Hashmi O. Rare but deadly: necrolytic migratory erythema and glucagonoma syndrome. JAAD Case Reports. 2017;76(6):AB213.
37. John AM, Schwartz RA. Glucagonoma syndrome: a review and update on treatment. J Eur Acad Dermatol Venereol. 2016;30(12):2016–22.
38. Castro PG, de Leon AM, Trancon JG, Martinez PA, Alvarez Perez JA, Fernandez Fernandez JC, et al. Glucagonoma syndrome: a case report. J Med Case Rep. 2011;5:402.
39. Rongioletti F, Rebora A. Updated classification of papular mucinosis, lichen myxedematosus, and scleromyxedema. J Am Acad Dermatol. 2001;44(2):273–81.
40. Cowper SE, Rabach M, Girardi M. Clinical and histological findings in nephrogenic systemic fibrosis. Eur J Radiol. 2008;66(2):191–9.
41. Rongioletti F, Merlo G, Carli C, Cribier B, Metze D, Calonje E, et al. Histopathologic characteristics of scleromyxedema: a study of a series of 34 cases. J Am Acad Dermatol. 2016;74(6):1194–200.
42. Sindhuphak W, Vibhagool A. Acquired hypertrichosis lanuginosa. Int J Dermatol. 1982;21(10):599–601.
43. Slee PH, van der Waal RI, Schagen van Leeuwen JH, Tupker RA, Timmer R, Seldenrijk CA, et al. Paraneoplastic hypertrichosis lanuginosa acquisita: uncommon or overlooked? Br J Dermatol. 2007;157(6):1087–92.
44. Mehregan DA, Winkelmann RK. Necrobiotic xanthogranuloma. Arch Dermatol. 1992;128(1):94–100.
45. Ugurlu S, Bartley GB, Gibson LE. Necrobiotic xanthogranuloma: long-term outcome of ocular and systemic involvement. Am J Ophthalmol. 2000;129(5):651–7.
46. Nambudiri VE, McLaughlin C, Lo TC, Zembowicz A, Moschella S. Successful multimodality treatment of recalcitrant necrobiotic xanthogranuloma using electron beam radiation and intravenous immunoglobulin. Clin Exp Dermatol. 2016;41(2):179–82.
47. Umbert I, Winkelmann RK. Necrobiotic xanthogranuloma with cardiac involvement. Br J Dermatol. 1995;133(3):438–43.
48. Heberton M, Azher T, Council ML, Khanna S. Metastatic cutaneous melanoma presenting with melanoma-associated retinopathy. Dermatol Surg. 2019;45(4):606–7.
49. Queen D, Gu Y, Lopez A, Chen D, Geskin LJ. Paraneoplastic cerebellar ataxia in Merkel cell carcinoma of unknown primary. JAAD Case Reports. 2019;5(5):398–400.
50. Kyle RA, Bayrd ED. Amyloidosis: review of 236 cases. Medicine (Baltimore). 1975;54(4):271–99.
51. Woollons A, Black MM. Nodular localized primary cutaneous amyloidosis: a long-term follow-up study. The British journal of dermatology. 2001;145(1):105–9.
52. Powell FC, Schroeter AL, Su WP, Perry HO. Pyoderma gangrenosum: a review of 86 patients. Q J Med. 1985;55(217):173–86.
53. Bennett ML, Jackson JM, Jorizzo JL, Fleischer AB, Jr., White WL, Callen JP. Pyoderma gangrenosum. A comparison of typical and atypical forms with an emphasis on time to remission. Case review of 86 patients from 2 institutions. Medicine (Baltimore). 2000;79(1):37–46.
54. Walling HW, Snipes CJ, Gerami P, Piette WW. The relationship between neutrophilic dermatosis of the dorsal hands and sweet syndrome: report of 9 cases and comparison to atypical pyoderma gangrenosum. Archives of dermatology. 2006;142(1):57–63.
55. Luz FB, Gaspar TAP, Kalil-Gaspar N, Ramos-e-Silva M. Multicentric reticulohistiocytosis. Journal of the European Academy of Dermatology and Venereology : JEADV. 2001;15(6):524–31.

56. Sigurgeirsson B, Lindelof B, Edhag O, Allander E. Risk of cancer in patients with dermatomyositis or polymyositis. A population-based study. The New England journal of medicine. 1992;326(6):363–7.
57. Frances C, el Rassi R, Laporte JL, Rybojad M, Papo T, Piette JC. Dermatologic manifestations of relapsing polychondritis. A study of 200 cases at a single center. Medicine (Baltimore). 2001;80(3):173–9.
58. Kalgaard OM, Seem E, Kvernebo K. Erythromelalgia: a clinical study of 87 cases. J Intern Med. 1997;242(3):191–7.
59. Ito M TT. Pseudo-ichthyosis acquise en taches circularis: Pityriasis circinata. Ann Dermatol Syph. 1960;87:26–37.
60. Sanchez-Guerrero J, Gutierrez-Urena S, Vidaller A, Reyes E, Iglesias A, Alarcon-Segovia D. Vasculitis as a paraneoplastic syndrome. Report of 11 cases and review of the literature. The Journal of rheumatology. 1990;17(11):1458–62.
61. Ota S, Haruyama T, Ishihara M, Natsume M, Fukasawa Y, Sakamoto T, et al. Paraneoplastic IgA Vasculitis in an Adult with Lung Adenocarcinoma. Internal medicine (Tokyo, Japan). 2018;57(9):1273–6.
62. Abdel Kader El Tal, Zeina Tannous, (2008) Cutaneous Vascular Disorders Associated with Internal Malignancy. Dermatologic Clinics 26 (1):45–57
63. James S.M. Yeh, Stephanie E. Munn, Tim A. Plunkett, Peter G. Harper, Deborah J. Hopster, Anthony W. du Vivier, (2000) Coexistence of acanthosis nigricans and the sign of Leser-Tr?lat in a patient with gastric adenocarcinoma: A case report and literature review. Journal of the American Academy of Dermatology 42 (2):357–62
64. Fett N, Haynes K, Propert KJ, Margolis DJ. Five-year malignancy incidence in patients with chronic pruritus: a population-based cohort study aimed at limiting unnecessary screening practices. Journal of the American Academy of Dermatology. 2014;70(4):651–8.
65. Larson VA, Tang O, Stander S, Kang S, Kwatra SG. Association between itch and cancer in 16,925 patients with pruritus: Experience at a tertiary care center. Journal of the American Academy of Dermatology. 2019;80(4):931–7.

Genodermatoses with Malignant Potential

3

Melia Hernandez Holt, Vincent Liu, and Jon Dyer

Abbreviations

APC	Adenomatous polyposis coli
BCC	Basal cell carcinoma
BCCNS	Basal cell carcinoma nevus syndrome
FAP	Familial adenomatous polyposis
HLRCC	Hereditary leiomyomatosis and renal cell cancer
MEN	Multiple endocrine neoplasia
MPNST	Malignant peripheral nerve sheath tumor
MTS	Muir-Torre syndrome
NF	Neurofibroma
TS	Tuberous sclerosis

M. Hernandez Holt (✉)
The Dermatology Group, Orlando, FL, USA

V. Liu
Department of Dermatology, University of Iowa Health Care, Iowa, IA, USA
e-mail: vincent-liu@uiowa.edu

J. Dyer
UPMB General Dermatology, University of Missouri, Columbia, MO, USA
e-mail: dyerja@health.missouri.edu

© Springer Nature Switzerland AG 2021
V. Liu (ed.), *Dermato-Oncology Study Guide*,
https://doi.org/10.1007/978-3-030-53437-0_3

Learning Objectives

1. Identify the cutaneous manifestations and their corresponding internal malignancies of commonly encountered genodermatoses with malignant potential.
2. Connect the underlying genetic abnormality with its corresponding genodermatoses with malignant potential.
3. Describe malignancy screening recommendations for patients with genodermatoses (Table 3.1).

1. **Nevoid Basal Cell Carcinoma Syndrome (Gorlin syndrome)**
   (a) Gene Mutation
      - PATCHED (PTCH)
         - PTCH is a sonic hedgehog membrane receptor which normally inhibits smoothened and thereby inhibits activation of Gli transcription factor. Inactivating mutations in PTCH result in constitutive activation of smoothened and Gli.
      - Autosomal dominant (50% are sporadic mutations)
   (b) Clinical Findings
      - Cutaneous:
         - Multiple basal cell carcinomas (Fig. 3.1a) (can mimic nevi due to pigmentation); palmar and plantar pits (Fig. 3.1b) (appear by early adulthood).

Table 3.1 Synopsis of genodermatoses with malignant potential

Genodermatosis		Genetic defect	Primary cutaneous features	Primary associated malignancy
1. Nevoid Basal Cell Carcinoma Syndrome		PATCHED (PTCH)	Numerous BCCs Palmar pits	Medulloblastoma
2. Birt-Hogg-Dubé Syndrome		Folliculin (FLCN)	Fibrofolliculomas trichodiscomas, angiofibromas	Renal cell carcinoma
3. Familial Colorectal Polyposis (Gardner syndrome)		Adenomatous polyposis coli (APC)	Epidermoid cysts (can present at birth), fibromas (back)	Colorectal carcinoma; Also, thyroid, hepatobiliary, pancreatic
4. Hereditary leiomyomatosis and renal cell cancer (HLRCC) (Reed Syndrome)		Fumarate hydratase	Leiomyomas	Renal cell carcinoma
5. Muir-Torre Syndrome (MTS)		MSH2, MLH1, and/or MSH6, also PMS2	Sebaceous neoplasms	Colorectal carcinoma Gentourinary carcinoma
6. Multiple Endocrine Neoplasia (MEN)	Type 1	MEN1	Angiofibromas (face) Collagenomas Hypopigmented macules	Gastrinomas Insulinomas Thymic/bronchial carcinomas
	Type 2A	RET	Lichen/macular amyloid	Thyroid carcinoma Pheochromocytoma
	Type2B	RET	Multiple muconsal neuromas	Thyroid carcinoma Pheochromocytoma

Genodermatosis	Genetic defect	Primary cutaneous features	Primary associated malignancy
7. Multiple Hamartoma Syndrome (Cowden Syndrome)	PTEN	Trichilemmomas Sclerotic fibromas Cobblestoning of oral mucosa	Breast carcinoma Thyroid carcinoma Endometrial carcinoma Genitourinary carcinoma
8. Neurofibromatosis type 1 (von Recklinghausen disease)	NF-1	Neurofibromas Café-au-lait macules	Juvenile myelomonocytic leukemia (MJL) MPNST (malignant peripheral nerve sheath tumor)
9. Peutz-Jeghers syndrome	STK11 (LKB1)	Mucosal melanotic macules	GI adenocarcinoma
10. Tuberous Sclerosis	TSC1 (hamartin), TSC2 (tuberin)	Angiofibromas Periungual fibromas Gingival fibromas Hypomelanotic macules Connective tissue nevi	CNS tumors Renal carcinoma
11. Familial Malignant Melanoma	CDKN2a, encoding p16 and p14	Melanocytic nevi Solar lentigines, ephelides	Melanoma Pancreatic cancer

Fig. 3.1 Nevoid BCC syndrome

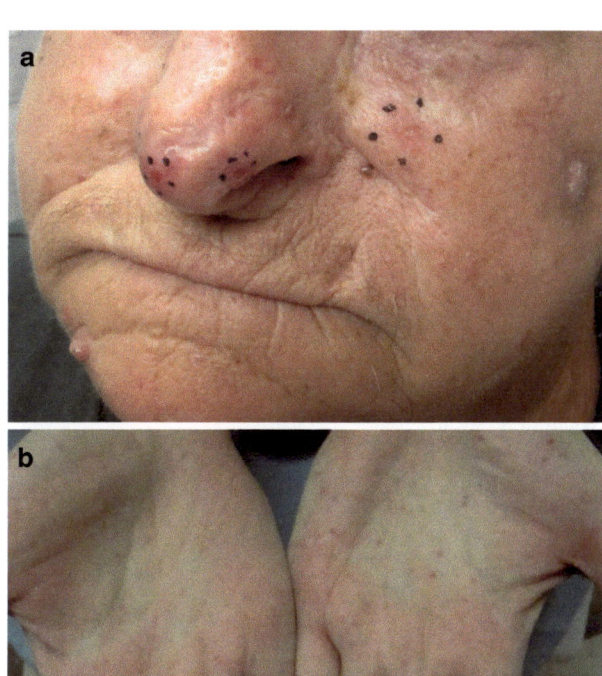

- Additional findings:
  - Odontogenic keratocysts (especially of the mandible), bifid ribs, calcification of falx cerebri, agenesis of corpus callosum, craniofacial anomalies (frontal bossing, hypertelorism, macrocephaly, widened nasal bridge), cardiac and ovarian fibromas, meningiomas.
- (c) Malignancy Association
  - Medulloblastoma—Desmoplastic (most present by age 2; 20× increased risk with SUFU mutations)
- (d) Screening Recommendations
  - Panoramic radiograph (orthopantomogram-digital) of the jaw performed annually starting at age 8 years
  - Cranial MRI to exclude medulloblastoma performed annually until age 8 years

Note: Individuals with mutations in the suppressor of fused gene (SUFU) can meet the clinical criteria for Gorlin syndrome and may be at an increased risk of medulloblastoma than those with a mutation in PATCHED [1].

2. **Birt-Hogg-Dubé Syndrome**
   - (a) Gene Mutation
     - Folliculin (FLCN), tumor suppressor
     - Autosomal dominant; variable within families
   - (b) Clinical Findings
     - Cutaneous:
       - Fibrofolliculomas (Fig. 3.2), trichodiscomas, angiofibromas, and acrochordons primarily on the head and neck; onset after age 20.
     - Additional findings:
       - Pulmonary blebs (basilar) which can result in spontaneous pneumothorax
   - (c) Malignancy Associations
     - Renal cell carcinomas (chromophobe and hybrid oncocytic variants)
   - (d) Screening Recommendations

Fig. 3.2 Birt-Hogg-Dubé syndrome clinical

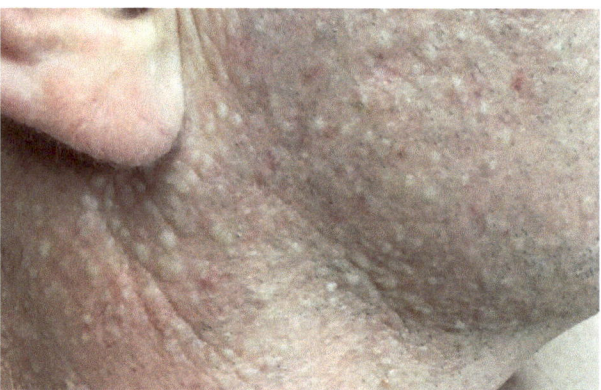

- Gadolinium enhanced abdominal MRI starting at age 20 years and repeating every 3–4 years to screen for renal tumors
- Baseline CT chest to screen for pulmonary blebs [2]

Note: Patients should be counseled against smoking and scuba diving due to risk of pneumothorax

3. **Familial Colorectal Polyposis (Gardner syndrome)**
   (a) Gene Mutation
       - Adenomatous polyposis coli (APC), tumor suppressor which downregulates Wnt/β-catenin pathway
       - Autosomal dominant
   (b) Clinical Findings
       - Cutaneous:
         – Epidermoid cysts (can present at birth), fibromas (back)
       - Additional findings:
         – Multiple GI adenomas
         – Osteomas (maxilla and mandible), fibrous tumors (skin, mesentery, retroperitoneum), bilateral congenital hypertrophy of the retinal pigment epithelium (CHRPE), dental anomalies (supernumerary teeth, odontomas, unerupted teeth), desmoid tumors (can be locally aggressive and frequently arise following surgery)
   (c) Malignancy Associations
       - Colorectal carcinoma is inevitable without surgical intervention
       - Thyroid (papillary) carcinoma, hepatoblastoma (children), pancreatic carcinoma, biliary carcinoma, adrenal carcinoma, sarcomas
   (d) Screening Recommendations
       - Annual sigmoidoscopy beginning at age 10 years until polyps are detected then transition to annual colonoscopy until colectomy [3]
       - Prophylactic colectomy when polyposis becomes evident (usually between age 15 and 25 years)
       - Upper endoscopy every 1–3 years
       - Liver ultrasound and serum alpha fetoprotein until age 5 (hepatoblastoma screening)

Note: Cutaneous findings frequently present before the polyposis begins. Gardner syndrome is considered a phenotypic variant of familial adenomatous polyposis (FAP) syndrome with prominent extraintestinal involvement. Turcot syndrome type 2 refers to patients with heterozygous APC mutations with brain tumors (often medulloblastomas) and adenomatous polyposis.

4. **Hereditary leiomyomatosis and renal cell cancer (HLRCC) (Reed Syndrome)**
   (a) Gene Mutation
       - Fumarate hydratase
       - Encodes *fumarase* which converts fumarate to malate in the mitochondrial Krebs cycle
       - Autosomal dominant

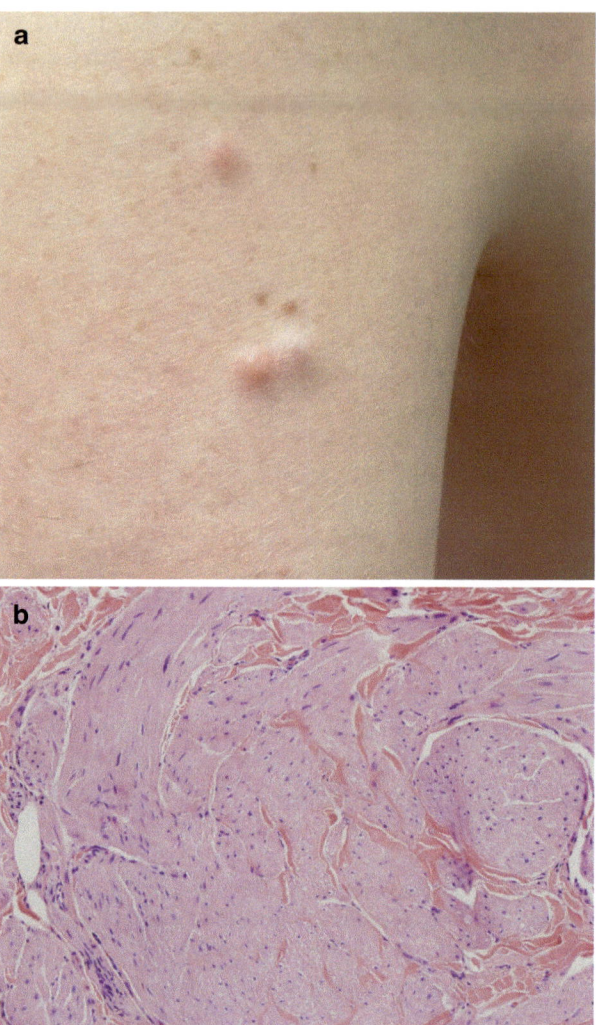

Fig. 3.3 Hereditary leiomyomatosis. (**a**) Clinical (**b**) Leiomyoma

(b) Clinical Findings
- Cutaneous:
  - Cutaneous leiomyomas (piloleiomyomas > genital leiomyomas and angioleiomyomas) in second to fourth decade of life (Fig. 3.3)
- Additional findings:
  - Uterine leiomyomas (>90% of affected women)

(c) Malignancy Associations
- Renal cell (type II papillary) carcinoma, early onset and aggressive
- Uterine leiomyomas with rare transformation to leiomyosarcomas

(d) Screening Recommendations
   - Baseline abdominal MRI starting at age 10, repeat annually to screen for renal cell carcinoma
   - Baseline pelvic ultrasound with annual pelvic exam [4]

5. **Muir-Torre Syndrome (MTS)**
   (a) Gene Mutation
      - MSH2, MLH1, and/or MSH6, as well as PMS2, which are DNA mismatch repair genes
      - Autosomal dominant heterozygous mutation
   (b) Clinical Findings
      - Cutaneous:
         - Sebaceous neoplasms (sebaceous adenomas (Fig. 3.4), sebaceomas, sebaceous carcinomas)
         - Basal cell carcinomas and keratoacanthomas with sebaceous differentiation
      Note: Cutaneous findings—mean age of presentation: 55
   (c) Malignancy Associations
      - Colorectal carcinoma (61%)
      - Genitourinary carcinoma (22%)
      - Breast carcinoma, hematologic malignancies, endometrial carcinoma, head and neck cancers
   (d) Screening Recommendations
      - Colonoscopy every 2 years beginning at age 20 years

Fig. 3.4 Muir-Torre syndrome

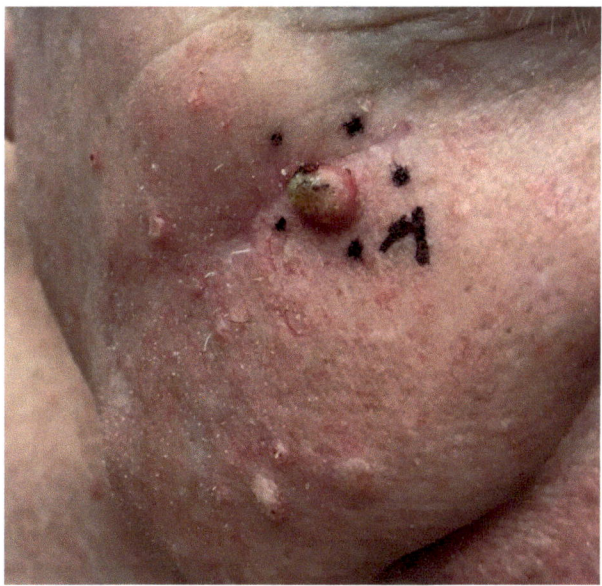

- Renal ultrasound, urinalysis, CBC annually
- Breast and pelvic exams with Pap smears annually beginning at age 30 years
- Consider: upper GI endoscopy, chest X-ray, cervical and urine cytology, occult blood testing, serial hemoglobin and liver function tests

Note: Muir-Torre syndrome (MTS) is considered a subtype of hereditary non-polyposis colorectal cancer (HNPCC) syndrome (Lynch syndrome). The MTS subtype is diagnosed due to the presence of at least one sebaceous tumor (or tumor with sebaceous differentiation) and a visceral malignancy in the absence of predisposing factors [3]

6. **Multiple Endocrine Neoplasia (MEN) Types 1, 2A, 2B**
   - **Type 1**
     (a) Gene Mutation
        - MEN1, Menin (a nuclear protein), often truncating mutations
        - Autosomal dominant
     (b) Clinical Findings
        - Cutaneous:
            - Facial angiofibromas (smaller, fewer, and more likely to be on upper lip compared to tuberous sclerosis), collagenomas, lipomas, hypopigmented macules
        - Additional findings:
            - Parathyroid: hyperparathyroidism (hyperplasia/adenoma)
            - Pituitary: prolactinoma, acromegaly, Cushing syndrome
            - Pancreas: islet cell hyperplasia, adenoma
     (c) Malignancy Associations
        - Gastrinoma (Zollinger-Ellison Syndrome), insulinomas, glucagonomas, thymic carcinoma, bronchial carcinoma
     (d) Screening Recommendations
        - Plasma calcium and PTH levels
        - Fasting gastrin, glucagon, vasoactive intestinal polypeptide (VIP), pancreatic polypeptide, chromogranin A, insulin with associated glucose level
        - Plasma prolactin and IGF-1 level annually with pituitary MRI every 3–5 years
        - Pancreatic and duodenal imaging (MRI, CT, or endoscopic ultrasound)
        - Abdominal (adrenal) imaging (MRI or CT) every 3 years
        - Chest imaging (CT or MRI) every 1–2 years [5, 6]
   - **Type 2A**
     (a) Gene Mutation
        - RET proto-oncogene (codon 634), tyrosine kinase receptor, point mutation
        - Autosomal dominant
     (b) Clinical Findings
        - Cutaneous:

- Lichen or macular amyloid, pruritus, notalgia paresthetica (onset prior to medullary thyroid carcinoma)
  - Additional findings:
    - Parathyroid: hyperparathyroidism (hyperplasia/adenoma)—20–30%
    - Thyroid: C-cell hyperplasia
  (c) Malignancy Associations
  - Thyroid carcinoma (medullary)—95%
  - Pheochromocytoma—50%
  (d) Screening Recommendations
  - Prophylactic thyroidectomy prior to age 10 years
  - Plasma free metanephrine or 24-h urinary metanephrine and vanillylmandelic acid levels to screen for pheochromocytoma
- **Type 2B**
  (a) Gene Mutation
  - RET proto-oncogene (Met918Thr), tyrosine kinase receptor, missense mutation
  - Autosomal dominant
  (b) Clinical Findings
  - Cutaneous:
    - Multiple mucosal neuromas (eyelids, lips, tongue, buccal, gingiva, palate, larynx), can be present at birth (Fig. 3.5)

Fig. 3.5 MEN-tongue neuromas

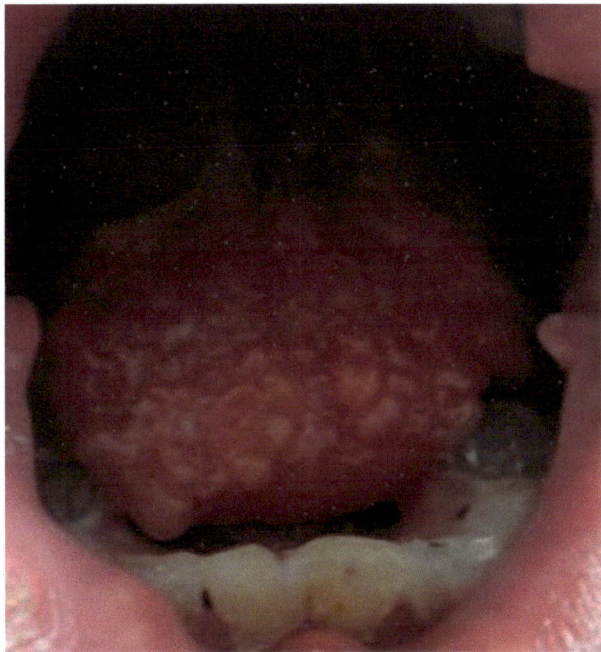

- Rarely cirumoral lentigines, hypertrichosis, synophrys,café-au lait macules reported, rare cutaneous neuromas
- Additional findings:
  - Ganglioneuromatosis (megacolon, diverticulosis), Marfanoid habitus, high-arched palate, flat nasal bridge, pectus excavatum, bilateral pes cavus, high patella, scoliosis, joint laxity; prominently thickened corneal nerves

(c) Malignancy Associations
- Medullary thyroid carcinoma (develops in most patients; usually in first second decade of life)
- Pheochromocytoma (50% of cases)

(d) Screening Recommendations
- Prophylactic thyroidectomy prior to 10 years of age; often in infancy
- Plasma free metanephrine or 24-h urinary metanephrine and vanillylmadelic acid levels to screen for pheochromocytoma

Note: Hirschsprung disease and familial medullary thyroid carcinoma are also due to mutations in RET.

7. **Multiple Hamartoma Syndrome (Cowden Syndrome)**
   (a) Gene Mutation
   - Protein-tyrosine phosphatase (PTEN), tumor suppressor
   - PTEN downregulates phosphatidylinositol 3-kinase (PI3K)/Akt/mammalian target of rapamycin (mTOR) pathway
   - Autosomal dominant
   (b) Clinical Findings
   - Cutaneous:
     - Trichilemmomas (Fig. 3.6), oral papillomas, acral keratosis, sclerotic fibromas, cobblestoning of the oral mucosa

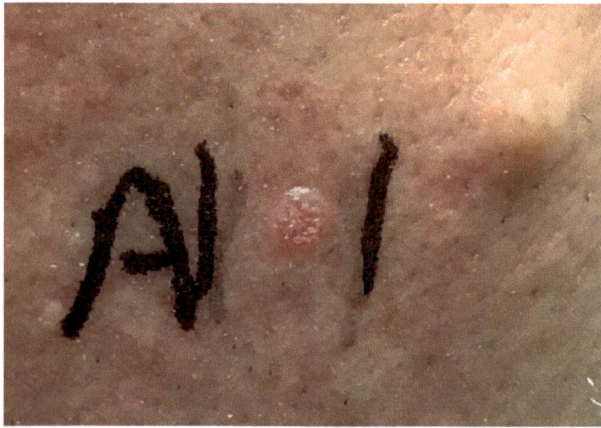

Fig. 3.6 Trichilemmoma, as seen in Cowden syndrome

- Lipomas, angiolipomas, vascular anomalies (fast flow), acrochordons, inverted follicular keratosis, apocrine hidrocystomas; possible increased risk for melanoma
  - Additional findings:
    - Thyroid (goiter and adenoma), breast (fibrocystic disease and fibroadenomas), ovarian cysts, uterine leiomyomas, urethral polyps, craniomegaly
    - Hamartomatous polyps can affect the entire GI tract (more common in the colon) usually asymptomatic with low risk of malignant degeneration (~9% lifetime)
- (c) Malignancy Associations
  - Breast carcinoma (25–50% of women; mean age of onset of 40 years)
  - Thyroid carcinoma (risk 5–10%; follicular > papillary)
  - Endometrial carcinoma (5–10% of women)
  - Others: teratomas, genitourinary carcinomas; >5% lifetime risk of melanoma
- (d) Screening Recommendations
  - Breast: monthly self-exams, biannual clinical exams (starting at age 25), annual mammogram and MRI (starting at age 30 or 5–10 years prior to the earliest breast cancer diagnosis in the family), consider prophylactic mastectomy
  - Thyroid: baseline thyroid function and ultrasound, annual ultrasound (starting at age 18)
  - Uterine: annual endometrial biopsies (starting at age 35 or 5 years prior to the earliest endometrial cancer diagnosis in the family)
  - Renal: annual renal ultrasound, urinalysis, urine cytology

Note: PTEN hamartoma tumor syndrome includes Cowden syndrome, Bannayan-Riley-Ruvalcaba syndrome, Lhermitte-Duclos disease, and autism/macrocephaly syndrome which are all caused by PTEN mutations. Bannayan-Riley-Ruvalcaba syndrome has been characterized by findings present around birth (such as genital lentigines, macrocephaly, and developmental delay). Lhermitte-Duclos disease is characterized by hamartomas of the cerebella ganglion cells which leads to ataxia and seizures.
[Germline mutations in succinate dehydrogenase B/C/D (SDHB/C/D) and KLLN promoter methylation epimutation (hypermethylation) (respectively 10% and ~30% of CS and CS-like syndrome patients without germline PTEN pathogenic variants) cause Cowden syndrome and Cowden syndrome-like phenotypes [7].

8. **Neurofibromatosis type 1 (von Recklinghausen disease)** (Table 3.2)
   - (a) Gene Mutation
     - NF-1, neurofibromin, tumor suppressor
     - Neurofibromin accelerates the hydrolysis of GTP to GDP, inhibiting the Ras signalling of the Ras-mitogen-activated protein kinase (MAPK) pathway
     - Autosomal dominant (30–50% spontaneous mutations)

Table 3.2 Diagnostic criteria for NF1

Presence of two or more:
• Six or more café-au-lait macules (>5 mm in prepubertal, >15 mm in postpubertal)
• Two or more neurofibromas or one plexiform neurofibroma
• Axillary or inguinal freckling
• Optic glioma
• Two or more iris hamartomas (Lisch nodules)
• Bone lesions (thinning of long bone cortex, sphenoid wing dysplasia)
First degree relative with NF1

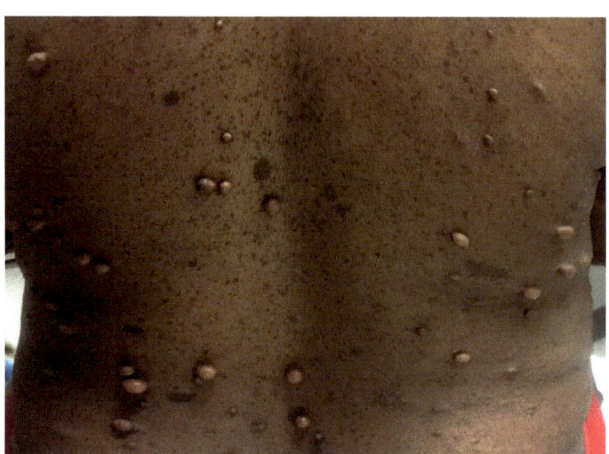

Fig. 3.7 Neurofibromatosis

(b) Clinical Findings:
- Cutaneous:
  - Neurofibromas (Fig. 3.7), Café-au-lait macules (CALMs), axillary/inguinal "freckling" (Crowe's sign), plexiform neurofibroma, juvenile xanthogranuloma (JXG); nevus anemicus
- Additional findings:
  - Ocular: Lisch nodule (iris hamartoma), optic glioma, hypertelorism, glaucoma
  - Skeletal: Macrocephaly, sphenoid wing dysplasia (manifesting as pulsating exophthalmos), scoliosis, thinning of long bone cortex (resulting in tibial bowing and pseudoarthrosis), spina bifida
  - Cardiovascular: hypertension (can be essential or secondary to renal artery stenosis or pheochromocytoma), pulmonary stenosis
  - Neurologic: unidentified bright objects on MRI, learning disabilities, seizures, hydrocephalus; sleep disturbance; headache

Note:
- Plexiform neurofibromas and skeletal defects (sphenoid wing dysplasia) are more likely to be congenital than other findings
- Café-au-lait macules are the earliest cutaneous manifestation

(c) Malignancy Associations
- Juvenile myelomonocytic leukemia (JML) (sometimes with associated juvenile xanthogranuloma (JXG)), pheochromocytoma, optic glioma, malignant peripheral nerve sheath tumor (neurosarcoma), breast carcinoma

Note: Rapid growth or pain within an existing plexiform neurofibroma is concerning for transformation into a malignant peripheral nerve sheath tumor (neurosarcoma)

(d) Screening Recommendations
- Blood pressure at each visit (if hypertension develops obtain renal arteriography and 24-h urine catecholamine and metanephrine measurement due to risk of renal artery stenosis and pheochromocytoma)
- Annual ophthalmology exam
- Annual screening for scoliosis
- Developmental/behavioral assessment and close surveillance for learning disabilities
- Annual dermatologic exam especially if plexiform neurofibroma is present (obtain PET/CT if concern for malignant transformation)
- Mammogram annually for women starting at age 40 years

Note:
- Localized mosaic (segmental) NF1 is characterized by café-au-lait macules, neurofibromas, and/or "freckling" in one or more body segments and is an example of mosaicism due to a postzygotic mutation in NF1.
- **Legius Syndrome** is due to a SPRED-1 mutation (inhibits Ras-Raf interaction inhibiting MAPK pathway). It presents with multiple café-au-lait macules, intertriginous "freckling", mild intellectual disability, and macrocephaly. Unlike NF1, patients do not have associated ocular findings.
- Constitutional mismatch repair deficiency is caused by homozygous or compound heterozygous mutations in the mismatch repair genes associated with Lynch syndrome. The resulting syndrome is similar to NF1 with some patients meeting criteria for NF1 especially with CALM. However they exhibit early onset of Lynch syndrome associated tumors.
- McCune-Albright syndrome is characterized by large CALM with irregular borders, polyostotic fibrous dysplasia, and precocious puberty [3].

9. **Peutz-Jeghers Syndrome**
   (a) Gene Mutation
   - STK11 (LKB1), serine threonine kinase, tumor suppressor
   - Autosomal dominant

(b) Clinical Findings
  • Cutaneous:
    – Mucosal melanotic macules, genital and acral melanotic macules
  • Additional findings:
    – Multiple hamartomatous polyps throughout the GI tract (can result in **intussusception** or hemorrhage)
(c) Malignancy Associations
  • GI adenocarcinoma
  • Ovarian, breast, testicular, uterine, thyroid, and pancreas carcinoma
(d) Screening Recommendations
  • Upper GI endoscopy (EGD) and colonoscopy at age 8, if normal repeat at age 18 then every 3 years [3]
Note: Laugier-Hunziker syndrome is a sporadic condition characterized by longitudinal melanonychia and oral melanotic macules without internal manifestations.

10. **Tuberous Sclerosis**
  (a) Gene Mutation
    • TSC1 (hamartin) and TSC2 (tuberin)
    • Hamartin and tuberin act to inhibit the mTOR pathway via the Rheb (GTPase) protein
    • Autosomal dominant (~75% are spontaneous mutations)
  (b) Clinical Findings (Table 3.3)
    • Cutaneous:
      – Angiofibromas (Fig. 3.8a), fibrous plaque of the forehead, periungual fibromas (Koenen tumors) (Fig. 3.8b), gingival fibromas, hypomelanotic macules (ash leaf (Fig. 3.8c) and confetti macules), connective tissue nevi (Shagreen patch), café-au-lait macules
    • Additional findings:
      – Ocular: retinal hamartomas and astrocytomas

Table 3.3 Diagnostic criteria for tuberous sclerosis: two major OR one major and two minor features

Major features	Minor features
Cutaneous • Facial angiofibromas or forehead plaque • Periungual fibromas • Hypomelanotic macules (3 or more, not including confetti macules) • Connective tissue nevi (Shagreen patch)	**Cutaneous** • "Confetti"hypomelanotic macules • Gingival fibromas
Extracutaneous • Retinal nodular hamartomas • Cortical tubers • Subependymal nodules • Subependymal giant cell astrocytoma • Renal angiomyolipoma • Cardiac rhabdomyoma • Lymphangioleiomyomatosis	**Extracutaneous** • Multiple randomly distributed dental enamel pits • Retinal achromic patch • Cerebral white matter radial migration lines • Multiple renal cysts • Bone cysts • Non-renal hamartoma

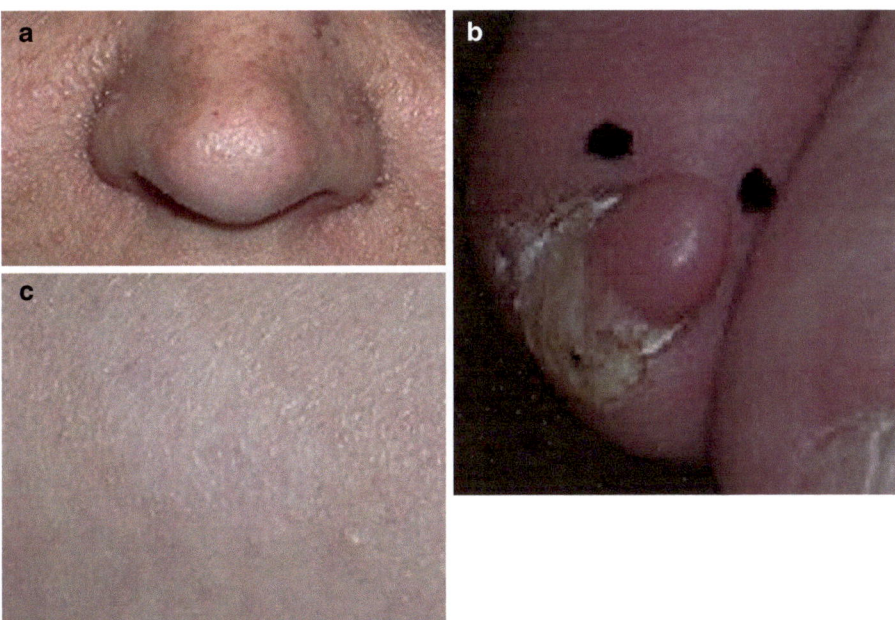

Fig. 3.8 Tuberous Sclerosis. (**a**) Perinasal angiofibromas, (**b**) Periungual fibroma, (**c**) Ash-leaf macule

  - – Skeletal: bone cysts
  - – Dental: enamel pits, gingival fibromas
  - – Pulmonary: lymphangioleiomyomatosis
  - – Renal: angiomyolipomas, cysts
  - – Endocrine: precocious puberty, hypothyroidism
  - – Neurologic: seizures, intellectual impairment, subependymal nodules, cortical tubers, infantile spasms, intracranial calcification
  - – Cardiac: myocardial rhabdomyoma, Wolf-Parkinson-White syndrome

Note: Hypopigmented macules are frequently the first cutaneous sign of tuberous sclerosis. Cardiac rhabdomyomas are frequently the first extracutaneous sign of tuberous sclerosis. Both can present at birth.

Rapamycin and other mTOR inhibitors can be used topically and systemically in the treatment of tuberous sclerosis.

  - (c) Malignancy Associations
    - • CNS tumors (including giant cell astrocytoma), renal carcinoma
  - (d) Screening Recommendations
    - • Neurodevelopmental and behavioral evaluation
    - • Ophthalmology exam
    - • Electrocardiography (ECG) at baseline and as needed
    - • Echocardiography (if <5 years, ECG abnormalities, or cardiac symptoms)
    - • Renal ultrasound (every 1–3 years if normal)
    - • Chest CT (in adult women only to screen for lymphangioleiomyomatosis)

- Cranial CT or MRI (every 1–3 years until adolescence)
- EEG (if history of seizures) [3]

11. **Familial Malignant Melanoma**
    (a) Gene Mutation
        - CDKN2A, encodes p16 and p14
            - p16 inhibits cyclin-dependent kinase (CDK) 4 and CDK 6 preventing phosphorylation of retinoblastoma (Rb) protein preventing G1 to S phase transition in the cell cycle
            - p14 prevents ubiquination of p53 preventing proteasomal degradation and p14 which regulate cell cycle progression via and p53, respectively
        Note: CDKN2A mutations are the most common mutations found in familial melanoma patients. Mutations in melanocortin 1 receptor (MC1R), tyrosinase (TYR), tyrosinase related protein 1 (TYRP1), and SLC45A2 also confer a high risk of developing familial melanoma.
    (b) Clinical Findings
        - Melanocytic nevi (>50), multiple atypical melanocytic nevi, solar lentigines, and ephelides
        - Additional risk factors include a personal history of at least 3 melanomas, at least 1 first or second degree relative with melanoma, relative with pancreatic cancer (Fig. 3.9)
    (c) Malignancy Associations
        - Melanoma
        - Pancreatic cancer
    (d) Screening Recommendations
        - Regular skin exam (at least every 6 months)
        - Endoscopic ultrasonography, endoscopic retrograde cholangiopancreatography (ERCP), +/− MRI to screen for pancreatic cancer [8]

Fig. 3.9 Familial malignant melanoma

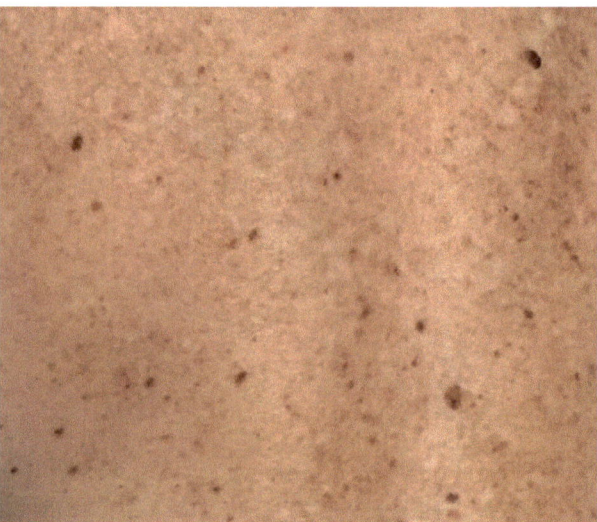

Review Questions and Answers

1. A patient with a history of a spontaneous pneumothorax presents with multiple fibrofolliculomas on the head and neck. The patient should undergo screening for which internal malignancy?
   (a) Colorectal carcinoma
   (b) Medullary thyroid carcinoma
   (c) Medulloblastoma
   (d) Renal carcinoma

   Answer: (d) Renal carcinoma. This patient's clinical picture is consistent with Birt-Hogg-Dubé syndrome. They are at an increased risk of renal carcinomas, specifically chromophobe and hybrid oncocytic types, and should undergo a gadolinium enhanced abdominal MRI starting at age 20 and repeating every 3–4 years.

2. What are the three most common malignancies reported in women with multiple hamartoma syndrome (Cowden syndrome)?
   (a) Breast, endometrial, colon
   (b) Breast, thyroid, endometrial
   (c) Breast, thyroid, colon
   (d) Thyroid, endometrial, colon

   Answer: (b) Breast, thyroid, and endometrial carcinoma are the three most common malignancies found in women with Cowden syndrome. GI hamartomas can affect the entire GI tract but they have a relatively low risk of malignant potential.

3. An infant with a history of a cardiac rhabdomyoma at birth presents with five hypopigmented macules and one café-au-lait macule. You suspect the infant has which genodermatosis?
   (a) Multiple endocrine neoplasia type 1 (MEN1)
   (b) Multiple hamartoma syndrome (Cowden syndrome)
   (c) Neurofibromatosis type 1 (NF1)
   (d) Tuberous sclerosis

   Answer: (d) Tuberous sclerosis. Cardiac rhabdomyomas and hypopigmented macules (3 or more) are the earliest signs of tuberous sclerosis and they are both considered major criteria for diagnosis. Hypopigmented macules and café-au-lait macule are both common in the general population. Hypopigmented macules can also be seen in patients with MEN1. Café-au-lait macules can be seen in tuberous sclerosis but are more common in NF1.

4. A patient presents with multiple café-au-lait macules, axillary "freckling", macrocephaly, intellectual disability, and macrocephaly. NF-1 testing is within normal limits. The patient likely has a mutation in which gene?
   (a) Hamartin

(b) MEN1
(c) MSH2
(d) SPRED1

Answer: (d) SPRED1. Legius syndrome is due to a SPRED1 mutation. It presents with multiple café-au-lait macules, intertriginous "freckling", mild intellectual disability, and macrocephaly. Unlike patients with neurofibromatosis type 1 (NF1), patients with Legius syndrome do not have associated ocular findings.

References

1. Smith MJ, Beetz C, Williams SG, et al. Germline mutations in SUFU cause Gorlin syndrome-associated childhood medulloblastoma and redefine the risk associated with PTCH1 mutations. J Clin Oncol. 2014;32(36):4155–61. Epub 2014 Nov 17.
2. Gupta N, Sunwoo BY, Kotloff RM. Birt-Hogg-Dube syndrome. Clin Chest Med. 2016;37(3):475–86.
3. Holman JD, Dyer JA. Genodermatoses with malignant potential. Curr Opin Pediatr. 2007;19(4):446–54.
4. Patel VM, Handler MZ, Schwartz RA, Lambert WC. Hereditary leiomyomatosis and renal cell cancer syndrome: an update and review. JAAD. 2017;77(1):149–58.
5. Norton JA, Krampitz G, Jensen RT. Mulitple endocrine neoplasia: genetics and clinical management. Surg Oncol Clin N Am. 2015;24(4):795–832.
6. Thakker RV, Newey PJ, Walls GV, et al. Clinical practice guidelines for multiple endocrine neoplasia type 1 (MEN1). J Clin Endocrinol Metab. 2012;97(9):2990–3011. https://doi.org/10.1210/jc.2012-1230. (Screening).
7. Mahdi J, Mester JL, Nizialek EA, et al. Germline PTEN, SDHB-D, and KLLN alterations in endometrial cancer patients Cowden and Cowden-like syndromes: an international, multicenter, prospective study. Cancer. 2015;121(5):688–96. https://doi.org/10.1002/cncr.29106. Epub 2014 Nov 5.
8. Soura E, Eliades PJ, Shannon K, et al. Hereditary melanoma: update on syndromes and management: genetics of familial atypical multiple mole melanoma syndrome. J Am Acad Dermatol. 2016;74(3):395–407.

Systemic Implications of Melanoma

4

Ryan M. Svoboda, Giselle Prado, and Darrell S. Rigel

Abbreviations

AAD   American Academy of Dermatology
ASCO   American Society of Clinical Oncology
cSCC   Cutaneous squamous cell carcinoma
CT   Computed tomography
FAMM   Familial atypical multiple mole
FISH   Fluorescent in situ hybridization
IBD   Inflammatory bowel disease
KA   Keratoacanthoma
OR   Odds Ratio
PET   Positron emission tomography
PUVA   Psoralen plus ultraviolet A
RR   Relative Risk
SDDI   Sequential digital dermoscopy imaging
SPF   Sun protection factor
SSO   Society of Surgical Oncology
UV   Ultraviolet
UVB   Ultraviolet B

R. M. Svoboda (✉)
Department of Dermatology, Penn State Health, Hershey, PA, USA

G. Prado
Hough Graduate School of Business, University of Florida, Gainesville, FL, USA

D. S. Rigel
Clinical Professor of Dermatology, New York University Medical Center, New York, NY, USA

© Springer Nature Switzerland AG 2021
V. Liu (ed.), *Dermato-Oncology Study Guide*,
https://doi.org/10.1007/978-3-030-53437-0_4

Learning Objectives

1. To understand how melanoma as a cutaneous disease can have systemic impact.
2. To recognize risk factors for development of melanoma and list prevention strategies.
3. To articulate the appropriate steps for diagnosis, staging, and management of melanoma.
4. To describe the benefits as well as the cutaneous adverse effects of systemic therapeutics for melanoma.

Epidemiology

Incidence and Mortality

- Incidence of melanoma in the United States continues to rise dramatically; systemic involvement (regional and/or distant metastases) is apparent at diagnosis in approximately 15% of cases [1].
  - Estimates predict 91,270 newly diagnosed cases of invasive melanoma and an additional 87,290 cases of melanoma *in situ* in 2018 [2]. This represents a considerable increase from 2017, which saw an estimated 87,110 new cases of invasive melanoma and 74,680 cases of *in situ* disease [3].
  - Overall lifetime risk of developing invasive melanoma for an individual living in the United States in the year 2018: 1/47, up from 1/150 in 1985 [4].
  - Overall lifetime risk of developing invasive **or** *in situ* melanoma for an individual living in the United States in the year 2018: 1/24 [4] (Fig. 4.1).
  - *2020 lifetime risk represents an estimated value
  - Highest risk: non-Hispanic white individuals
    1/27 (men) and 1/45 (women) chance of developing invasive melanoma [5].
  - These numbers may significantly underestimate the true incidence of the disease, as multiple studies have demonstrated significant underreporting of new melanoma diagnoses to centralized cancer registries by physicians [6–8].
- Similar trends seen worldwide in recent decades.
  - Developed nations and those with populations comprised primarily of non-Hispanic white individuals have seen the largest increases.
  - However, some recent data suggests that rates may be stabilizing in cohorts of younger individuals, particularly in Australia, New Zealand, the United States, and Canada [9].
- Increased incidence may partially be related to increased detection of existing disease due to promotion of secondary prevention efforts [10].
  - However, incidence of tumors of all thicknesses has increased—indicating a true rise in melanoma incidence to at least some extent [11].
- Morbidity and mortality related to melanoma is substantial.
  - Number of deaths related to melanoma exceeds that of all other skin cancers combined.
  - Directly related to extent of disease spread.

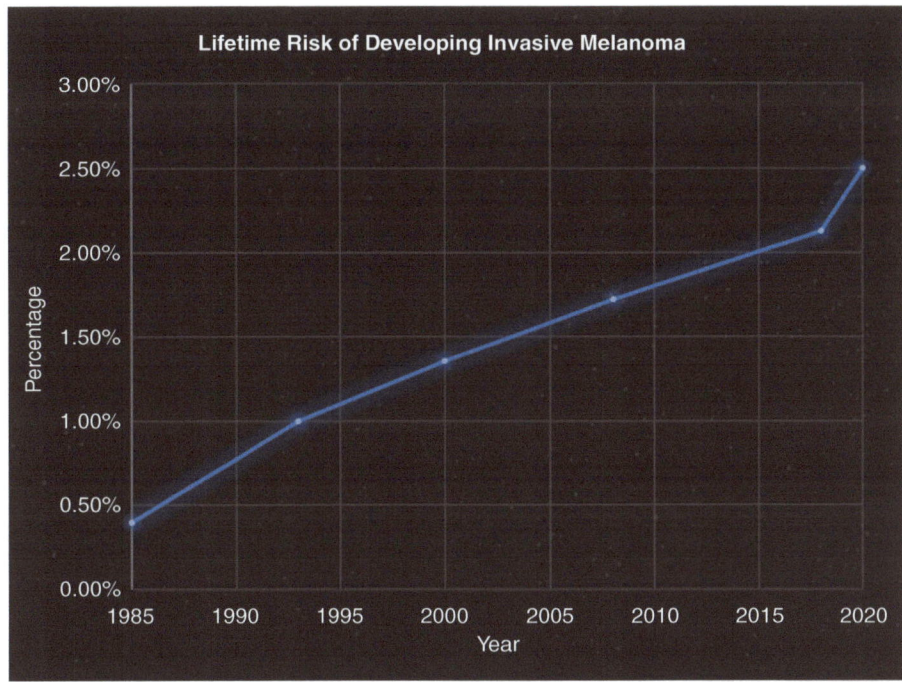

Fig. 4.1 Line graph depicting lifetime risk of invasive melanoma between 1985 and 2020

> Long-term survival rates ~99% for localized disease versus 63% in those with evidence of metastasis to regional lymph nodes and 20% in patients with distant metastasis [3].

- After steadily increasing for many years, melanoma-related mortality in the United States has leveled out and is starting to demonstrate a slight decline [12].

  > Annual melanoma-related deaths peaked at 10,130 in 2016.

  > Expected to reach a 5-year low of 9320 in 2018 (roughly one death attributable to melanoma occurs every hour).

- Five-year survival rates have steadily increased from 81.8% in 1975 to 93.1% in 2009 and continue to improve [3, 13].

  > Related to enhanced secondary prevention efforts and evolving treatment options for advanced melanoma.

Risk Factors

- Ultraviolet radiation
  - Single most important risk factor for the development of melanoma.

    > History of intense, intermittent sun exposure (especially with blistering sunburn) more strongly associated with melanoma risk than overall cumulative UV exposure (OR 1.71 vs 1.18) [14].

    > History of indoor tanning associated with dose-dependent increase in risk of melanoma development [15].

- Age
  - Incidence rises with increasing age, as does the risk of melanoma-related death.
  - Mean age at diagnosis = 64 years [13].
  - Despite an overall predilection for older individuals, melanoma is the most common cancer in men under the age of 50 and third most common in women of this age group [16, 17].
- Gender
  - Slightly more common in men than women overall [13].
    Gender predilection more pronounced in Caucasian individuals.
    However, in younger individuals (under the age of 40), incidence is higher in women.
    - Likely **partially** related to higher rate of indoor tanning among this demographic [18].
    - Recent evidence demonstrating this trend in individuals of all skin types suggests that tanning practices do not account entirely for this finding [19].
- Ethnicity
  - Incidence varies across ethnic groups.
  - Highest incidence rates seen in Caucasians, followed by Hispanics, Asians, and black individuals.
    Despite lower incidence, 5-year survival rates in Hispanic and black populations have consistently been shown to be lower than in Caucasians.
  - Distribution of melanoma subtypes also varies by ethnicity.
    For example, acral lentiginous melanoma is the least common subtype in Caucasians and the most common subtype in black individuals [20].
  - Origin of many malignant melanomas in certain ethnic groups (e.g. African Americans and Asians) may be unrelated to UV exposure, as evidenced by a propensity to develop lesions in non-sun-exposed anatomic locations [21].
- Family History
  - Most cases of melanoma are sporadic, but approximately 1/10 occur in patients with a family history of melanoma [22].
  - Familial atypical multiple mole (FAMMM) syndrome represents a subset of familial cases that is truly hereditary in nature (as opposed to being related to similar exposures).
    Caused by a single germline mutation in one of several genes, most commonly *CDKN2A* or *CDK4*, leading to increased UV susceptibility [23].
    Certain FAMMM-associated gene mutations also carry an increase risk of developing other internal malignancies.
    Patients require close dermatologic follow-up
    - More likely to develop melanoma at a younger age
    - Have a greater chance of developing multiple lesions throughout their life

- Genetic testing should be considered for certain populations:
    - Patients with 3 or more invasive melanomas, particularly for those arising before age 45 years old
    - Patients with 3 or more blood relatives on one side of the family with history of melanoma or pancreatic cancer
- Skin Phenotype
    - Fair-skinned individuals and those more prone to sunburn (e.g. Fitzpatrick Skin Type I and II) display a higher risk of melanoma.
    - Number of nevi appears to correlate with risk of developing melanoma.
        Individuals with over 100 nevi have a substantially increased risk of developing melanoma compared to those with fewer than 15 nevi (RR 6.89, 95% CI 4.63–10.25) [24].
        Individuals with 5 or more atypical nevi have a significantly increased risk of developing melanoma, and are more likely to present with more advanced lesions [25].
- Inflammatory Bowel Disease
    - Incidence of melanoma appears increased in patients with IBD.
    - Meta-analysis of 172,837 patients with IBD from 1940 to 2009: 37% increased relative risk of developing melanoma compared to the general population [26].
    - Risk slightly higher in the subset of patients with Crohn's disease versus those with ulcerative colitis.
- Immunosuppression
    - Solid organ transplant patients have an approximately 2.5 times increased risk of developing melanoma when compared to the general population [27, 28].
    - Melanoma accounts for greater proportion of subsequent skin malignancies in pediatric transplant patients compared to adult transplant patients [29].
    - Donor-derived metastatic melanoma is rare but has been reported on multiple occasions [30], underscoring the importance of careful donor selection and exclusion of malignant melanoma prior to transplantation.
- Phototherapy
    - Treatment with phototherapy (both PUVA and narrow-band UVB therapy) is associated with an increased risk of developing melanoma in patients with psoriasis [31].
        Magnitude of increased risk is less than that for non-melanoma skin cancers [31].
    - Careful evaluation of risks and benefits is esseential prior to initiation of phototherapy for the treatment of psoriasis.
    - Close monitoring for the development of skin cancer during and after completion of therapy is imperative.
- Parkinson's Disease
    - Associated with an overall decreased risk of malignancy (all types in aggregate), but a higher risk of malignant melanoma.
    - Etiology: unclear and likely multifactorial in nature, involving both genetic and environmental risk factors [32].

Prevention

Primary Prevention on an Individual Level

- Regular sunscreen use
  - Effective in preventing the development of malignant melanoma.
  - Regular, daily use is more beneficial than discretionary use [33].
  - Estimates suggest that a 5% annual increase in sunscreen usage over the next 10 years would lead to 231,053 fewer melanomas being diagnosed in the Caucasian population of the United States over a period of 15 years [34].
- The impact of SPF
  - Traditionally, it has been felt that since sunscreens with SPF 30 block approximately 96% of UVB rays, usage of higher SPF products would offer little in terms of added benefit.
  - However, these numbers come from laboratory studies in which the density of sunscreen application was higher than that routinely utilized by consumers in real-world settings.
  - In actuality, consumers typically apply sunscreen at densities far less than that used to determine the SPF value listed on the product label. Additionally, reapplication data confirms that consumers typically reapply sunscreen far less frequently than the intervals recommended on sunscreen labels [35].
  - In actual use conditions, SPF 100+ sunscreen has been shown to be significantly more effective than SPF 50+ sunscreen in the prevention of sunburn [35]. Although long-term data tracking melanoma risk are not currently available, it can be inferred that higher SPF sunscreens likely confer an advantage in terms of melanoma risk reduction (given the established epidemiologic association between sunburn and subsequent melanoma development).

Public Health Initiatives

- ABCD(E) Campaign and Skin-Self Examination
  - ABCDs of melanoma: originally developed in 1985 by the New York University Melanoma Cooperative Group.
  - Purpose: to improve early detection of melanoma by creating an organized framework that could be applied not only by healthcare providers but also by patients and their loved ones, who initally detect up to 65% of melanomas using skin-self examination [36, 37].
  - Algorithm uses a simple pneumonic to help individuals evaluate their pigmented skin lesions for the common morphologic features of melanoma:
    - A-Asymmetry
    - B-Border irregularity
    - C-Color variegation
    - D-Diameter >6 mm
  - The ABCD framework, with the later addition of "E" for evolution, has allowed for earlier identification of lesions of malignant melanoma while still in a curable phase [38].

- Public Screening Programs
  - May serve to increase access to secondary prevention services.
  - One such program, the SPOTme® Campaign sponsored by the American Academy of Dermatology, has identified over 28,500 malignant melanomas since its inception in 1985 [39].
- Public health programs targeted at secondary prevention are synergistic with primary preventive efforts such as proper sun hygiene practices.

Work-up and Evaluation

Clinical Skin Examination

- Screening guidelines vary, but the American Academy of Dermatology endorses regular total body clinical skin examination for high-risk patients [40].
  - Personal history of skin cancer
  - Melanoma in two or more first-degree relatives
  - Increased number of melanocytic nevi (>100)
- Entire skin surface should be examined, including often-overlooked areas such as the scalp, interdigital folds, and perianal area.
- Photography may be useful for monitoring for evolution of pigmented lesions overtime, as ~20% of melanomas arise from preexisting nevi [41].

Biopsy Considerations

- Cytologic analysis alone is insufficient in the assessment of pigmented lesions concerning for melanoma.
  - Tumor thickness is important in determining stage and has implications on subsequent evaluation and treatment.
  - Choice of biopsy technique should take this into consideration; curettage biopsy should be avoided.
- Both the American Academy of Dermatology and the National Comprehensive Cancer Network recommend narrow-excision biopsy (excisional biopsy) with at least 0.5 cm margins for pigmented lesions concerning for melanoma [42, 43].
- Actual management varies significantly from published recommendations, with shave biopsy (35%) being employed most commonly by dermatologists, followed by narrow excision (31%), saucerization/scoop shave (12%), punch (11%), and wide-local excision (3%) [44] (Fig. 4.2).
  - Of these techniques, punch biopsy appears to have highest rate of false negative diagnosis (23%), followed by shave (4.5%) and excisional biopsy (1.7%) [45].
  - In 95% of cases, the thickness from shave biopsy is predictive of the final stage.
- Ultimately, approach to biopsy should take into account lesion and patient-specific characteristics, but the goal should be to ensure highest degree of diagnostic accuracy so that early treatment can be offered, decreasing the risk of subsequent systemic involvement.

Fig. 4.2 US
Dermatologists' preferred
biopsy technique for
pigmented lesions
suspicious for melanoma

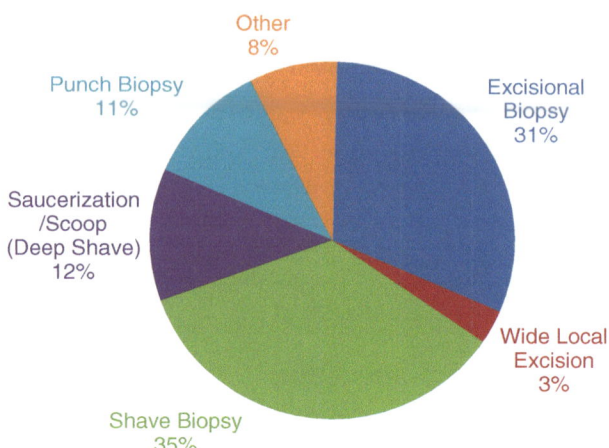

Adjunctive Diagnostic Tests

- Dermoscopy
  - Shown to improve detection of early melanoma compared to visual examination alone (potentially allowing for definitive treatment prior to systemic involvement) [46].
  - Sequential digital dermoscopy imaging (SSDI) may further enhance early detection.
- Advanced imaging modalities
  - Reflectance confocal microscopy
      High sensitivity (>90%) for the detection of melanoma [47].
      Requires significant training; results are user-dependent.
  - Optical coherence tomography
      High sensitivity (~80%) and specificity (>90%) for the diagnosis of melanoma [48].
      Like reflectance confocal microscopy, requires significant skill.
      May be confined with reflectance confocal microscopy to further enhance diagnosis.
- Molecular assays (Table 4.1)
  - Pigmented lesion assay
      Non-invasive 2-gene molecular test.
      Sample obtained by clinician via an adhesive patch used to strip tissue from a pigmented lesion.
      Results are meant to aid in the decision to biopsy; may be particularly useful in cosmetically sensitive areas.
      High sensitivity (~95%) and low-moderate specificity (30–50%) [49].
  - Fluorescent in situ hybridization (FISH) [50]
      Used as an adjuvant to histopathology in equivocal cases.

Table 4.1 Commercially available genomic testing for melanoma

Clinical question	Test
Decision to biopsy?	2 GEP
Diagnostically challenging or ambiguous histopathology?	FISH
	23 GEP
Prognostic risk for local recurrence/distant metastases?	31 GEP[a]

[a]See discussion below—Adjunctive Molecular Tests for Prognosis

- 23-gene expression profile (23-GEP) test [51]
  Quantifies expression level of 23 genes from the primary tumor.
  Like FISH, meant for use to aid in diagnosis of melanoma in cases with challenging histopathologic features.

Management

Staging (Tables 4.2 and 4.3)

- Survival by AJCC Stage
  - Both 5-year and 10-year survival show a clear, inverse relationship with AJCC stage.
    Stage I melanoma: 10-year survival exceeds 80%.
    Stage IV melanoma: 10-year survival approximately 20% [52].
  - Proper staging allows for communication of prognostic information and also has implications on treatment strategies.

Staging Work-Up
- All patients with newly-diagnosed melanoma require a complete physical examination that includes palpation of all superficial lymph node basins [42].
  - Patients with palpable lymphadenopathy should be offered therapeutic lymph node dissection.
- A comprehensive review of systems should be aimed at uncovering symptoms of distant metastasis (e.g. new onset neurologic symptoms should prompt imaging for brain metastases).

Role of Imaging Studies
- Asymptomatic individuals without suggestion of regional or distant metastases based on comprehensive history and physical exam generally do not require further bloodwork or imaging studies based on current AAD recommendations [42].
  - High rate of false-positives in this population.
  - Not cost-effective.
  - Risks of ionizing radiation (e.g. x-ray, CT, PET).
- Patients with evidence of systemic involvement
  - Patients with micrometastatic regional lymph node involvement.
    Low rate of distant metastases, but CT or PET could be considered as visceral involvement would alter available treatment options [42].

Table 4.2 AJCC T-classification eighth edition [52]

T-classification	Breslow depth (mm)	Ulceration
In situ	N/A	N/A
T1a	<0.8	–
T1b	<0.8	+
	0.8–1.0	+/–
T2a	1.1–2.0	–
T2b	1.1–2.0	+
T3a	2.1–4.0	–
T3b	2.1–4.0	+
T4a	>4.0	–
T4b	>4.0	+

Table 4.3 AJCC staging system corresponding to TNM classification [52]

Stage	T	N	M
0	In situ	0	0
IA	1a-1b	0	0
IB	2a	0	0
IIA	2b-3a	0	0
IIB	3b-4a	0	0
IIC	4b	0	0
IIIA	1a/b-2a	1a-2a	0
IIIB	1a-2a	1b-2b	0
	2b-3a	1a-2b	0
IIIC	1a-3a	2c-3c	0
	3b-4a	1a-3c	0
	4b	1a-2c	0
IIID	4b	3a-3c	0
IV	1a-4b	0-3c	1

- Patients with clinically palpable regional lymph nodes.

  Higher detection rate of occult distant metastases; advanced imaging modalities should be considered.

Role of Sentinel Lymph Node Biopsy and Completion Lymph Node Dissection

- The goal of sentinel lymph node biopsy (SLNBx) is to detect micrometastatic regional disease in patients without clinically apparent lymph node involvement (palpable lymphadenopathy).
- Traditionally, patients with positive SLNBx were offered completion lymph node dissection.
- However, the role of and indications for sentinel lymph node biopsy (SLNBx) and subsequent recommendations for clinical lymph node dissection in malignant melanoma are changing as more data on its impact become available.
  - Multicenter Selective Lymphadenectomy I (MSLT-I) Trial [53]

    Multicenter, prospective trial randomizing melanoma patients without clinically apparent lymphadenopathy to wide local excision and SLNBx (with clinical lymph node dissection if positive) versus wide local excision with nodal observation (with therapeutic lymph node dissection for development of clinically apparent lymphadenopathy).

SLNBx with biopsy-based management improved disease-free survival but not melanoma-specific survival in the overall study group. Among patients with tumors 1.2 mm or thicker and positive nodal metastases, biopsy-based management did lead to improved melanoma-specific survival.

– Multicenter Selective Lymphadenectomy II (MSLT-II) Trial [54]

Multicenter, prospective trial randomzing patients with positive SLNBx to immediate clinical lymph node dissection versus frequent ultrasonographic nodal observation (with therapeutic lymph node dissection for development of lymphadenopathy detectable by ultraound).

Immediate clinical lymph node dissection provided prognostic information and was associated with an increased rate of regional disease control but did not improve melanoma-specific survival.

Complications (lymphedema) were 4× higher in the nodal dissection group.

– DeCOG-SLT [55]

Multi-center trial randomizing patients with SLN positive, ≥1 mm thick melanoma of the torso or extremities to immediate completion lymph node dissection versus observation.

Immediate completion lymph node dissection was not associated with lower overall or distant-metastasis-free survival. However, sample size was not adequate to providie the satatsitcal power to demonstrate a statistically significant difference between the two groups.

High rate of adverse events was observed in the dissection group.

- Current American Society of Clinical Oncology (ASCO) and Society of Surgical Oncology (SSO) Joint Guidelines [56]

– SLNBx (for patients with no clinically apparent nodal disease)

Thin lesions

- T1a (non-ulcerated, <0.8 mm).
  – Routine SLNBx is *not* recommended.
- T1b (0.8–1.0 mm with or without ulceration, <0.8 mm with ulceration).
  – SLNBx may be considered on a patient-specific basis after thorough discussion of the risks/benefits/alternatives.

Intermediate-thickness lesions

- T2–T3 (1.0–4.0 mm).
  – SLNBx is recommended.

Thick lesions

- T4 (>4.0 mm)
  – SLNBx may be recommended to patient after a thorough discussion of the potential risks and benefits of the procedure, as well as the alternatives.

– Completion lymph node dissection

For patients with positive SLNBx with low-risk features, careful observation or completion lymph node dissection can be offered. Patients with high-risk clinical features should generally warrant completion lymph node dissection; observation should only be offered after thorough discussion of the risks and benefits of this strategy.

High-risk features: immunosuppression, extracapsular extension, concomitant microsatellitosis of the primary lesion, micrometastasis to >3 lymph nodes, and micrometastasis in >2 nodal basins.

Adjunctive Molecular Testing for Prognosis

- 31 Gene Expression Profile (31-GEP) Test
  - Assesses prognosis in patients with Stage I or II invasive melanoma

    Test performed on formalin-fixed, paraffin-embedded tissue.

    Quantifies expression level of 31 genes from the primary tumor.

    Applies a validated algorithm to dichotomize patients into low-risk (Class 1) and high-risk (Class 2) for the subsequent development of metastatic disease.

    The results of the test have been shown to accurately differentiate patients based on 5-year recurrence-free and distant-metastasis-free survival rates in multiple, independent studies [57–61].

    - 5 year disease-free survival
      - Class 1 ~97%
      - Class 2 ~37%

    May allow for more individualized approach to follow-up and surveillance for metastasis.

    Results are synergistic with information obtained from AJCC staging and SLNBx.

    - When 31-GEP results are combined with AJCC Individualized Melanoma Patient Outcome Prediction Tool, identification of patients with high-risk melanomas is improved above either individual tool [62].

Treatment

Surgical Treatment

- Surgical management (wide local excision) remains the mainstay of therapy for localized melanoma.
- Goals of wide local excision in the treatment of cutaneous melanoma:
  - To completely excise all malignant tissue, preventing systemic spread of disease.
  - To minimize morbidity, cosmetic disfigurement, functional impairment, and cost.
- Surgical considerations
  - Timing of surgical excision is of paramount importance, particluarly in early-stage disease.

    On multivariate analysis stratified by stage, increased time from biopsy (over 30 days) has been associated with increased risk of melanoma-related mortality in patients with Stage I disease [63].

Table 4.4 Currently recommended surgical margins for malignant melanoma

Breslow thickness	Surgical margins
In-situ	5 mm
<1 mm	1 cm
1–2 mm	1–2 cm
>2 mm	2 cm

    Ensuring definitive treatment of Stage I melanoma patients within 30 days of diagnosis improves patient-oriented outcomes.

    Prevention of progression to systemic disease is of paramount importance.

- Recommended margins for excision vary depending on the thickness of the tumor (Table 4.4).

    Principle of treatment: ensure complete eradication of disease while allowing for a cosmetically acceptable response and minimum morbidity.

- Surgery for metastatic melanoma
  - Surgical removal of metastases (metastasectomy) may be indicated in appropriately selected patients with isolated metastases [64].
  - Combinatation of metastasectomy and systemic medical therapy may further enhance survival.
  - Despite recommendations, available data suggest that 5 mm margins for melanoma *in situ* may be inadequate when compared to more conservative margins of excision ranging from 6 to 9 mm [65].
  - Controversy also exists regarding the necessary margin of excision for patients with relatively thick primary lesions (> 2 mm).

    Prior data suggested an increased frecquency of local recurrence in patients undergoing excision with narrow margins (1 cm).

    Recent studies have attempted to elucidate whether a long-term survival advantage exists when wide margins are employed.
    - Multicenter trial of 900 patients with Breslow thickness > 2 mm aimed to investigate if narrower margins (1 cm) could produce similar results [66].
      - Patients randomized to narrow (1 cm) versus wide (3 cm) margins.
    - Patients in the narrow margin group demonstrated a statistically significant reduction in melanoma-specific, but not overall survival, out to 12 years.
    - Results suggest that 1 cm excision margins may be inadequate for cutaneous melanoma with Breslow thickness greater than 2 mm.

    More studies are needed to investigate the adequacy of 1 cm excision margins for thinner melanomas (<2 mm) that exhibit other poor prognostic factors.

Medical Therapies for Systemic Melanoma

- Traditional chemotherapy regimens have not yielded desirable results in the treatment of melanoma, with low response rates, no significant improvement in overall survival, and a high incidence of toxicity [67].

Table 4.5 Molecular targets and corresponding therapies in melanoma

Immune checkpoint inhibition		Targeted antitumor therapy	
PD-1	*CTLA-4*	*BRAF*	*MEK*
Nivolumab	Ipilumab	Vemurafenib	Trametinib
Pembrolizumab		Dabrafenib	Cobimetinib

- Isolated limb perfusion with regional administration of chemotherapy may be a viable option for appropriate patients with regional lymph node metastases [43].
- Other traditional systemic therapies, such as high-dose interferon have shown variable results but likely lead to some measurable improvement in disease-free survival [43].
- An evolving understanding of the derangement of cell-signaling processes involved in the malignant transformation of melanocytes has led to multiple therapeutic targets.
- These targets provide options for systemic treatment of advanced, metastatic (Stage IV) melanoma when complete surgical excision of all cancerous cells is not possible.
- In addition to the growing number of molecular targets, genetics is playing an increasing role.
  - The presence of and ability to test for certain "driver" mutations allows for an individualized treatment approach.
- Current targeted therapeutic options for advanced melanoma fall under two categories (Table 4.5):
  - Immune checkpoint inhibition
  - Targeted antitumor therapy

Immune Checkpoint Inhibitors (Immunotherapy)
- Melanoma cells undergo predictable changes in the expression of cell surface ligands that limits the body's immune response against the cancer.
- Due to the nature of the mechanism of action of immunotherapy, it is not unusual for there to be a delay in the time from onset of therapy to achievement of an objective response when compared to classic chemotherapy.
- Occasionally, patients will experience transient worsening of disease upon initiation of therapy.
  - Particularly important to take into account when considering early discontinuation of therapy in patients who experience worsening of lesions or development of new lesions shortly after beginning therapy with immune checkpoint inhibitors.
- Age >65 independent predictor (in univariate and multivariate regression models) of improved progression-free and overall survival following checkpoint inhibitor therapy [68].
- Anti-programmed cell death receptor 1 (PD-1) antibodies
  - PD-1: receptor expressed on the surface of activated T cells. Binds to corresponding ligands on melanoma cells, leading to T cell to recognize the tumor cell as "self," thereby allowing it to evade destruction by the immune system.

- Anti PD-1 antibodies block the binding of PD-1 to its ligands, thereby allowing the activated T cells to recognize the melanoma cells as "foreign," leading to a targeted immune response.
- In patients with PD-L1-negative tumors, the combination of PD-1 and CTLA-4 blockade has been shown to be more effective than either agent alone [69].
- PD-1 blockade may be particularly useful in the treatment of advanced desmoplastic melanoma, a rare subtype of melanoma characterized by blunted response to most traditional chemotherapy regimens and a low incidence of targetable driver mutations [70].

   In one study of 60 patients with desmoplastic malignant melanoma treated with anti-PD-1 therapy, 42 (70%) experienced an objective response with 19 (32%) experiencing a complete response [70].

   The benefit likely results from a high mutational burden in these tumors and frequent pre-existing adaptive immune response that is limited by melanoma cells expression of PD-L1.
- The makeup of the gut microbiome appears to influence the efficacy of PD-1-based immunotherapy, with certain bacteria being associated with a positive response to therapy. Likewise, non-responders appear to have an imbalance of GI flora, possibly leading to impaired cellular immunity [71].

   Maintenance of favorable GI flora could help ensure maximum efficacy of anti-PD-1 therapy.

   These data suggest a potential role for fecal transplant in a subset of patients prior to onset of anti-PD-1 therapy, although further studies are needed to investigate this possibility [72].
- Nivolumab

   Fully human IgG4 monoclonal antibody that blocks the PD-1 receptor.

   Indications for use of nivolumab in advanced melanoma:
   - BRAF V600 wild-type unresectable or metastatic melanoma, as a single agent.
   - BRAF V600 mutation-positive unresectable or metastatic melanoma, as a single agent.
   - Unresectable or metastatic melanoma, in combination with ipilimumab.

   Improves progression-free survival in a significant subset of patients, with the majority of those who respond to therapy demonstrating a sustained response beyond 1 year [73].

   Drug related adverse events occur in approximately 71% of patients treated. Most common adverse events experienced are fatigue (25%). Pruritus (17%), diarrhea (13%), and rash (13%) [74].

   Drug-related serious adverse events occur in over 10% of those treated. Serious adverse events that have been associated with nivolumab therapy include pneumonitis, vitiligo, colitis, hepatitis, hypophysitis, and thyroiditis.
- Pembrolizumab

   IgG4 monoclonal antibody and PD-1 inhibitor indicated for use in advanced melanoma.

Common adverse events: were fatigue (21%), rash (15%), pruritus (14%), asthenia (12%), and hypothyroidism (10%) [69].

In clinical trials, pembrolizumab led to significant improvement in both progression-free and overall survival with less high-grade toxicity when compared to ipilimumab [69].

- Anti-CTLA-4 antibodies
  - Cytotoxic T lymphocyte associated antigen 4 is an immune checkpoint receptor that transmits an inhibitory signal to T cells in order to downregulate the immune response [75]. The gene encoding CTLA-4 is constitutively activated in certain cancers.
  - Ipilimumab is a human monoclonal antibody directed against CTLA-4. By antagonizing CTLA-4, ipilimumab activates proliferation of tumor-specific T cells, leading to immune-mediated tumor regression.
  - Ipilimumab has been shown to improve overall survival in patients with advanced melanoma [76]. Both response rate and adverse event rate are dose-dependent [77].
  - The most common immune-related adverse events in patients treated with ipilimumab are hepatotoxicity, dermatitis, diarrhea, hypophysitis, and uveitis [78]. Other dermatologic adverse events include pruritus and vitiligo.
  - Combination of radiation therapy and CTLA-4 inhibition may improve progression-free survival, but resistance related to upregulation of PD-L1 by melanoma cells occurs [79].

    Resistance may be overcome through addition of anti-PD-1 agent.
- Combination checkpoint inhibitor therapy with anti-PD-1 and anti-CTLA-4 therapy
  - Survival outcomes for patients with unresectable, Stage III or IV melanoma treated with a combination of nivolumab and ipilimumab are encouraging, with 3-year survival rates exceeding 60% in patients treated with escalating doses [80].
  - Among previously untreated patients with metastatic melanoma with PD-L1 negative tumors, the combination of PD-1 and CTLA-4 blockade appears to be more effective than either agent alone [81].
  - Grade 3 and 4 treatment-related adverse events are common in patients treated with this combination (59%) [80].

Targeted Antitumor Therapy

- Our evolving understanding of the common oncogene mutations involved in the development of melanoma has led to the elucidation of specific molecular targets for which systemic therapy can be directed.
- The RAS-RAF-MEK-ERK pathway is an important cell-signaling pathway that involves a cascade of signaling initiated by a transmembrane receptor tyrosine kinase [82].

- This leads to the activation of a G-protein, RAS, followed by the subsequent activation of three sequential protein kinases, RAF, MEK, and ERK. Following the activation of ERK, the protein translocates to the nucleus, where it serves as an activator of nuclear transcription factors.
- Gene mutations resulting in constitutive activation of this pathway leads to the overexpression of genes that allow transformed cells to escape the cell cycle.

  For example, a single valine→glutamic acid amino acid substitution of *BRAF*, the gene encoding the RAF protein (V600E mutation) leads to constitutive activation of the RAS-RAF-MEK-ERK pathway and is seen in 41–53% of melanomas [83, 84].

  A glutamate→arginine or glutamate→lysine substitution at codon 61 of *NRAS*, the gene encoding RAS, also leads to constitutive activation independent of external inputs, and is seen in 17–29% of melanomas [83, 84].

  Therapies inhibiting these proteins allow for targeted treatment in patients with driver mutations leading to constitutive activation of the RAS-RAF-MEK-ERK pathway.

  The frequencies in which mutations in *BRAF* and *NRAS* (the gene encoding RAS) are observed appear to vary depending on the histologic subtype of melanoma and the degree of sun exposure, indicating that melanoma is a heterogenous disease and that personalized treatment approaches based on the genetic signature of an individual's malignancy is feasible [84].

- Development of resistance to inhibitors of the RAS-RAF-MEK-ERK pathway is common and may be delayed by the use of combination therapy targeting multiple steps in the pathway (i.e. BRAF and MEK inhibitors).
- BRAF Inhibitors
  - BRAF inhibitor therapy is efficacious in melanomas with a positive BRAF driver mutation.
  - Not FDA-approved at this time for treatment of wild-type BRAF lesions (absence of BRAF mutation).

    Mutation tested for via testing performed on a formalin-fixed, paraffin-embedded pathology specimen.
  - Vemurafenib

    BRAF inhibitor approved for the treatment of unresectable or metastatic melanoma with a proven BRAF V600E mutation.
    - ~50% with BRAF V600E mutation respond to therapy [85].

    Superior 6-months overall (84% vs 64%, $p < 0.001$) and progression-free survival rates compared to dacarbazine in Phase 3 randomized trial [86].

    Median progression-free survival >7 months; median overall survival = 16 months [87, 88].

    Most common adverse events: arthralgia (34%), development of keratoacanthoma (KA) or squamous cell carcinoma (cSCC) (31%), rash (25%), nausea (15%), photosensitivity (15%), fatigue (12%), pruritus (12%), and lymphopenia (6%) [86].

- In Phase 3 trials, 38% of patients required dose reduction secondary to treatment-related adverse events [86].
- The vast majority of vemurafenib-associated keratinocyte neoplasms are keratoacanthomas, with only ~10% representing invasive cSCC [88].
  - Development of KAs or SCCs typically occurs early in the treatment course (median 8 weeks) [88].
  - Higher incidence of NRAS mutations (60%) in KAs/cSCCs in patients treated with vemurafinib compared to patients with non-vemurafenib-associated lesions [89].
    Pre-existing keratinocytes with NRAS mutations may be unmasked by vemurafinib.
- Photosensitivity: lowers UVA minimal erythemal dose
  - Broad-spectrum sunscreen and UV-protective clothing should be recommended to patients on vemurafenib [90].

Vemurafenib resistance
- Development of resistance is common and problematic.
- Typically occurs in 2–18 months [85].
- Related to bypass of RAS-RAF-MEK-ERK pathway by tumor cells [91].
- Heat shock protein-90 (HSP90) inhibition may be an effective strategy at combating multiple divergent resistance pathways [91].

- Dabrafenib

  BRAF inhibitor approved, in combination with trametinib (MEK inhibitor) for the treatment of unresectable or metastatic melanoma with demonstrated BRAF V600E or V600K mutations [92].

  Like vemurafenib, dabrafenib has demonstrated improved progression-free survival compared to dacarbazine.

  Treatment-related adverse events noted in ~50% of patients [92].
  - Most common adverse events: skin-related toxic effects, fever, fatigue, arthralgia, and headache [92].

- The BRAF inhibitor paradox

  BRAF inhibitors lead to inhibition of the RAS-RAF-MEK-ERK pathway in cells harboring a BRAF mutation, but paradoxically activate the signaling pathway in cells void of a mutation [93].

  Preexisting benign pigmented lesions are at risk for malignant transformation in the setting of BRAF inhibitor therapy [94].

  Risk of developing a second primary melanoma while on BRAF inhibitor therapy does not appear to correlate with the length of exposure to therapy [95].

Total body photography and nevus monitoring with dermoscopy may be a useful secondary prevention strategy in patients on BRAF inhibitors [95].
- MEK inhibitors
  - Inhibit mitogen-activated protein kinase enzymes MEK1 and/or MEK2.
  - Used in the treatment of *BRAF*-mutated advanced melanoma.
  - Trametinib

    MEK1/MEK2 inhibitor approved for use as a single agent or in combination with dabrafenib in patients with unresectable or metastatic melanoma with a proven BRAF V600E or BRAF V600K mutation [96]. Also approved in combination with dabrafenib as an adjuvant treatment for Stage III melanoma with a proven BRAF V600E mutation following wide local excision of the primary lesion.

    Demonstrates improved overall and progression-free survival versus traditional chemotherapy [97].

    Median progression-free survival in monotherapy ~4.8–5.7 months versus 11 months in combination with trametinib [97, 98].

    Most common adverse events are dermatologic (rash, acneiform dermatitis).
  - Cobimetinib

    MEK inhibitor approved for the treatment of unresectable or metastatic advanced melanoma with proven BRAF V600E or V600K mutation, in combination with vemurafinib.

    When added to vemurafenib therapy, associated with significant improvement in progression-free survival compared to vemurafenib monotherapy (9.9 months versus 6.2 months), with a non-significant increase in the incidence of grade 3 and 4 adverse events [99].

Conclusions

- Melanoma remains a significant worldwide public health problem.
- Long-term outcomes in early stage disease are excellent; progression to systemic involvement is associated with dismal intermediate and long-term survival.
- An optimal approach to treatment of melanoma takes into account the systemic implications of widespread disease. Thus, work-up and treatment (even of early stage disease) should be conducted in the context of prevention (if possible) and early detection of systemic involvement.

Review Questions and Answers

1. In which of the following patients would genetic testing for Familial atypical multiple mole (FAMMM) syndrome be appropriate?
   (a) A patient with a single invasive cutaneous melanoma diagnosed at age 32
   (b) A patient with two prior distinct cutaneous melanomas diagnosed at age 67 and age 72
   (c) A 42 year-old patient with a single invasive cutaneous melanoma, who also has a strong family history of melanoma (mother, maternal uncle, and maternal grandfather) and pancreatic cancer (maternal uncle and maternal grandfather)
   (d) A 41 year-old patient with no history of melanoma but with a family history of melanoma (mother and father, both after age 50).
   (e) A patient with a family history of breast cancer (mother) and colon cancer (paternal grandfather).

   Answer: (c) A 42 year-old patient with a single invasive cutaneous melanoma, who also has a strong family history of melanoma (mother, maternal uncle, and maternal grandfather) and pancreatic cancer (maternal uncle and maternal grandfather)

2. Which of the following biopsy methods would NOT be appropriate for a 4 mm asymmetric, irregular pigmented lesion concerning for potential melanoma?
   (a) Shave biopsy with saucerization
   (b) 5 mm punch biopsy
   (c) Excisional biopsy narrow margins
   (d) Curettage biopsy

   Answer: (d) Curettage biopsy

3. Which of the following genetic tests could be considered as an adjunct in order to add prognostic information regarding the risk of local recurrence/distant metastasis in a 42 year old patient diagnosed with a 0.7 mm Breslow Depth malignant melanoma (pT1a)?
   (a) 2 gene expression profile test
   (b) 23 gene expression profile test
   (c) 31 gene expression profile test
   (d) Fluorescence in situ hybridization (FISH)
   (e) Enzyme linked immunosorbent assay (ELISA)

   Answer: (c) 31 gene expression profile test

4. For which of the following patients would **sentinel lymph node biopsy** be recommended?
   (a) A patient with a 0.4 mm Breslow depth, non-ulcerated melanoma of the left chest with no clinically palpable lymph nodes

(b) A patient with a 2.3 mm Breslow depth, non-ulcerated melanoma of the left chest with a clinically palpable, firm/fixed 2 cm left axillary lymph node

(c) A patient with a 12.0 mm Breslow depth desmoplastic melanoma of the left chest with no clinically palpable lymph nodes

(d) A patient with a lentigo maligna (melanoma *in situ*) of the right cheek

(e) A patient with a 2.3 mm Breslow depth, non-ulcerated melanoma of the left chest with no clinically palpable lymph nodes

Answer: (e) A patient with a 2.3 mm Breslow depth, non-ulcerated melanoma of the left chest with no clinically palpable lymph nodes

5. Which of the following systemic treatment options for melanoma is considered to be a targeted antitumor therapy?

(a) Vemurafenib

(b) Ipilumab

(c) Nivolumab

(d) Pembrolizumab

(e) Interferon alpha 2a

Answer: (a) Vemurafenib

References

1. Siegel RL, Miller KD, Jemal A. Cancer statistics, 2015. CA Cancer J Clin. 2015;65(1):5–29.
2. American Cancer Society. Key statistics for melanoma skin cancer 2018. https://www.cancer.org/cancer/melanoma-skin-cancer/about/key-statistics.html.
3. Siegel RL, Miller KD, Jemal A. Cancer statistics, 2017. CA Cancer J Clin. 2017;67(1):7–30.
4. DS R. NYU melanoma cooperative group statistical report. 2018.
5. Siegel RL, Miller KD, Jemal A. Cancer statistics, 2016. CA Cancer J Clin. 2016;66(1):7–30.
6. Svoboda RM, Glazer AM, Farberg AS, Cowdrey MCE, Rigel DS. Melanoma reporting practices of United States dermatologists. Dermatol Surg. 2018;44(11):1391–5.
7. Cartee TV, Kini SP, Chen SC. Melanoma reporting to central cancer registries by US dermatologists: an analysis of the persistent knowledge and practice gap. J Am Acad Dermatol. 2011;65(5 Suppl 1):S124–32.
8. Heuring E, Chen SC. Melanoma underreporting among US dermatopathologists: a pilot study. J Cutan Pathol. 2018;45(7):550–1.
9. Erdmann F, Lortet-Tieulent J, Schuz J, Zeeb H, Greinert R, Breitbart EW, et al. International trends in the incidence of malignant melanoma 1953-2008—are recent generations at higher or lower risk? Int J Cancer. 2013;132(2):385–400.
10. Apalla Z, Nashan D, Weller RB, Castellsague X. Skin cancer: epidemiology, disease burden, pathophysiology, diagnosis, and therapeutic approaches. Dermatol Ther. 2017;7(Suppl 1):5–19.
11. Shaikh WR, Dusza SW, Weinstock MA, Oliveria SA, Geller AC, Halpern AC. Melanoma thickness and survival trends in the United States, 1989–2009. J Natl Cancer Inst. 2016;108(1):djv294.
12. Cronin KA, Lake AJ, Scott S, Sherman RL, Noone AM, Howlader N, et al. Annual report to the nation on the status of cancer, part I: national cancer statistics. Cancer. 2018;124(13):2785–800.
13. National Cancer Institute. Surveillance, epidemiology, and end results program 2017. https://seer.cancer.gov/statfacts/html/melan.html.
14. Elwood JM, Jopson J. Melanoma and sun exposure: an overview of published studies. Int J Cancer. 1997;73(2):198–203.

15. Colantonio S, Bracken MB, Beecker J. The association of indoor tanning and melanoma in adults: systematic review and meta-analysis. J Am Acad Dermatol. 2014;70(5):847–57.e1–18.
16. American Cancer Society. Cancer facts and figures 2018. https://www.cancer.org/content/dam/cancer-org/research/cancer-facts-and-statistics/annual-cancer-facts-and-figures/2018/cancer-facts-and-figures-2018.pdf.
17. Skin Cancer Foundation. Skin cancer facts & statistics 2018. https://www.skincancer.org/skin-cancer-information/skin-cancer-facts.
18. Matthews NH, Li WQ, Qureshi AA, Weinstock MA, Cho E. Epidemiology of melanoma. In: Ward WH, Farma JM, editors. Cutaneous melanoma: etiology and therapy. Brisbane: Codon Publications The Authors; 2017.
19. Yuan TA, Meyskens F, Liu-Smith F. A cancer registry-based analysis on the non-white populations reveals a critical role of the female sex in early-onset melanoma. Cancer Causes Control. 2018;29:405–15.
20. Feng Z, Zhang Z, Wu XC. Lifetime risks of cutaneous melanoma by histological subtype and race/ethnicity in the United States. J Louisiana State Med Soc. 2013;165(4):201–8.
21. Agbai ON, Buster K, Sanchez M, Hernandez C, Kundu RV, Chiu M, et al. Skin cancer and photoprotection in people of color: a review and recommendations for physicians and the public. J Am Acad Dermatol. 2014;70(4):748–62.
22. Nabil R, Plasmeijer E, van Doorn R, Bergman W, Kukutsch NA. Unscheduled visits of patients with familial melanoma to a pigmented lesion clinic: evaluation of patients' characteristics and suspicious lesions. Acta Derm Venereol. 2018;98:667–70.
23. Soura E, Eliades PJ, Shannon K, Stratigos AJ, Tsao H. Hereditary melanoma: update on syndromes and management: genetics of familial atypical multiple mole melanoma syndrome. J Am Acad Dermatol. 2016;74(3):395–407; quiz 8–10.
24. Gandini S, Sera F, Cattaruzza MS, Pasquini P, Abeni D, Boyle P, et al. Meta-analysis of risk factors for cutaneous melanoma: I. Common and atypical naevi. Eur J Cancer. 2005;41(1):28–44.
25. Geller AC, Mayer JE, Sober AJ, Miller DR, Argenziano G, Johnson TM, et al. Total nevi, atypical nevi, and melanoma thickness: an analysis of 566 patients at 2 US centers. JAMA Dermatol. 2016;152(4):413–8.
26. Singh S, Nagpal SJ, Murad MH, Yadav S, Kane SV, Pardi DS, et al. Inflammatory bowel disease is associated with an increased risk of melanoma: a systematic review and meta-analysis. Clin Gastroenterol Hepatol. 2014;12(2):210–8.
27. Engels EA, Pfeiffer RM, Fraumeni JF Jr, Kasiske BL, Israni AK, Snyder JJ, et al. Spectrum of cancer risk among US solid organ transplant recipients. JAMA. 2011;306(17):1891–901.
28. Vajdic CM, McDonald SP, McCredie MR, van Leeuwen MT, Stewart JH, Law M, et al. Cancer incidence before and after kidney transplantation. JAMA. 2006;296(23):2823–31.
29. Penn I. De novo malignances in pediatric organ transplant recipients. Pediatr Transplant. 1998;2(1):56–63.
30. Ali FR, Lear JT. Melanoma in organ transplant recipients: incidence, outcomes and management considerations. J Skin Cancer. 2012;2012:404615.
31. Maiorino A, De Simone C, Perino F, Caldarola G, Peris K. Melanoma and non-melanoma skin cancer in psoriatic patients treated with high-dose phototherapy. J Dermatolog Treat. 2016;27(5):443–7.
32. Disse M, Reich H, Lee PK, Schram SS. A review of the association between Parkinson disease and malignant melanoma. Dermatol Surg. 2016;42(2):141–6.
33. Green AC, Williams GM, Logan V, Strutton GM. Reduced melanoma after regular sunscreen use: randomized trial follow-up. J Clin Oncol. 2011;29(3):257–63.
34. Olsen CM, Wilson LF, Green AC, Biswas N, Loyalka J, Whiteman DC. How many melanomas might be prevented if more people applied sunscreen regularly? Br J Dermatol. 2018;178(1):140–7.
35. Williams JD, Maitra P, Atillasoy E, Wu MM, Farberg AS, Rigel DS. SPF 100+ sunscreen is more protective against sunburn than SPF 50+ in actual use: results of a randomized, double-blind, split-face, natural sunlight exposure clinical trial. J Am Acad Dermatol. 2018;78(5):902–10.e2.
36. Koh HK, Miller DR, Geller AC, Clapp RW, Mercer MB, Lew RA. Who discovers melanoma? Patterns from a population-based survey. J Am Acad Dermatol. 1992;26(6):914–9.

37. Brady MS, Oliveria SA, Christos PJ, Berwick M, Coit DG, Katz J, et al. Patterns of detection in patients with cutaneous melanoma. Cancer. 2000;89(2):342–7.
38. Rigel DS, Friedman RJ. The rationale of the ABCDs of early melanoma. J Am Acad Dermatol. 1993;29(6):1060–1.
39. AAD statement on USPSTF recommendation on skin cance screening [press release]. 2016.
40. Dermatology AAo. How to perform a total body skin exam 2015.
41. Tsao H, Pehamberger H, Sober A. Precursor lesions and markers of increased risk for melanoma. In: Cutaneous melanoma. St Louis: Quality Medical Publishing Inc.; 1998. p. 65–79.
42. Bichakjian CK, Halpern AC, Johnson TM, Foote Hood A, Grichnik JM, Swetter SM, et al. Guidelines of care for the management of primary cutaneous melanoma. American Academy of Dermatology. J Am Acad Dermatol. 2011;65(5):1032–47.
43. Coit DG, Andtbacka R, Bichakjian CK, Dilawari RA, Dimaio D, Guild V, et al. Melanoma. J Natl Comp Cancer Netw. 2009;7(3):250–75.
44. Farberg AS, Rigel DS. A comparison of current practice patterns of US dermatologists versus published guidelines for the biopsy, initial management, and follow up of patients with primary cutaneous melanoma. J Am Acad Dermatol. 2016;75(6):1193–7.e1.
45. Ng JC, Swain S, Dowling JP, Wolfe R, Simpson P, Kelly JW. The impact of partial biopsy on histopathologic diagnosis of cutaneous melanoma: experience of an Australian tertiary referral service. Arch Dermatol. 2010;146(3):234–9.
46. Kittler H, Pehamberger H, Wolff K, Binder M. Diagnostic accuracy of dermoscopy. Lancet Oncol. 2002;3(3):159–65.
47. Winkelmann RR, Farberg AS, Glazer AM, Cockerell CJ, Sober AJ, Siegel DM, et al. Integrating skin cancer-related technologies into clinical practice. Dermatol Clin. 2017;35(4):565–76.
48. Xiong YQ, Mo Y, Wen YQ, Cheng MJ, Huo ST, Chen XJ, et al. Optical coherence tomography for the diagnosis of malignant skin tumors: a meta-analysis. J Biomed Opt. 2018;23(2):1–10.
49. Ferris LK, Jansen B, Ho J, Busam KJ, Gross K, Hansen DD, et al. Utility of a noninvasive 2-gene molecular assay for cutaneous melanoma and effect on the decision to biopsy. JAMA Dermatol. 2017;153(7):675–80.
50. Reimann JDR, Salim S, Velazquez EF, Wang L, Williams KM, Flejter WL, et al. Comparison of melanoma gene expression score with histopathology, fluorescence in situ hybridization, and SNP array for the classification of melanocytic neoplasms. Mod Pathol. 2018;31(11):1733–43.
51. Ko JS, Matharoo-Ball B, Billings SD, Thomson BJ, Tang JY, Sarin KY, et al. Diagnostic distinction of malignant melanoma and benign nevi by a gene expression signature and correlation to clinical outcomes. Cancer Epidemiol Biomark Prev. 2017;26(7):1107–13.
52. Gershenwald JE, Scloyler RA. Melanoma staging: American Joint Committee on Cancer (AJCC) 8th edition and beyond. Ann Surg Oncol. 2018;25:2105–10.
53. Morton DL, Thompson JF, Cochran AJ, Mozzillo N, Nieweg OE, Roses DF, et al. Final trial report of sentinel-node biopsy versus nodal observation in melanoma. N Engl J Med. 2014;370(7):599–609.
54. Faries MB, Thompson JF, Cochran AJ, Andtbacka RH, Mozzillo N, Zager JS, et al. Completion dissection or observation for sentinel-node metastasis in melanoma. N Engl J Med. 2017;376(23):2211–22.
55. Leiter U, Stadler R, Mauch C, Hohenberger W, Brockmeyer N, Berking C, et al. Complete lymph node dissection versus no dissection in patients with sentinel lymph node biopsy positive melanoma (DeCOG-SLT): a multicentre, randomised, phase 3 trial. Lancet Oncol. 2016;17(6):757–67.
56. Wong SL, Faries MB, Kennedy EB, Agarwala SS, Akhurst TJ, Ariyan C, et al. Sentinel lymph node biopsy and management of regional lymph nodes in melanoma: American Society of Clinical Oncology and Society of Surgical Oncology clinical practice guideline update. J Clin Oncol. 2018;36(4):399–413.
57. Gerami P, Cook RW, Russell MC, Wilkinson J, Amaria RN, Gonzalez R, et al. Gene expression profiling for molecular staging of cutaneous melanoma in patients undergoing sentinel lymph node biopsy. J Am Acad Dermatol. 2015;72(5):780–5.e3.
58. Gerami P, Cook RW, Wilkinson J, Russell MC, Dhillon N, Amaria RN, et al. Development of a prognostic genetic signature to predict the metastatic risk associated with cutaneous melanoma. Clin Cancer Res. 2015;21(1):175–83.

59. Hsueh EC, DeBloom JR, Lee J, Sussman JJ, Covington KR, Middlebrook B, et al. Interim analysis of survival in a prospective, multi-center registry cohort of cutaneous melanoma tested with a prognostic 31-gene expression profile test. J Hematol Oncol. 2017;10(1):152.
60. Hsueh EC, Schwartz TL, Lizalek JM, Hunborg PS, Hurley MY, Hinyard LJ. Prospective validation of gene expression profiling in primary cutaneous melanoma. American Society of Clinical Oncology; 2016.
61. Zager JS, Gastman BR, Leachman S, Gonzalez RC, Fleming MD, Ferris LK, et al. Performance of a prognostic 31-gene expression profile in an independent cohort of 523 cutaneous melanoma patients. BMC Cancer. 2018;18(1):130.
62. Ferris LK, Farberg AS, Middlebrook B, Johnson CE, Lassen N, Oelschlager KM, et al. Identification of high-risk cutaneous melanoma tumors is improved when combining the online American Joint Committee on cancer individualized melanoma patient outcome prediction tool with a 31-gene expression profile-based classification. J Am Acad Dermatol. 2017;76(5):818–25.e3.
63. Conic RZ, Cabrera CI, Khorana AA, Gastman BR. Determination of the impact of melanoma surgical timing on survival using the National Cancer Database. J Am Acad Dermatol. 2018;78(1):40–6.e7.
64. Deutsch GB, Kirchoff DD, Faries MB. Metastasectomy for stage IV melanoma. Surg Oncol Clin N Am. 2015;24(2):279–98.
65. Kunishige JH, Brodland DG, Zitelli JA. Surgical margins for melanoma in situ. J Am Acad Dermatol. 2012;66(3):438–44.
66. Hayes AJ, Maynard L, Coombes G, Newton-Bishop J, Timmons M, Cook M, et al. Wide versus narrow excision margins for high-risk, primary cutaneous melanomas: long-term follow-up of survival in a randomised trial. Lancet Oncol. 2016;17(2):184–92.
67. Wilson MA, Schuchter LM. Chemotherapy for melanoma. Cancer Treat Res. 2016;167:209–29.
68. Perier-Muzet M, Gatt E, Peron J, Falandry C, Amini-Adle M, Thomas L, et al. Association of immunotherapy with overall survival in elderly patients with melanoma. JAMA Dermatol. 2018;154(1):82–7.
69. Robert C, Schachter J, Long GV, Arance A, Grob JJ, Mortier L, et al. Pembrolizumab versus ipilimumab in advanced melanoma. N Engl J Med. 2015;372(26):2521–32.
70. Eroglu Z, Zaretsky JM, Hu-Lieskovan S, Kim DW, Algazi A, Johnson DB, et al. High response rate to PD-1 blockade in desmoplastic melanomas. Nature. 2018;553(7688):347–50.
71. Routy B, Le Chatelier E, Derosa L, Duong CPM, Alou MT, Daillere R, et al. Gut microbiome influences efficacy of PD-1-based immunotherapy against epithelial tumors. Science (New York, NY). 2018;359(6371):91–7.
72. Matson V, Fessler J, Bao R, Chongsuwat T, Zha Y, Alegre ML, et al. The commensal microbiome is associated with anti-PD-1 efficacy in metastatic melanoma patients. Science (New York, NY). 2018;359(6371):104–8.
73. Topalian SL, Hodi FS, Brahmer JR, Gettinger SN, Smith DC, McDermott DF, et al. Safety, activity, and immune correlates of anti-PD-1 antibody in cancer. N Engl J Med. 2012;366(26):2443–54.
74. Weber JS, Hodi FS, Wolchok JD, Topalian SL, Schadendorf D, Larkin J, et al. Safety profile of Nivolumab monotherapy: a pooled analysis of patients with advanced melanoma. J Clin Oncol. 2017;35(7):785–92.
75. McCoy KD, Le Gros G. The role of CTLA-4 in the regulation of T cell immune responses. Immunol Cell Biol. 1999;77(1):1–10.
76. Lens M, Testori A, Ferucci PF. Ipilimumab targeting CD28-CTLA-4 axis: new hope in the treatment of melanoma. Curr Top Med Chem. 2012;12(1):61–6.
77. Wolchok JD, Neyns B, Linette G, Negrier S, Lutzky J, Thomas L, et al. Ipilimumab monotherapy in patients with pretreated advanced melanoma: a randomised, double-blind, multicentre, phase 2, dose-ranging study. Lancet Oncol. 2010;11(2):155–64.
78. Horvat TZ, Adel NG, Dang TO, Momtaz P, Postow MA, Callahan MK, et al. Immune-related adverse events, need for systemic immunosuppression, and effects on survival and time to treatment failure in patients with melanoma treated with ipilimumab at memorial Sloan Kettering Cancer center. J Clin Oncol. 2015;33(28):3193–8.

79. Ngiow SF, McArthur GA, Smyth MJ. Radiotherapy complements immune checkpoint block-ade. Cancer Cell. 2015;27(4):437–8.
80. Callahan MK, Kluger H, Postow MA, Segal NH, Lesokhin A, Atkins MB, et al. Nivolumab plus ipilimumab in patients with advanced melanoma: updated survival, response, and safety data in a phase I dose-escalation study. J Clin Oncol. 2018;36(4):391–8.
81. Larkin J, Chiarion-Sileni V, Gonzalez R, Grob JJ, Cowey CL, Lao CD, et al. Combined nivolumab and ipilimumab or monotherapy in untreated melanoma. N Engl J Med. 2015;373(1):23–34.
82. McCain J. The MAPK (ERK) pathway: investigational combinations for the treatment of BRAF-mutated metastatic melanoma. P&T. 2013;38(2):96–108.
83. Kono M, Dunn IS, Durda PJ, Butera D, Rose LB, Haggerty TJ, et al. Role of the mitogen-activated protein kinase signaling pathway in the regulation of human melanocytic antigen expression. Mol Cancer Res. 2006;4(10):779–92.
84. Lee JH, Choi JW, Kim YS. Frequencies of BRAF and NRAS mutations are different in his-tological types and sites of origin of cutaneous melanoma: a meta-analysis. Br J Dermatol. 2011;164(4):776–84.
85. Tuma RS. Getting around PLX4032: studies turn up unusual mechanisms of resistance to melanoma drug. J Natl Cancer Inst. 2011;103(3):170–1, 177.
86. Chapman PB, Hauschild A, Robert C, Haanen JB, Ascierto P, Larkin J, et al. Improved survival with vemurafenib in melanoma with BRAF V600E mutation. N Engl J Med. 2011;364(26):2507–16.
87. Flaherty KT, Puzanov I, Kim KB, Ribas A, McArthur GA, Sosman JA, et al. Inhibition of mutated, activated BRAF in metastatic melanoma. N Engl J Med. 2010;363(9):809–19.
88. Sosman JA, Kim KB, Schuchter L, Gonzalez R, Pavlick AC, Weber JS, et al. Survival in BRAF V600-mutant advanced melanoma treated with vemurafenib. N Engl J Med. 2012;366(8):707–14.
89. Su F, Viros A, Milagre C, Trunzer K, Bollag G, Spleiss O, et al. RAS mutations in cuta-neous squamous-cell carcinomas in patients treated with BRAF inhibitors. N Engl J Med. 2012;366(3):207–15.
90. Dummer R, Rinderknecht J, Goldinger SM. Ultraviolet A and photosensitivity during vemu-rafenib therapy. N Engl J Med. 2012;366(5):480–1.
91. Paraiso KH, Haarberg HE, Wood E, Rebecca VW, Chen YA, Xiang Y, et al. The HSP90 inhibi-tor XL888 overcomes BRAF inhibitor resistance mediated through diverse mechanisms. Clin Cancer Res. 2012;18(9):2502–14.
92. Hauschild A, Grob JJ, Demidov LV, Jouary T, Gutzmer R, Millward M, et al. Dabrafenib in BRAF-mutated metastatic melanoma: a multicentre, open-label, phase 3 randomised con-trolled trial. Lancet (London, England). 2012;380(9839):358–65.
93. Sullivan RJ, Flaherty KT. Resistance to BRAF-targeted therapy in melanoma. Eur J Cancer. 2013;49(6):1297–304.
94. Debarbieux S, Dalle S, Depaepe L, Poulalhon N, Balme B, Thomas L. Second primary mela-nomas treated with BRAF blockers: study by reflectance confocal microscopy. Br J Dermatol. 2013;168(6):1230–5.
95. Yagerman S, Flores E, Busam K, Lacouture M, Marghoob AA. Overview photography and short-term mole monitoring in patients taking a BRAF inhibitor. JAMA Dermatol. 2014;150(9):1010–1.
96. Administration FaD. 2014.
97. Flaherty KT, Robert C, Hersey P, Nathan P, Garbe C, Milhem M, et al. Improved survival with MEK inhibition in BRAF-mutated melanoma. N Engl J Med. 2012;367(2):107–14.
98. Long GV, Stroyakovskiy D, Gogas H, Levchenko E, de Braud F, Larkin J, et al. Dabrafenib and trametinib versus dabrafenib and placebo for Val600 BRAF-mutant melanoma: a mul-ticentre, double-blind, phase 3 randomised controlled trial. Lancet (London, England). 2015;386(9992):444–51.
99. Larkin J, Ascierto PA, Dreno B, Atkinson V, Liszkay G, Maio M, et al. Combined vemurafenib and cobimetinib in BRAF-mutated melanoma. N Engl J Med. 2014;371(20):1867–76.

Surgical Approach to Cutaneous Malignancy with Systemic Implications

5

Patricia Richey and Nkanyezi Ferguson

Abbreviations

ADPA	Aggressive digital papillary adenocarcinoma
AUC	Appropriate use criteria
BCC	Basal cell carcinoma
CLL	Chronic lymphocytic leukemia
CNB	Core needle biopsy
cSCC	Cutaneous squamous cell carcinoma
CT	Computed tomography
DFSP	Dermatofibrosarcoma protuberans
ED&C	Electrodesiccation and curettage
EMPD	Extramammary paget's disease
EPC	Eccrine porocarcinoma
FNA	Fine needle aspiration
HPV	Human papillomavirus
IHC	Immunohistochemical
LM	Lentigo maligna
LN	Lymph node
MCC	Merkel cell carcinoma
MMS	Mohs micrographic surgery

P. Richey · N. Ferguson (✉)
University of Iowa Health Care, Iowa City, IA, USA
e-mail: Patricia-richey@uiowa.edu; Nkanyezi-ferguson@uiowa.edu

© Springer Nature Switzerland AG 2021
V. Liu (ed.), *Dermato-Oncology Study Guide*,
https://doi.org/10.1007/978-3-030-53437-0_5

MRI	Magnetic resonance imaging
NMSC	Non-melanoma skin cancer
PET	Positron emission tomography
PNI	Perineural invasion
RT	Radiotherapy
SCC	Squamous cell carcinoma
SLNB	Sentinel lymph node biopsy
US	Ultrasonography
UVA	Ultraviolet A radiation
UVB	Ultraviolet B radiation
VEGF	Vascular endothelial growth factor
WLE	Wide local excision

Learning Objectives

1. Identify correct clinical settings in which a specific type of biopsy should be performed, and understand the technique behind each biopsy type.
2. In treating skin cancer, identify scenarios in which surgery must be performed, as well as when surgery can be avoided.
3. Identify appropriate treatment of melanoma and non-melanoma skin cancers, with particular focus given to surgical considerations in these conditions.

Surgical Issues of Systemic Malignancy

Introduction

- Treatment of cutaneous malignancies is a difficult and complex undertaking, with a multitude of opinions, evidence-based and experience-based, available in the literature.
- While surgical management is required in many cases, it is important to note that treatment of many of these malignancies often require a multidisciplinary team-based approach.
- The performance of a skin biopsy is an intrinsic part of the initial management of a patient suspected of having a skin cancer; however, when considering options for treating a skin cancer once it has been diagnosed, it is important to know both when surgery must be performed, as well as when surgery can be avoided.
- In this chapter, we address treatment of melanoma and non-melanoma skin cancers, with particular focus given to surgical considerations in these conditions.

Diagnostic and Procedural Considerations in Dermatooncology

Introduction [1–3]

- Skin biopsies are essential in diagnosing dermatooncologic conditions and selection of an appropriate biopsy modality has important clinical implications. General principles when considering biopsy type and location should be taken into consideration.
  - Identify a primary lesion (i.e. a well-formed new lesion)
  - Lesions with overlying secondary changes should be avoided as these changes may disrupt the primary histopathology
  - Biopsy the thickest portion and avoid necrotic tissue
  - Partial or superficial biopsies may not be representative of the entire process or allow for evaluation of possible dermal involvement

- For large lesions multiple biopsies may be needed with focus on sampling the most atypical portion
- Ulcers may have secondary changes underneath the ulcer bed resulting in potential for misdiagnosis therefore biopsy should include the ulcer edge, ulcer, and adjacent inflamed skin
- If clinical differential diagnosis includes melanoma or other tumor where assessment of the base of the tumor is important and transection should be avoided, biopsies of adequate depth (e.g. excisional biopsy, punch biopsy or saucerization biopsy) should be employed

Shave Biopsy [1–3, 9]

- Most common biopsy technique
- Consists of the epidermis and papillary with variable amount of superficial reticular dermis
- Used when pathologic process is primarily epidermal (e.g. actinic keratosis, squamous cell carcinoma in situ, superficial basal cell carcinoma, etc.)
- Useful procedure for diagnosing superficial carcinomas
- Technique
  - Wheal is created from anesthesia to elevate area to be biopsied
  - Skin is stabilized with the non-dominant hand
  - Razor or blade is held in a semi-curved/concave position parallel to the skin
  - Back and forth sawing motion is utilized to remove the specimen (Fig. 5.1)

Saucerization Biopsy [1–3, 7, 9]

- Biopsy is intentionally deeper
- Consists of the epidermis, papillary dermis, and reticular dermis

Fig. 5.1 Shave biopsy

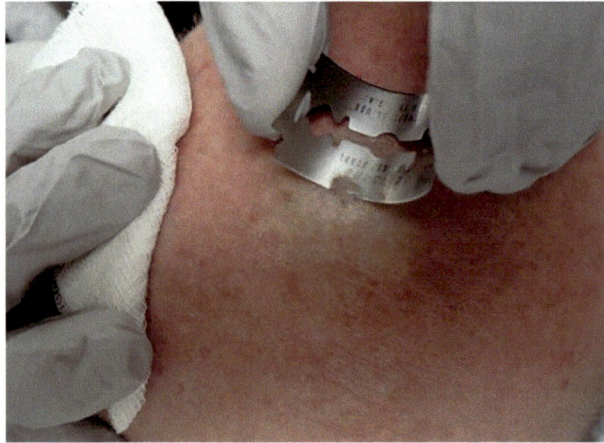

- Used to biopsy neoplasms (e.g. squamous cell carcinoma, basal cell carcinoma, dysplastic nevi, etc.)
- When performed appropriately, the diagnostic value is comparable to that of most incisional/excisional biopsies and do not statistically affect survival rates for malignant lesions
  - Advantages: Easy to perform, less invasive, quick, avoids use of sutures, inexpensive, reduces sampling error in larger lesions when compared to a punch biopsy
  - Disadvantages: Risk of transection of melanoma, cosmesis and effectiveness depends on performing providers experience (Fig. 5.2a, b)

Punch Biopsy [1–3, 9]

- The punch biopsy samples a cylindrical shaped specimen
- Consists of epidermis, dermis, and subcutaneous fat
- Diameter of the metal "barrel" varies from 2 to 10 mm
- Helpful for examining processes within the dermis and superficial subcutaneous fat
- Technique
  - Skin is stabilized with the non-dominant hand
  - Firm and constant downward pressure is applied with a circular twisting motion
  - A 'give' is felt once subcutaneous tissue is reached indicating a full-thickness cut has been made
  - Biopsy is elevated and handled with care in order to avoid 'crush artifact' from forceps
     Lymphocytes are susceptible to crush artifact
  - Iris scissors are utilized to excise the biopsy specimen at the base
  - Firm pressure is applied for hemostasis
  - Biopsy sites <3 mm can heal by secondary intention, however biopsies >3 mm are usually closed with interrupted epidermal sutures for superior cosmesis (Fig. 5.3a–c)

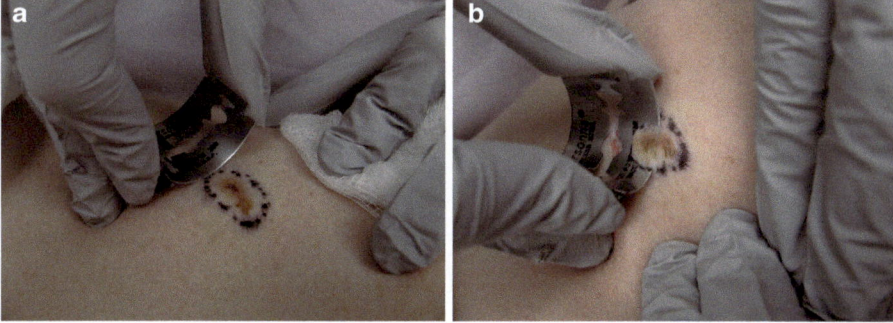

Fig. 5.2 This image displays a saucerization biopsy, where the intended depth of the biopsy (encompassing the epidermis, papillary dermis and reticular dermis) is deeper than an average shave biopsy

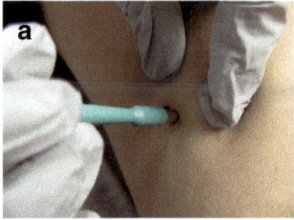 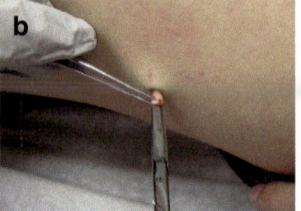 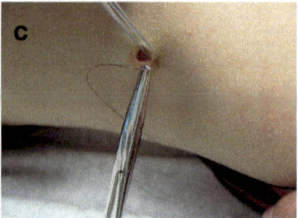

Fig. 5.3 This image displays a punch biopsy, where a cylindrical specimen is obtained in order to visualize the epidermis, dermis and subcutaneous fat

Incisional Biopsy [1–3]

- An incisional biopsy involves removal of a wedge of tissue from the center or edge of a lesion
- Optimal option for sampling deep subcutaneous fat or fascia for pathologic examination
- Can also be used to sample a significant portion of large-sized tumors
- Specimen should include the thickest and most clinically representative area
- Imbedding and specimen sectioning should occur in the long axis

Excisional Biopsy [1–3]

- Excisional biopsy removes the entire lesion
- Includes epidermis, dermis, and subcutaneous fat
- Biopsy of choice for a suspected invasive cutaneous melanoma
- Indications
    - Pigmented lesions suspicious for melanoma
    - Suspected metastases
    - When presumed pathology involves deep dermis or subcutaneous fat
- Technique
    - An ellipse is drawn around the lesion
    - A margin of normal skin around the lesion is included in the biopsy
        Approximately 1–5 mm margins for malignant lesions
    - Technique is similar to an elliptical excision (see below for further details)

Wide Local Excision [1–3]

- Wide local excision (WLE) is used to surgically remove a biopsy proven malignancy
- The surgical margin taken during WLE excision varies and depends on the specific malignancy
- Technique
    - Excision is started at the apex of the ellipse with a scalpel blade (#15 or #10) held at 90° (perpendicular to the skin)

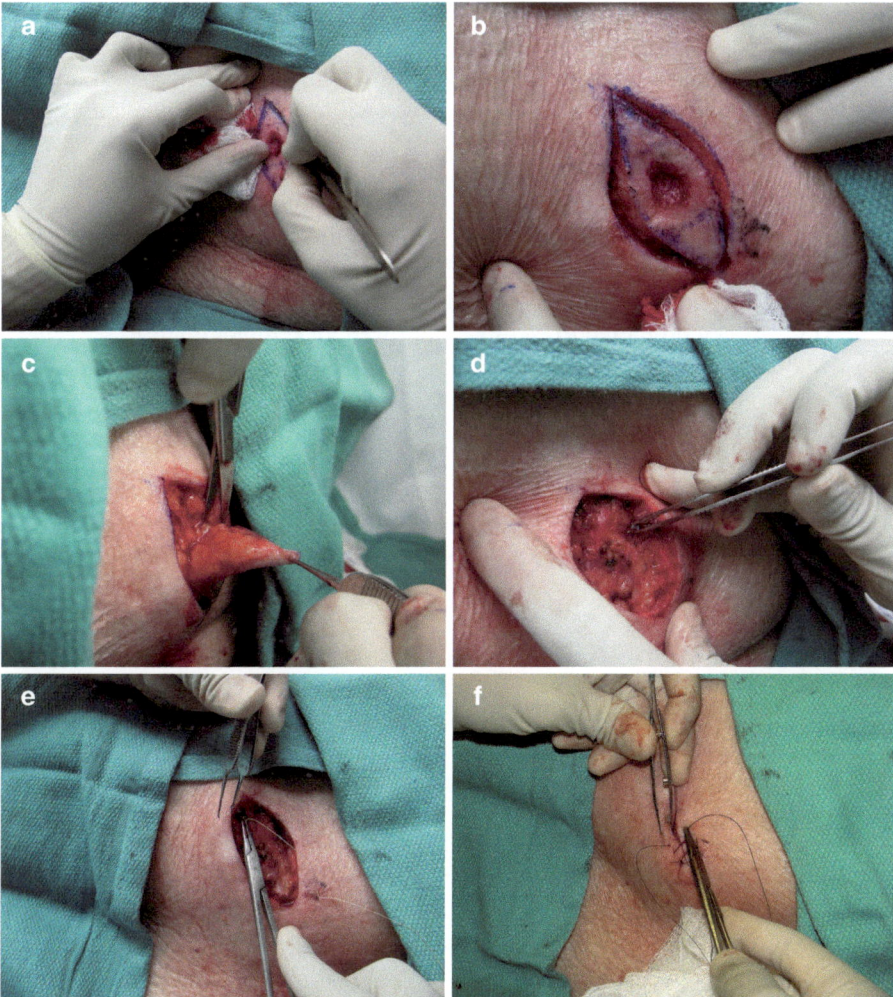

Fig. 5.4 Wide local excision, pictured here, is used to remove a biopsy-proven malignancy, making sure to obtain the proper surgical margin

- The scalpel is held in the dominant hand and the non-dominant hand used to apply opposing tension
- Blade should go completely through the dermis to the subcutaneous fat
- Specimen is carefully lifted at the edge with forceps
- Iris scissors are used to remove tissue and depth at the base and tips kept equal in order to avoid boating effect at closure
- Hemostasis is obtained with meticulous spot cautery
- Wound is closed with buried absorbable and non-absorbable epidermal sutures placed in an interrupted or running fashion (Fig. 5.4a–f)

Mohs Micrographic Surgery (MMS) [1–3]

- MMS is a combined surgical and pathologic technique utilized to remove complex and ill-defined cutaneous malignancies with histologic evaluation of 100% of the tissue margins
- Procedure offers tissue sparing potential in locations such as the head and neck where wide local excision (WLE) may not be feasible
- Frozen tissue specimens are sectioned horizontally or tangentially into serial sections. All major tissue borders, including the peripheral and deep margins, are evaluated histologically on the same day as surgery
- MMS has been promoted to facilitate improved local tumor control
  - Major components:
      Surgical excision
      Histopathologic examination
      Precise mapping
      Wound management
- For certain tumors, MMS offers a better alternative to standard excision, with increased cure rates and tissue conservation (Fig. 5.5a–c)

Fine Needle Aspiration (FNA) [4–6]

- FNA utilizes a thin, hollow needle connected to a syringe to sample a small amount of fluid and tissue cells from the suspected lesion for diagnostic purposes
- The cytological specimen is then evaluated under microscopy
- Superficial lesions can be palpated to facilitate correct placement of needle
- Deeper lesions may require needle guidance via an imaging modality such as ultrasonography (US) or computed tomography (CT)
- Advantages:
  - Minimally invasive, safe and cost-effective technique
  - Sensitive and specific for the diagnosis of malignancy with diagnostic accuracy ranging from 90 to 99%
  - Requires little equipment and provides a rapid report of results

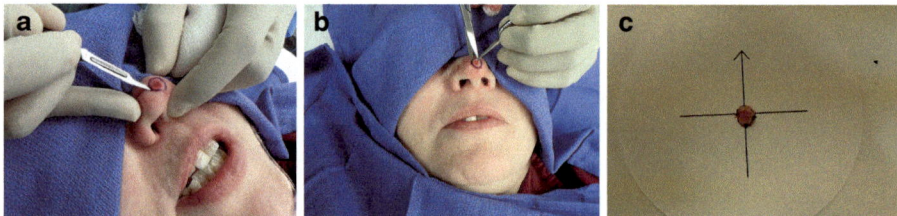

Fig. 5.5 In Mohs micrographic surgery, a surgical specimen (or "layer") is obtained (**a**, **b**), and a map is developed (**c**), so that precise tumor location can be identified during histopathologic review

- Disadvantages:
  - Aspiration technique and interpretation relies on the experience of the cytopathologist
  - A portion of aspirates are unsatisfactory
  - Inability toassess tissue architecture limits diagnostic assessment
- Indications
  - Mass lesion with a clinical suspicion of malignant tumor (palpable or deep-seated)

Core Needle Biopsy (CNB) [4–6]

- CNB needles remove a small cylinder of tissue and are slightly larger than those utilized in fine needle aspiration (FNA)
- Core needle biopsies can serve as an alternative or adjunct to FNA cytology and open biopsies for obtaining histological diagnoses.
  - Image-guided CNB is increasingly replacing excisional lymph node biopsy in the diagnosis and sub-classification of malignant lymphadenopathies and enlarged lymph nodes
  - A vacuum-assisted device is sometimes used
- Advantages:
  - Minimally invasive, well-tolerated
  - Advantage over more invasive surgical excisional techniques in relation to reduced costs, lower rates of post-procedural complications, and fewer delays in the diagnosis
  - CNB is preferred in cases where more solid specimens are needed.
- Disadvantages:
  - Compared to FNA requires a larger needle, is more expensive and requires longer wait for results

Surgical Management of Cutaneous Malignancies with High Rate of Systemic Spread

Melanoma [10–24]

- Clinical Examination
  - Heterogenous color
  - Irregular borders (Fig. 5.6)
- Dermoscopy
  - Utilizes optical magnification to evaluate pigmented lesions
  - Useful tool to help with early recognition of malignant melanoma
    Heterogeneous and asymmetric colors and structures
    Irregular pigment network and borders
    Blue-white veil (indistinct blue coloration with overlying white haze)

Fig. 5.6 Superficial
spreading malignant
melanoma

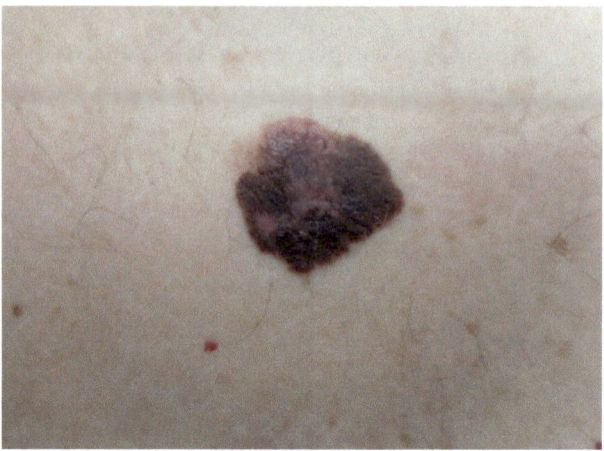

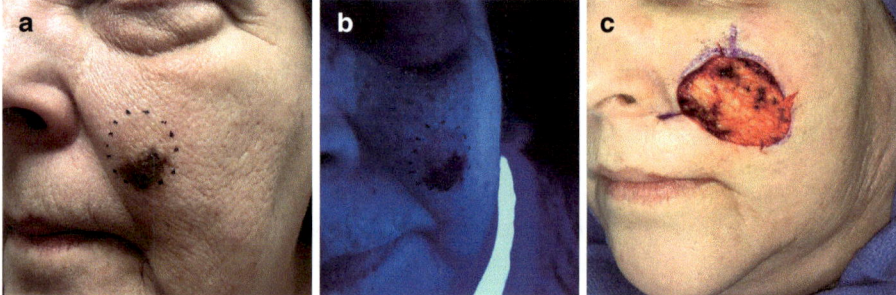

Fig. 5.7 Lentigo maligna. This figure demonstrates the correlation between Wood's lamp margins (**b**) and true margins of the tumor (**c**). Note the ill-defined clinical margins in (**a**)

      Structure-less areas lacking discernable pigment network/globules

      Regression pattern (white, scar-like areas and pepper-like blue-grey granules)

      Dotted or irregular vessels

- Wood's lamp
  - The true margins of lentigo maligna are often difficult to accurately delineate and may extend beyond the apparent clinical margins
  - Wood's lamp light (wavelength peak at 365 nm) can be used to improve margin delineation
  - Shorter wavelength of light from the Wood's lamp allows for improved visualization of contrast between superficial pigmented lesions and unaffected epidermis (without scatter interference from longer wavelengths)
  - Can also be utilized to evaluate for recurrence in scars from previous lentigo maligna excisions (Fig. 5.7a–c)
- Biopsy
  - Appropriate biopsy technique for melanoma is imperative in order to provide an adequate and representative sample for dermatopathology interpretation as

this provides prognostic information and has implications for staging and prognostication
- Biopsy technique preferred for melanoma is an excisional or complete biopsy of the entire clinically apparent lesion with a margin of adjacent normal-appearing skin
   Longitudinal/axial positioning of ellipse on the extremities is optimal for subsequent wide excision and sentinel lymph node biopsy if needed
- Incisional biopsy or partial sampling may be indicated for larger lesions or lesions in anatomically sensitive areas (e.g. face, acral region)
   The most atypical areas, identified clinically or under dermoscopy, should be selected for biopsy
   Caution as partial sampling may not represent true Breslow depth
- A deep saucerization biopsy may be satisfactory in experienced hands when the lesion is flat and the suspicion of melanoma is not high
- Sentinel lymph node biopsy
   - Sentinel lymph node biopsy (SLNB) is used for staging of melanoma patients at risk for occult nodal metastases and has a low complication rate (<5%)
   - The risk of SLNB metastasis is uncommon in patients with T1a melanomas (<5%); however this risk rises with increasing primary melanoma Breslow thickness (5–12% of T1b melanomas and >50% of T4b melanomas)
      SLNB is not recommended for T1a melanomas
      SLNB can be considered and discussed with patient for T1b melanomas
      SLNB should be discussed and offered for T2, T3, and T4 melanomas
   - Recommendation that SNLB is performed immediately prior to WLE in order to reduce disruption of lymphatic channels
- Surgical management
   - Surgical wide local excision (WLE) is the treatment modality of choice for local control of cutaneous melanoma without occult metastasis
      Surgical margin recommendations are based on prospective randomized controlled studies and consensus discussion
      Aim to remove the clinically apparent tumor and a surrounding margin of normal tissue
   - Melanoma in situ (excluding lentigo maligna)
      Excision with at least 0.5 cm margins is recommended
   - Lentigo maligna
      Lentigo maligna (LM) has unpredictable subclinical extension and WLE with 0.5-cm margins may be inadequate for tumor extirpation.
      Staged excision
      - 'Square method', 'perimeter technique' or 'spaghetti technique', 'slow Mohs', staged radial sections, and staged 'mapped' excisions
      - Removes the clinically apparent lesion and uses rushed permanent sections to identify the excised tissue for margin control
      - If margins are positive, additional tissue is surgically removed
      - Final closure of the wound bed is delayed until permanent sections show clear margins

Mohs micrographic surgery (MMS)
- Indicated for LM located on the face, ears or scalp
- Relies on comprehensive pathological assessment of the peripheral margins using frozen sections
- Technique is usually combined with immunohistochemical (IHC) stains to better delineate surgical margins
    (HMB45, MEL-5, MITF, MART-1, SOX10)
- Central debulk specimen is submitted for permanent section evaluation to ensure adequate staging of melanoma
- Mohs appropriate use criteria (AUC) can be used to help identify cases appropriate for Mohs surgery

– T1 Melanoma (Breslow ≤1 mm)

WLE with 1 cm margins is recommended

Efficacy of narrow excision (excision with 1-cm margins) for primary melanomas compared to 3 cm margins showed:
- No difference in development of metastatic disease involving regional nodes and distant organs
- No difference in disease-free survival rates and overall survival rates

– T2 Melanoma (Breslow >1–2 mm)

WLE with at least 1–2 cm margins is recommended

Location of tumor with consideration of functional and cosmetic factors may play a role in determining surgical margins for this group of patients

– T3 Melanoma (Breslow >2–4 mm) and T4 Melanoma (>4 cm)

2-cm excision margin is recommended for melanomas with tumor thickness greater than 2 mm in most clinical guidelines
- Margins greater than 2 cm do not show any significant differences in important outcome parameters such as local or distant metastases and overall survival

- Metastases
  – Local and regional metastases represent lymphatic or angiotropic extension (reflected by the "N" status in TNM staging)
  – Local recurrence is defined as regrowth within 2 cm of the surgical scar following definitive excision of a primary melanoma with appropriate surgical margins
  – Microsatellites

    Microscopic foci of discontinuous tumor located adjacent or deep to the primary tumor
  – Satellite metastases:

    Clinically evident skin or subcutaneous tumor within 2 cm of the primary tumor
  – In-transit metastases:

    Skin or subcutaneous metastases that are more than 2 cm from the primary lesion but are not beyond the regional nodal basin

    Typically located between the primary site and the regional nodes

    May extend in a direction opposite to that of the closest regional nodal basin in patients with extensive disease

Tumor blockage of lymphatic channels and altered patterns of lymphatic flow
– Management for in transit and satellite metastases
In the absence of extensive metastatic disease, resection of single or multiple in transit metastases is the treatment of choice when complete resection can be achieved with an acceptable level of morbidity
A histologically negative margin is adequate

High Risk Non-melanoma Skin Cancer [25–34]

Cutaneous Squamous Cell Carcinoma (cSCC)
- Introduction
  – Second most common cutaneous malignancy in the general population
  – Most common skin cancer in dark-skinned patients
  – Association with HPV when located in the perianal region, external genitalia, and around the nailbed
  – Estimated incidence of lymph node metastasis 3.7–5.2%
  – Estimated incidence of disease-specific death are 1.5–2.1%
- Clinical and histologic factors associated with increased risk of local recurrence, lymph node metastasis, and disease-specific death include:
  – Tumor diameter of 2 cm or larger
    Maximum clinical diameter of the lesion prior to surgery (Fig. 5.8a, b)
  – Depth of tumor invasion
    Tumor thickness >2.0 mm (in particular, >6.0 mm) is a risk factor for local recurrence and metastasis
    Invasion of cSCC beyond the subcutaneous fat is associated with a significantly higher risk of disease-specific death
  – Perineural invasion (tumor cells within the nerve sheath)
    Small-caliber perineural invasion (nerves <0.1 mm in caliber) have no independent association with poor outcomes
    Large-caliber perineural invasion (0.1 mm diameter or greater) has greater risk of lymph node metastasis (Fig. 5.9)

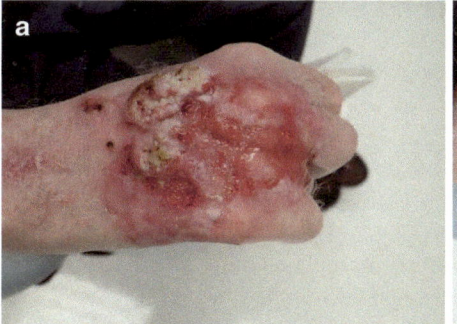 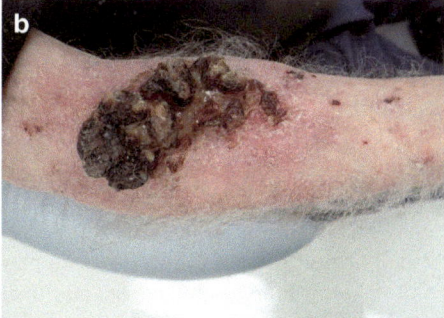

Fig. 5.8 This figure (**a, b**) displays two clinical photos of aggressive squamous cell carcinomas

Fig. 5.9 Histopathology of SCC with perineural invasion

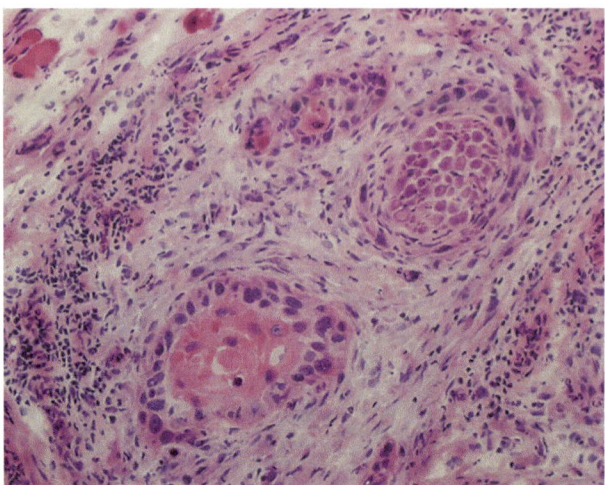

Fig. 5.10 Squamous cell carcinoma of the lip

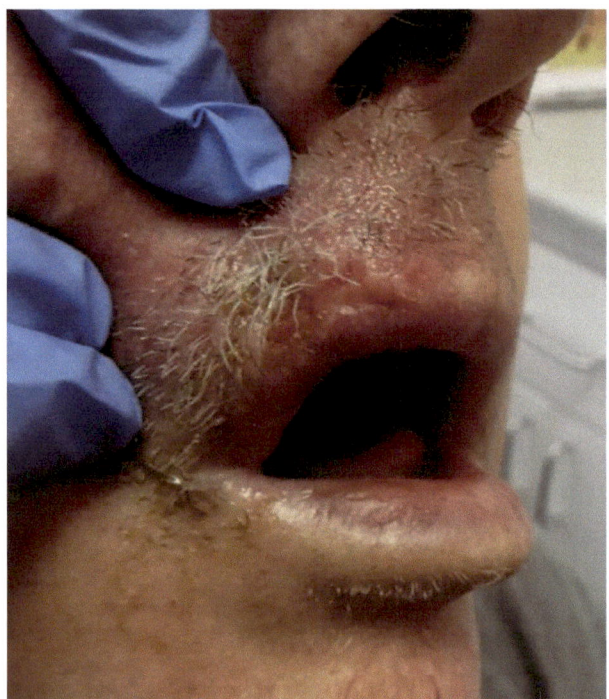

– Poorly differentiated histology
    Poor differentiation is independently associated with local recurrence, lymph node metastasis, and disease-specific death
– Location on ear or lip
    Particular anatomic locations such as lip (vermillion and hair-bearing) and ear demonstrate an increased risk of local recurrence and metastatic potential (Fig. 5.10)

Fig. 5.11 Histopathology of SCC lymphovascular invasion

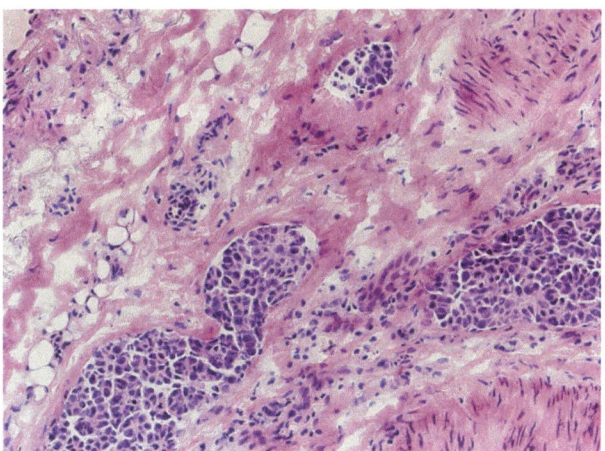

- Lymphovascular invasion (Fig. 5.11)
- Recurrent tumor
- Association with burns, ulcers, and radiation
- Human papillomavirus (HPV)
    Anatomic sites such as the penis, vulva, and perianal skin, tend to have HPV-related pathogenesis
- Immunocompromised host status
- Staging
  - Regional lymph node (LN) exam
    All patients should undergo a regional LN exam
  - Clinically enlarged nodes should be examined through fine needle aspiration (FNA) or core needle biopsy (CNB)
  - Early detection of nodal metastasis may improve outcomes
  - Tumor-positive lymph nodes are managed by aggressive surgical resection (local and regional disease)
  - Radiologic imaging
    Standard method to detect subclinical nodal spread
    Utilized for high-risk cutaneous SCC for nodal staging and for preoperative planning
    - CT
      - Evaluates for extracapsular nodal spread, skull-base invasion, central nodal necrosis, and cartilage involvement
    - MRI
      - Evaluates for perineural involvement (detects only advanced perineural spread) and definition of tissue planes
- Treatment
  - Surgical excision
    5 mm excision margin is recommended for excision of cSCC
  - Mohs micrographic surgery

Frozen tissue specimens are sectioned horizontally or tangentially into serial sections.

Peripheral and deep margins, are evaluated histologically on the same day as surgery

Preserves the maximum amount of normal tissue

Important from a functional, aesthetic, and reconstructive perspective

– Mohs Moat

For larger tumors, Mohs moat technique, under local or tumescent anesthesia, may be utilized to clear peripheral margin

A multidisciplinary approach is often utilized in these scenarios, combining dermatologic surgery with orthopedic oncology, plastic surgery, and/or otolaryngology for deep tumor resection

Due to the large amount of tissue requiring time for meticulous frozen tissue processing and review by the Mohs surgeon, a wider margin is obtained to allow for clearance within one to two resection layers

– Electrodessication and curettage

Electrodesiccation and curettage (ED&C) is a treatment option for squamous cell carcinoma in situ and for low-risk cutaneous SCC in the appropriate clinical setting

Retrospective studies show acceptably high cure rates with ED&C when compared to excision for low risk cutaneous SCC

Procedure involves three rounds of curettage and fulguration

Advantages:

Requires less time and resources than excision

Disadvantages:

No histologic confirmation of malignancy removal is performed

May provide inferior cosmesis in certain anatomic locations (Fig. 5.12a–c)

– Radiotherapy (RT)

Postoperative adjuvant RT is indicated in patients with locally advanced cSCC of the head and neck

After radical resection, indications for postoperative RT may include the following:

• Positive surgical margins
• Perineural invasion (PNI) of large caliber nerves

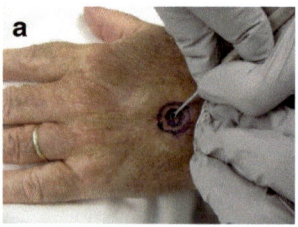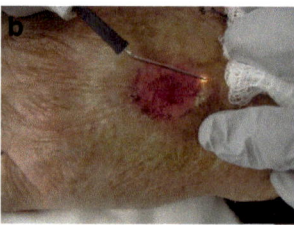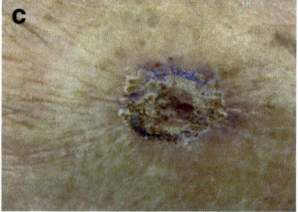

Fig. 5.12 Electrodesiccation and curettage (ED&C), pictures here, is a treatment option for basal cell carcinoma, squamous cell carcinoma in situ and for low-risk cutaneous SCC in the appropriate clinical setting

- Positive lymph nodes
- Invasion of bone or cartilage
- Extensive skeletal muscle infiltration

Primary RT can be considered for medically inoperable patients or in areas of poor surgical cosmetic outcome

Basal Cell Carcinoma

- Most common malignancy in the United States
  - Approximately 70–80% of all skin malignancies are basal cell carcinomas (BCC)
  - Commonly found in elderly male patients in sun-exposed areas of the body (Fig. 5.13)
- The 5-year cause-specific survival is 99% with low rates (<0.1%) of regional lymph node metastases or distant metastases and low 1% rate of perineural spread
- Aggressive forms histologically
  - Infiltrative
  - Basosquamous
  - Micronodular
  - Morpheaform/sclerosing
- If left untreated basal cell carcinomas can become locally invasive and destructive requiring multimodality therapy to achieve local control
- Risk factors for aggressive behavior
  - Perineural invasion (PNI)
  - Diameter greater than 2 cm
  - Long-standing presence or prior therapy (incomplete excision or prior RT)
  - History of immunosuppression
  - Poorly defined borders
- Treatment
  - Surgical excision (4 mm margins)
      4 mm excision margin is recommended for excision of BCCs

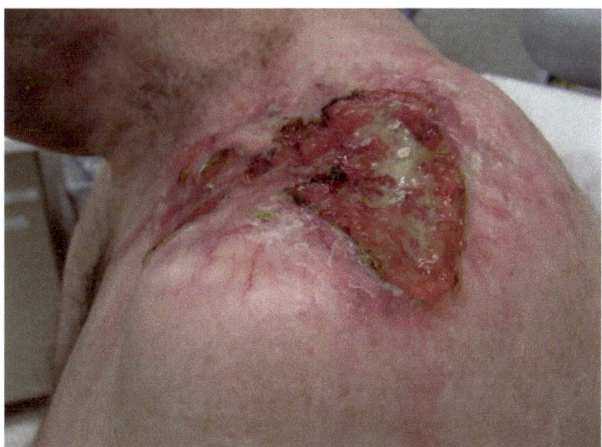

Fig. 5.13 Basal cell carcinoma

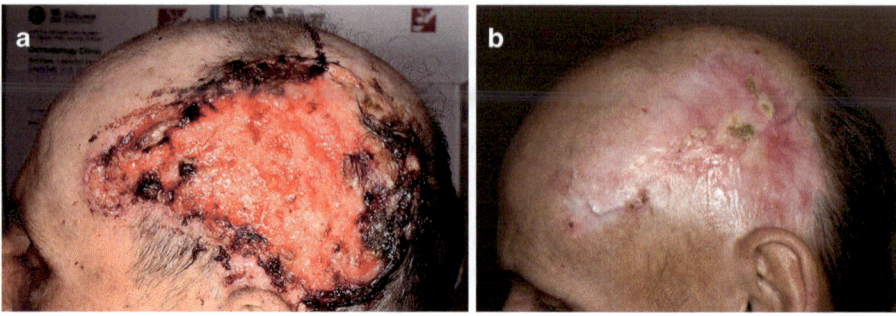

Fig. 5.14 This figure depicts an aggressive basal cell carcinoma before (**a**) and after (**b**) treatment with vismodegib, a Hedgehog pathway inhibitor

- Mohs micrographic surgery
    Frozen tissue specimens are sectioned horizontally or tangentially into serial sections.
    Peripheral and deep margins, are evaluated histologically on the same day as surgery
    Preserves the maximum amount of normal tissue
    Important from a functional, aesthetic and reconstructive perspective
- Electrodessication and curettage
    Electrodesiccation and curettage (ED&C) is a treatment option for superficial BCC and for low-risk BCC in the appropriate clinical setting
    Not appropriate for lesions with extension into deep dermis
- Radiotherapy (RT)
    Primary use: Consideration in cases where cosmetic or functional outcomes may be unacceptable with definitive surgical management
    Adjuvant use: Performed to the postoperative bed alone due to low risk of nodal or perineural spread
- Antineoplastic agents
    Most mutations in BCC result from loss of function of the patched homologue-1 (PTCH1)
    Small molecule inhibitors of the smoothened receptor of the hedgehog signaling pathway can be utilized for metastatic or locally advanced basal cell carcinomas (Fig. 5.14a, b)

Merkel Cell Carcinoma (MCC) [35–39]

- Introduction
    - Rare, very aggressive neuroendocrine tumor of the skin with a high frequency of local recurrence and metastasis and a high mortality rate
    - Development can be associated with integrated Merkel cell polyomavirus
    - Predisposing factors include ultraviolet light exposure and immunosuppression
    - Clinically presents as a red to violaceous papule or nodule located commonly on the head and neck > lower extremity and buttocks

- Immunohistologically the tumor demonstrates cytokeratin (CK)-20 staining in a perinuclear dot pattern
- Commonly presents as localized cutaneous lesion; however, MCC can quickly metastasize to involve regional draining lymph nodes or other distant sites (e.g. skin, lung, liver, bone, and central nervous system)
- Surgical Management
  - Surgical resection, sentinel lymph node biopsy (SLNB), and radiotherapy have been the gold standard of management in patients with localized disease. In more advanced disease, chemotherapy and now, immunotherapy, play a role in treatment.
  - Wide local excision (WLE)
    In localized MCC (TNM stage I and II) wide local excision is recommended with the goal of obtaining clear pathologic margins
    WLE margins of 1–2 cm is recommended with excision down to muscle fascia or periosteum where possible feasible
    High rate of local recurrence with WLE alone (range 25–40%)
  - Mohs micrographic surgery (MMS)
    MMS can offer tissue sparing potential in locations such as the head and neck where WLE may not be feasible
    Provides an exhaustive and comprehensive histologic assessment of the margins
    For the treatment of early stage MCC, MMS appears to be as effective as WLE
    Prospective studies are needed given current conflicting literature comparing WLE versus MMS for treatment of MCC
  - Sentinel lymph node biopsy (SLNB)
    Approximately 25–30% of patients with no clinical evidence of lymphadenopathy will have pathological tumor involvement of lymph node(s)
    SLNB should be considered in patients with MCC as it can provide vital prognostic information and helps direct management
    SLNB is usually performed immediately prior to WLE
    If the SLN is involved, further imaging with positron emission tomography/computed tomography (PET/CT) should be performed and complete nodal dissection and adjuvant therapy considered
  - Clinically apparent lymphadenopathy
    Fine-needle aspiration or core needle biopsy should be performed to work up clinically suspicious lymphadenopathy
    Further imaging with PET/CT should be performed to evaluate for distant metastatic disease
    If no distant metastases then complete nodal dissection should be performed with consideration of adjuvant therapy
- Adjuvant therapy
  - Radiotherapy (RT)
    RT to the primary excision site and regional lymph node basin is most commonly used as adjuvant therapy after surgery in early-stage disease

Decision to use adjuvant RT may also be influenced by high risk patients (e.g. immunosuppressed) or high risk tumor characteristics (e.g. >1 cm, lymphatic or intravascular invasion)
- Chemotherapy in MCC is mainly reserved for patients with metastatic disease or as palliative therapy in symptomatic patients
    Adjuvant chemotherapy has not been shown to improve outcomes and responses tend not to be durable with poor side effect profile
- Immunotherapy and emerging therapeutics
    Current research focus on the creation of new immunotherapeutic strategies such as immune checkpoint inhibitors, and identifying predictive biomarkers is evolving in order to increase treatment efficacy in refractory patients

Dermatofibrosarcoma Protuberans [40–43]

- Introduction
    - Slow growing, indolent tumor
    - DFSP may demonstrate fibrosarcomatous changes with risk for metastasis to the lungs
    - Strong clinical evidence for treatment is limited given the rare incidence of dermatofibrosarcoma protuberans (DFSP) (Fig. 5.15)
- Surgical management
    - High local recurrence given extension of spindle cells along the facial plane
    - Magnetic resonance imaging (MRI) can aid in pre-operative evaluation of extent of tumor
    - Surgical excision with comprehensive margin evaluation is critical for management
    - Mohs micrographic surgery (MMS)

Fig. 5.15 Dermatofibrosarcoma protuberans

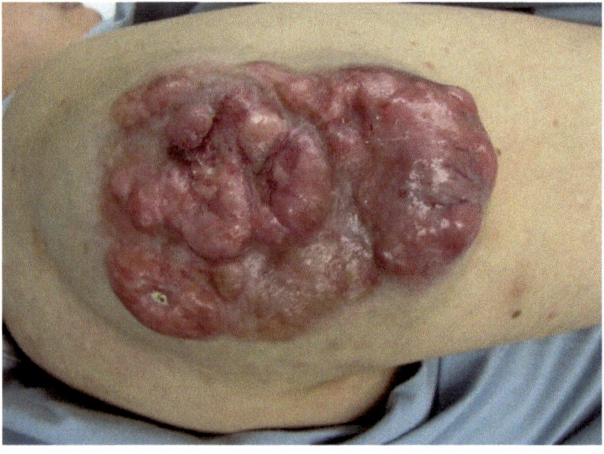

Frozen tissue specimens are sectioned horizontally or tangentially into serial sections for complete margin evaluation on the same day as surgery
- Wide local excision followed by complete circumferential peripheral and deep margin assessment
  Specimen is excised with wide margins (ranging from 2 to 4 cm)
  Specimen permanently fixed and then undergoes tangential sectioning at its borders and horizontal sectioning of the central portion at the base
  Closure can be delayed until margins are assessed
- Wide excision with traditional vertical sectioning results in only a fraction of the surgical margin examined
  Higher risk of false negative interpretation and of local recurrence from residual tumor cells
• Adjuvant therapy
  - Radiation therapy
    Postsurgical adjuvant when positive margins are present
    In cases of unresectable tumor
  - Targeted tyrosine kinase inhibitors
    Imatinib can block tumor progression in DFSP that contain aberrant COL1A1PDGFB fusion gene

Vascular Neoplasms (Angiosarcoma) [44, 45]

• Introduction
  - Rare and aggressive vascular neoplasm accounting for about 1–2% of all sarcomas of the skin
  - Neoplastic cells are characterized by overexpression of vascular endothelial growth factor (VEGF)
  - Local recurrence and metastasis to lymph nodes and lung are common
  - Overall prognosis is poor
• Surgical Management
  - Histopathologically clear surgical margins are of critical importance
  - Wide local excision
    Wide margins are recommended; however, no clear guidelines exist on the specific width of surgical margins in angiosarcoma
    Due to extensive microscopic extension beyond clinically apparent margins, surgical resection of the primary tumor has resulted in primary site cure rates that are universally disappointing for local control
  - Mohs micrographic surgery (MMS)
    Has been used for margin evaluation with several case reports describing good long-term control at the primary site, but no significant patient series have been performed
    Therefore, the role of MMS in the treatment of angiosarcoma has not been evaluated or defined and is not currently the preferred treatment for this commonly fatal neoplasm
  - Because of the local aggressiveness and seemingly multifocal nature, wide excision with or without MMS control of the excision is currently recommended

- Adjuvant therapy
  - Radiotherapy
    Combination of surgery and radiation is the mainstay of treatment for angiosarcoma
    Radiation alone may be used for palliation and does not improve overall survival
  - Chemotherapy
    Data regarding chemotherapy in angiosarcoma are inconsistent.
    While some studies show benefits of adjuvant chemotherapy, others report no increase in overall survival
    Nevertheless, chemotherapy is widely used in metastatic angiosarcoma and lesions that cannot be entirely resected

Adnexal Malignant Tumors

Malignant Sebaceous Tumors

Sebaceous Carcinoma [47–49]
- Introduction
  - Aggressive tumor arising from the sebaceous glands
  - Majority are located in the head and neck region with a predilection for the periocular area
  - May occur sporadically or as part of Muir-Torre syndrome
- Surgical Management
  - Wide local excision (WLE)
    Historically has been the standard of care for sebaceous carcinoma
    High rate of local recurrence is noted and recurrent disease can be treated with surgical re-excision
    Periorbital sebaceous carcinoma can lead to orbital invasion necessitating exenteration
    Sentinel lymph node biopsy (SLNB) has been advocated for both periorbital and extraorbital sebaceous carcinoma
  - Mohs micrographic surgery (MMS)
    Additional data are needed to allow for true comparison of the long-term outcomes of MMS versus WLE
    May be advantageous for the treatment of sebaceous carcinoma given the propensity for tumors to arise in cosmetically sensitive areas such as the eyelids and face
    MMS has been associated with good outcomes and low local recurrence rates for periorbital sebaceous carcinoma
- Adjuvant therapy
  - Radiotherapy

Adjuvant radiotherapy can be considered in patients with recurrent disease to reduce further recurrence risk

Regional lymph node metastasis may be treated with adjuvant radiation after regional lymph node dissection

Radiotherapy for primary treatment sebaceous carcinoma should be restricted to recurrent lesions, metastatic disease, or palliative treatment in patients who are not candidates for surgical excision

- Chemotherapy

Reports detailing the use of chemotherapy in sebaceous carcinoma are limited to a small number of case reports

Malignant Apocrine Tumors

Extramammary Paget's Disease (EMPD) [50, 51]

- Introduction
  - Paget's disease typically affects the breast but can often manifest in the genital area and, rarely, in the axilla of the elderly. The latter is called extramammary Paget's disease and accounts for 6% of cutaneous malignancies
- Surgical Management
  - The clinical margin of EMPD is often ambiguous, and therefore a standard treatment for EMPD has not yet been established. A conclusion regarding the most appropriate surgical margin has yet to be reached
  - Wide local excision

    Standard therapy is complete surgical excision with 2–5 cm safety margin or Mohs surgery
    - Wider margins recommended based on presence of skip lesions, as well as presence of disease past the clinically affected area

    Recurrence rates after standard surgical excision range from 33 to 62.5% versus 12 to 28% after Mohs surgery
  - Mohs micrographic surgery (MMS)

    Mohs surgery allows maximal tissue sparing in critical anatomic structures

    Efficacy of MMS for EMPD remains controversial. Some data suggest that EMPD has an irregular and multifocal growth pattern, with lesions histologically extending further than the clinically evident margin.

    Scouting biopsies prior to definitive surgical management may be beneficial for surgical planning

    Given the rarity of this tumor, the number of cases included in most studies is insufficient to definitively prove the efficacy of MMS
- Adjuvant therapy
  - Radiotherapy

    The role of radiation therapy following resection is uncertain but should be considered for those patients with close or positive resection margins or bulky disease
  - Chemotherapy

Chemotherapy has been recommended for locally advanced and metastatic disease but the optimal regimen for metastatic EMPD remains to be defined

Malignant Eccrine Tumors

Aggressive Digital Papillary Adenocarcinoma (ADPA) [52–55]
- Introduction
  - Aggressive digital papillary adenocarcinoma is a rare cancer
  - Slow growing and usually painless isolated nodule
  - Commonly involving the volar surface of the digit in close proximity to the distal interphalangeal joint or periungual region
  - Regional and distant metastases have been described
- Surgical Management
  - Surgical excision
    Although surgical excision is the accepted standard for management, there are no definitive guidelines and variation remains in the exact procedure or margins recommended
    - Management varies in the literature from wide local excision to digit amputation
    - Further comparative studies are needed to look at long term follow up, recurrence rates, and regional or distant spread
  - Mohs micrographic surgery (MMS)
    MMS has been successfully used as a digit-sparing alternative with evidence of prolonged disease-free survival in several cases
  - Other studies
    The role of lymph node (LN) analysis, including sentinel lymph node biopsy and LN dissection, is being refined for ADPA
- Adjuvant therapy
  - Radiotherapy
    The role of radiation therapy following resection is uncertain but should be considered for those patients with close or positive resection margins or bulky disease
  - Chemotherapy
    Chemotherapy has been recommended for locally advanced and metastatic disease but the optimal regimen for metastatic EMPD remains to be defined

Eccrine Porocarcinoma (EPC) [56–58]
- Introduction
  - Rare tumor thought to arise from the acrosyringium of the sweat gland of the skin
  - EPC most commonly develops on the lower extremities followed by the trunk and head and neck region
  - Most prevalent in adults aged 50–80 years

- Surgical Management
  - Wide local excision (WLE)
    - Most common therapy utilized
    - Local recurrence and metastasis have been reported
  - Mohs micrographic surgery (MMS)
    - In the most recent literature, MMS has been associated with fewer cases of metastasis when compared to excision
  - Other studies
    - Sentinel lymph node biopsy can be considered if patients possess high-risk features
    - Palpable lymphadenopathy should be further worked up with fine needle aspiration (FNA)
- Adjuvant therapy
  - Other treatments including chemotherapy, single-agent or multiagent chemotherapy
  - The development of checkpoint inhibitors may offer promise in the future

General Surgical Considerations in Patients with Systemic Malignancy

Immunosuppression [59–65, 68–70]

- Organ transplant recipients
  - Non-melanoma skin cancer (NMSC) accounts for 95% of all skin cancers in organ transplant recipients (OTR)
  - The risk of developing skin cancer in immunosuppressed is 10- to 250-fold higher than in immunocompetent patients
  - SCC to BCC ratio of 3 or 4:1
  - Appear after a mean interval of 8–10 years after transplantation
  - Incidence of NMSC increases steadily with time after transplantation
  - Risk factors
    - Fair color of skin, hair and eyes
    - Male sex
    - Cumulative ultraviolet radiation exposure
    - History of skin cancer prior to transplantation
    - Age at transplantation (50 years or older)
    - Duration and degree of immunosuppression
    - Organ transplanted (heart transplant associated with higher risk)
    - Infection with HPV
    - Low CD4 count
  - Role of immunosuppressive agents
    - Skin cancers result both from a decrease in immunosurveillance and from the direct oncogenic effects

- Cutaneous squamous cell carcinoma in organ transplant recipients
  65-fold increased incidence of SCC
  Human papilloma virus poses additional risks
  SCC are more aggressive, grow rapidly, and recur locally and metastasize
  Managing cSCC in the immunosuppressed
  - Sun protection
    - Sunscreen (broad-spectrum protection (protects against UVA and UVB rays), Sun Protection Factor (SPF) 30 or higher, water resistance) and sun avoidance (seek shade, protective clothing, caution near water and sand, avoid tanning beds)
  - Regular skin checks
  - Field therapy for actinic keratoses
    - Topical 5-fluorouracil (5-FU)
    - Topical imiquimod
    - Photodynamic therapy
  - Surgical clearance
    - Mainstay of treatment for high-risk SCC is complete surgical clearance of the lesion with histologically clear margins
      Excision
      Mohs micrographic surgery
  - Chemoprophylaxis
    - Retinoids
      Decrease cell growth
      Inhibit malignant transformation by downregulating proto-oncogenes and influencing growth factors
      Decreases new cancer formation, but does not alter the course of an existing tumor
      Chemopreventive properties within 2–3 months of the initiation of therapy
      Loss of the protective chemoprophylactic effect within 2–3 months after discontinuation of therapy
      Indications on a case by case basis however following guidelines can be considered:
      - Numerous SCC per year (5–10 per year)
      - Innumerable actinic keratosis and multiple NMSC
      - Accelerating frequency of SCC
      - Eruptive keratoacanthomas
      - High risk NMSC

- Nicotinamide (vitamin B3)
    Enhances repair of photodamaged DNA
    Prevents the inhibitory effects of ultraviolet radiation on the immune system without altering baseline immune responses
    Nicotinamide 500 mg twice daily
- Assessment of immune status
    Reduction in immunosuppression has been associated with a decrease in new SCC formation and improved outcomes in those with established aggressive disease
    Single-agent therapy appears to carry a lower risk than multi-agent immunosuppression
    Newer immunosuppressive agents (e.g. sirolimus) are associated with a lower incidence of SCC development when compared to older calcineurin inhibitors (CNIs)
    A balance must be struck between risks posed by skin cancer and risks of a new immunosuppressive regimen
    Determination of impact of multiple skin cancers on the patient's quality of life and estimates of morbidity and mortality the patient may suffer from skin cancer
- Radiation therapy
  - High-risk SCC is best approached with surgical excision, combined with adjuvant radiation therapy if indicated
  - Radiation therapy can be used as a primary treatment option for SCC however:
      Outcomes are generally inferior to surgery
      Tumors can recur quickly after treatment
      Iatrogenic carcinogenesis can occur years later in the radiated area

Autoimmune Cytopenias and CLL [66, 67]

- Patients with chronic lymphocytic leukemia (CLL) are at risk for more frequent and aggressive skin cancers
  - Increased rates of local recurrence, regional metastasis, and death
- Both melanoma and NMSC are reported as some of the most common second cancer associations in this population

Review Questions and Answers

1. Which cytopenia is associated with more frequent and more aggressive skin cancers?
   (a) AML
   (b) CML
   (c) CLL
   (d) ALL

   Answer: (c) CLL

2. Options for chemoprophylaxis and prevention of further skin cancer development include all but which of the following?
   (a) Retinoids
   (b) Nicotinamide
   (c) Sirolimus
   (d) Mycophenolate mofetil

   Answer: (d) Mycophenolate mofetil

3. Identify the standard accepted treatment of a superficial spreading melanoma with a Breslow of 2.1 mm.
   (a) Wide local excision with 0.5 cm margins. No sentinel lymph node biopsy.
   (b) Wide local excision with 1.0 cm margins, with sentinel lymph node biopsy.
   (c) Wide local excision with 1.0 cm margins. No sentinel lymph node biopsy.
   (d) Wide local excision with 2.0 cm margins, with sentinel lymph node biopsy.

   Answer: (d) Wide local excision with 2.0 cm margins, with sentinel lymph node biopsy

References

1. Ad Hoc Task Force, Connolly SM, Baker DR, Coldiron BM, Fazio MJ, Storrs PA, Vidimos AT, Zalla MJ, Brewer JD, Smith Begolka W, Ratings Panel, Berger TG, Bigby M, Bolognia JL, Brodland DG, Collins S, Cronin TA Jr, Dahl MV, Grant-Kels JM, Hanke CW, Hruza GJ, James WD, Lober CW, EI MB, Norton SA, Roenigk RK, Wheeland RG, Wisco OJ. AAD/ACMS/ASDSA/ASMS 2012 appropriate use criteria for Mohs micrographic surgery: a report of the American Academy of Dermatology, American College of Mohs Surgery, American Society for Dermatologic Surgery Association, and the American Society for Mohs Surgery. J Am Acad Dermatol. 2012;67(4):531–50.
2. American Medical Association. Mohs micrographic surgery. CPT Assist. 2006;16:1–7. Google Scholar.
3. Bolognia J, Jorizzo J, Schaffer JV. Dermatology. 2nd ed. Philadelphia: Elsevier Saunders; 2012.
4. Kalra A, Prucher GM, Hodges S. The role of core needle biopsies in the management of neck lumps. Ann R Coll Surg Engl. 2019;101:193–6.
5. Kocjan G. Fine needle aspiration cytology: diagnostic principles and dilemmas. 1st ed. Berlin: Springer; 2006.

6. Lin F, Liu H, Zhang J, Kremitske D. Handbook of practical fine needle aspiration and small tissue biopsies. 1st ed. Heidelberg: Springer; 2017.
7. Mendese G, Maloney M, Bordeaux J. To scoop or not to scoop: the diagnostic and therapeutic utility of the scoop-shave biopsy for pigmented lesions. Dermatol Surg. 2014;40(10):1077–83.
8. Özel D, Aydın T. A clinical compilation of lymph node pathologies comparing the diagnostic performance of biopsy methods. J Ultrasound. 2019;22(1):59–64.
9. Pickett H. Shave and punch biopsy for skin lesions. Am Fam Physician. 2011;84(9):995–1002.
10. Amin MB, Edge SB, Greene FL, et al., editors. AJCC cancer staging manual. 8th ed. New York: Springer; 2017.
11. Austin E, Mamalis A, Ho D, Jagdeo J. Laser and light-based therapy for cutaneous and soft-tissue metastases of malignant melanoma: a systematic review. Arch Dermatol Res. 2017;309(4):229–42.
12. Braun RP, Rabinovitz HS, Oliviero M, Kopf AW, Saurat JH. Dermoscopy of pigmented skin lesions. J Am Acad Dermatol. 2005;52:109–21.
13. Campos de Castro BA, JCS F, Pedrosa MS, Raquel de Almeida Marques D, Gonçalves VP, JMM P. Surg Cosmet Dermatol. 2015;7(1):65–7.
14. Gershenwald JE, Scolyer RA, Hess KR, Sondak VK, Long GV, Ross MI, Lazar AJ, Faries MB, Kirkwood JM, McArthur GA, Haydu LE, Eggermont AMM, Flaherty KT, Balch CM, Thompson JF, for members of the American Joint Committee on Cancer Melanoma Expert Panel and the International Melanoma Database and Discovery Platform. Melanoma staging: Evidence-based changes in the American Joint Committee on Cancer eighth edition cancer staging manual. CA Cancer J Clin. 2017;67(6):472–92.
15. Hayes AJ, Maynard L, Coombes G, et al. Wide versus narrow excision margins for high-risk, primary cutaneous melanomas: long-term follow-up of survival in a randomised trial. Lancet Oncol. 2016;17:184–92. Higgins HW 2nd, Lee KC, Galan A, Leffell DJ.Melanoma in situ: Part II. Histopathology, treatment, and clinical management.J Am Acad Dermatol. 2015 Aug;73(2):193–203.
16. Hunger RE, Angermeier S, Seyed Jafari SM, Ochsenbein A, Shafighi M. A retrospective study of 1- versus 2-cm excision margins for cutaneous malignant melanomas thicker than 2 mm. J Am Acad Dermatol. 2015;72:1054–9.
17. Kunishige JH, Brodland DG, Zitelli JA. Surgical margins for melanoma in situ: When 5-mm margins are really 9 mm. J Am Acad Dermatol. 2015;72(4):745.
18. Kasprzak JM, Xu Y. Diagnosis and management of lentigo maligna: a review. Drugs Context. 2015;4:212–81.
19. Paraskevas LR, Halpern AC, Marghoob AA. Utility of the Wood's light: five cases from a pigmented lesion clinic. Br J Dermatol. 2005;152(5):1039–44.
20. Robinson JK. Margin control for lentigo maligna. J Am Acad Dermatol. 1994;31:79–85.
21. Sladden MJ, Balch C, Barzilai DA, Berg D, Freiman A, Handiside T, Hollis S, Lens MB, Thompson JF. Surgical excision margins for primary cutaneous melanoma. Cochrane Database Syst Rev. 2009;4:CD004835.
22. Swanson NA, Lee KK, Gorman A, Lee HN. Biopsy techniques. Diagnosis of melanoma. Dermatol Clin. 2002;20:677–80.
23. Thomas JM, Newton-Bishop J, A'Hern R, Coombes G, Timmons M, Evans J, Cook M, Theaker J, Fallowfield M, O'Neill T, Ruka W, Bliss JM, United Kingdom Melanoma Study Group; British Association of Plastic Surgeons; Scottish Cancer Therapy Network. Excision margins in high-risk malignant melanoma. N Engl J Med. 2004;350(8):757–66.
24. Veronesi U, Cascinelli N, Adamus J, Balch C, Bandiera D, Barchuk A, Bufalino R, Craig P, De Marsillac J, Durand JC, et al. Thin stage I primary cutaneous malignant melanoma. Comparison of excision with margins of 1 or 3 cm. N Engl J Med. 1988;318(18):1159–62.
25. Delgado Jiménez Y, Camarero-Mulas C, Sanmartín-Jiménez O, et al. Differences of Mohs micrographic surgery in basal cell carcinoma versus squamous cell carcinoma. Int J Dermatol. 2018;57(11):1375–81.
26. Jennings L, Schmults CD. Management of high-risk cutaneous squamous cell carcinoma. J Clin Aesthet Dermatol. 2010;3(4):39–48.

27. Karia PS, Jambusaria-Pahlajani A, Harrington DP, et al. Evaluation of American Joint Committee on Cancer, International Union Against Cancer, and Brigham and Women's Hospital tumor staging for cutaneous squamous cell carcinoma. J Clin Oncol. 2014;32(4):327–34.
28. Mikhail GR, Mohs FE. Mohs micrographic surgery. Philadelphia: W.B. Saunders; 1991. p. 7–13.
29. Mierzwa ML. Radiotherapy for skin cancers of the face, head, and neck. Facial Plast Surg Clin North Am. 2019;27(1):131–8.
30. Motaparthi K, Kapil JP, Velazquez EF. Cutaneous squamous cell carcinoma: review of the eighth edition of the American Joint Committee on Cancer Staging Guidelines, prognostic factors, and histopathologic variants. Adv Anat Pathol. 2017;24(4):171–94.
31. Reschly MJ, Shenefelt PD. Controversies in skin surgery: electrodessication and curettage versus excision for low-risk, small, well-differentiated squamous cell carcinomas. J Drugs Dermatol. 2010;9(7):773–6.
32. Seth R, Revenaugh PC, Vidimos AT, Scharpf J, Somani AK, Fritz MA. Simultaneous intraoperative Mohs clearance and reconstruction for advanced cutaneous malignancies. Arch Facial Plast Surg. 2011;13(6):404–10.
33. Vance KK, Pytynia KB, Antony AK, Krunic AL. Mohs moat: peripheral cutaneous margin clearance in a collaborative approach for aggressive and deeply invasive basal cell carcinoma. Australas J Dermatol. 2014;55(3):198–200.
34. Wiznia LE, Federman DG. Treatment of basal cell carcinoma in the elderly: what nondermatologists need to know. Am J Med. 2016;129(7):655–60.
35. Femia D, Prinzi N, Anichini A, et al. Treatment of advanced merkel cell carcinoma: current therapeutic options and novel immunotherapy approaches. Target Oncol. 2018;13:567.
36. Kline L, Coldiron B. Mohs micrographic surgery for the treatment of merkel cell carcinoma. Dermatol Surg. 2016;42(8):945–51.
37. Singh B, Qureshi MM, Truong MT, Sahni D. Demographics and outcomes of stage I and II Merkel cell carcinoma treated with Mohs micrographic surgery compared with wide local excision in the National Cancer Database. J Am Acad Dermatol. 2018;79(1):126–134.e3.
38. Tai P. A practical update of surgical management of merkel cell carcinoma of the skin. ISRN Surg. 2013;2013:850797.
39. Tello TL, Coggshall K, Yom SS, Yu SS. Merkel cell carcinoma: an update and review: current and future therapy. J Am Acad Dermatol. 2018;78(3):445–54.
40. Bogucki B, Neuhaus I, Hurst EA. Dermatofibrosarcoma protuberans: a review of the literature. Dermatol Surg. 2012;38:537–51.
41. Criscito MC, Martires KJ, Stein JA. Prognostic factors, treatment, and survival in dermatofibrosarcoma protuberans. JAMA Dermatol. 2016;152(12):1365–71.
42. Lowe GC, Onajin O, Baum CL, Otley CC, Arpey CJ, Roenigk RK, Brewer JD. A comparison of Mohs micrographic surgery and wide local excision for treatment of dermatofibrosarcoma protuberans with long-term follow-up: the mayo clinic experience. Dermatol Surg. 2017;43(1):98–106.
43. Trofymenko O, Bordeaux JS, Zeitouni NC. Survival in patients with primary dermatofibrosarcoma protuberans: National Cancer Database analysis. J Am Acad Dermatol. 2018;78(6):1125–34.
44. Guadagnolo NA, Zagars GK, Araujo D, et al. Outcomes after definitive treatment for cutaneous angiosarcoma of the face and scalp. Head Neck. 2011;33(5):661–7.
45. Shustef E, Kazlouskaya V, Prieto VG, et al. Cutaneous angiosarcoma: a current update. J Clin Pathol. 2017;70(11):917–25.
46. Stein JM, Hrabovsky S, Schuller DE, et al. Mohs micrographic surgery and the otolaryngologist. Am J Otolaryngol. 2004;25(6):385–93.
47. Haber R, Battistella M, Pages C, et al. Sebaceous carcinomas of the skin: 24 cases and a literature review. Acta Derm Venereol. 2017;97(8):959–61.
48. Ad Hoc Task Force, Connolly SM, Baker DR, Coldiron BM, et al. AAD/ACMS/ASDSA/ ASMS 2012 appropriate use criteria for Mohs micrographic surgery: a report of the American Academy of Dermatology, American College of Mohs Surgery, American Society for

Dermatologic Surgery Association, and the American Society for Mohs Surgery. J Am Acad Dermatol. 2012;67:531–50.

49. Kyllo RL, Brady KL, Hurst EA. Sebaceous carcinoma: review of the literature. Dermatol Surg. 2015;41(1):1–15.
50. Bae JM, Choi YY, Kim H, et al. Mohs micrographic surgery for extramammary Paget disease: a pooled analysis of individual patient data. J Am Acad Dermatol. 2013;68(4):632–7.
51. Kyriazanos ID, Stamos NP, Miliadis L, Noussis G, Stoidis CN. Extra-mammary Paget's disease of the perianal region: a review of the literature emphasizing the operative management technique. Surg Oncol. 2011;20:e61–71.
52. Rismiller K, Knackstedt TJ. Aggressive digital papillary adenocarcinoma: population-based analysis of incidence, demographics, treatment, and outcomes. Dermatol Surg. 2018;44(7):911–7.
53. Duke WH, Sherrod TT, Lupton GP. Aggressive digital papillary adenocarcinoma (aggressive digital papillary adenoma and adenocarcinoma revisited). Am J Surg Pathol. 2000;24:775–84.
54. Haynes D, Thompson C, Leitenberger J, Vetto J. Mohs micrographic surgery as a digit-sparing treatment for aggressive digital papillary adenocarcinoma. Dermatol Surg. 2017;43:1487–9.
55. Knackstedt RW, Knackstedt TJ, Findley AB, Piliang M, et al. Aggressive digital papillary adenocarcinoma: treatment with Mohs micrographic surgery and an update of the literature. Int J Dermatol. 2017;56:1061–4.
56. Goon PKC, Gurung P, Levell NJ, et al. Eccrine porocarcinoma of the skin is rising in incidence in the east of England. Acta Derm Venereol. 2018;98(10):991–2.
57. Nazemi A, Higgins S, Swift R, et al. Eccrine porocarcinoma: new insights and a systematic review of the literature. Dermatol Surg. 2018;44(10):1247–61.
58. Song SS, Wu Lee W, Hamman MS, et al. Mohs micrographic surgery for eccrine porocarcinoma: an update and review of the literature. Dermatol Surg. 2015;41(3):301–6.
59. Berg D, Otley CC. Skin cancer in organ transplant recipients: epidemiology, pathogenesis, and management. J Am Acad Dermatol. 2002;47(1):1–17.
60. Brin L, Zubair AS, Brewer JD. Optimal management of skin cancer in immunosuppressed patients. Clin Dermatol. 2014;15(4):339–56.
61. Chen AC, Martin AJ, Choy B, et al. A phase 3 randomized trial of nicotinamide for skin-cancer chemoprevention. N Engl J Med. 2015;373:1618–26.
62. Euvrard S, Morelon E, Rostaing L, et al. Sirolimus and secondary skin-cancer prevention in kidney transplantation. N Engl J Med. 2012;367(4):329–39.
63. Garrett GL, Blanc PD, Boscardin J, et al. Incidence of and risk factors for skin cancer in organ transplant recipients in the United States. JAMA Dermatol. 2017;153(3):296–303.
64. Harwood CA, Leedham-Green M, Leigh IM, et al. Low-dose retinoids in the prevention of cutaneous squamous cell carcinomas in organ transplant recipients: a 16-year retrospective study. Arch Dermatol. 2005;141(4):456–64.
65. Jennings L, Schmults CD. Management of high-risk cutaneous squamous cell carcinoma. Am J Clin Aesthet Dermatol. 2010;3(4):39–48.
66. Mulcahy A, Mulligan SP, Shumack SP. Recommendations for skin cancer monitoring for patients with chronic lymphocytic leukemia. Leuk Lymphoma. 2018;59(3):578–82.
67. Onajin O, Brewer JD. Skin cancer in patients with chronic lymphocytic leukemia and non-Hodgkin lymphoma. Clin Adv Hematol Oncol. 2012;10:571–6.
68. Otley CC, Stasko T, Tope WD, et al. Chemoprevention of nonmelanoma skin cancer with systemic retinoids: practical dosing and management of adverse effects. Dermatol Surg. 2006;32(4):562–8.
69. Ulrich C, Kanitakis J, Stockfleth E, et al. Skin cancer in organ transplant recipients--where do we stand today? Am J Transplant. 2008;8(11):2192–8.
70. Zwald FO, Brown M. Skin cancer in solid organ transplant recipients: advances in therapy and management: part I. Epidemiology of skin cancer in solid organ transplant recipients. J Am Acad Dermatol. 2011;65(2):253–61.

Malignancy Risks of Dermatologic Therapies

6

Kevin K. Wu and April W. Armstrong

Abbreviations

AIDS	Acquired immunodeficiency syndrome
BB-UVB	Broad-band ultraviolet B
BCC	Basal cell carcinoma
CD	Cluster of differentiation
CI	Confidence interval
CMM	Cutaneous malignant melanoma
CsA	Cyclosporine
CTCL	Cutaneous T-cell lymphoma
EBV	Ebstein-Barr virus
ECG	Electrocardiogram
FDA	Food and Drug Administration
HAART	Highly active antiretroviral therapy
HHV	Human Herpesvirus
HIV	Human immunodeficiency virus
HPV	Human papilloma virus
HR	Hazard ratio
IARC	International Agency for Research on Cancer
IBD	Inflammatory bowel disease
Ig	Immunoglobulin
IL	Interleukin
IRR	Incidence rate ratio

K. K. Wu (✉)
Department of Medicine, University of California, Irvine, Orange, CA, USA

A. W. Armstrong
Department of Dermatology, University of Southern California Keck School of Medicine, Los Angeles, CA, USA

© Springer Nature Switzerland AG 2021
V. Liu (ed.), *Dermato-Oncology Study Guide*,
https://doi.org/10.1007/978-3-030-53437-0_6

IVIG	Intravenous immunoglobulin
MTX	Methotrexate
nbDMARDs	Nonbiologic disease modifying antirheumatic drugs
NB-UVB	Narrow-band ultraviolet B
NHL	Non-Hodgkin's lymphoma
NIH	National Institutes of Health
NMSC	Nonmelanoma skin cancer
OR	Odds ratio
PDE	Phosphodiesterase
PSOLAR	Psoriasis Longitudinal Assessment and Registry
PUVA	Psoralen-ultraviolet A
PY	Person years
RA	Rheumatoid arthritis
RCT	Randomized clinical trial
RR	Risk ratio
SCC	Squamous cell carcinoma
SEER	Surveillance, Epidemiology, and End Results Program
SIR	Standardized incidence rate
SLE	Systemic lupus erythematosus
TNF	Tumor necrosis factor
UVA	Ultraviolet A
UVB	Ultraviolet B

Learning Objectives

1. To understand an approach to ascertaining the risk of malignancy from a given therapeutic agent.
2. To quantify and compare the risk of cancer from therapies used to treat skin conditions.

Introduction

Patients with inflammatory skin diseases such as psoriasis and atopic dermatitis have higher risks of malignancies compared to healthy individuals regardless of which therapies they use [1, 2]. In these populations with inflammatory skin diseases, it is important to determine whether dermatologic therapies have an effect on the risk of malignancies. However, assessing the association between dermatologic therapies and malignancies is challenging because clinical trials lack adequate population sizes and long-term follow-up to detect clinically important differences in malignancy rates in those receiving the medication compared to those receiving placebo. In this chapter, we synthesize and interpret the literature on malignancy risks with dermatologic therapies with a focus on long-term follow-up data.

Definitions

- Person years (PY) = number of individuals × years on the drug
- Standardized incidence rates (SIR) = ratio of observed number of cancers to the expected number of cancers
- Odds ratio (OR) = ratio of the odds of cancer with a given exposure to the odds of cancer without the exposure
- Risk ratio (RR) = ratio of the probability of cancer with a given exposure to the probability of cancer without the exposure
- Hazard ratio (HR) = ratio of how often cancer events occur with a given exposure to how often cancer events occur without the exposure over time
- Incidence rate ratio (IRR) = ratio of the incidence rate with a given exposure to the incidence rate without the exposure

Interpreting Data on Rare Adverse Events

- The adequacy of medication safety databases for interpretation of malignancy risk is reliant upon the following:
  - *Number of patients*
    Are there enough patients in the trial to detect rare events? (see **Rule of 3's**)
  - *Length of exposure*
    Is the length of exposure to the drug during the trial long enough to detect rare events?
    - Post-market data may be required to assess long-term malignancy risk because the length of the clinical trial is likely too short to detect malignancy events.

- *Type of patients*
    Are the patients who participated in the clinical development program of the investigational drug representative of the general population?
    - If the population is not representative, the data may not be generalizable.
- *Comparison group*
    What is the reference population against which the trial/study population is being compared to, *the population with the disease regardless of therapy* or *the general population*?
    - Studies comparing the study population to the general population may be less informative because these studies are unable to control for the disease itself.
- **Rule of 3's** for rare, adverse drug-related events:
    - If the incidence of the event is 1 in *n* patients per year, at least 3*n* trial subjects per year are needed to see 1 case at 95% confidence interval (CI)
        Example: true incidence of toxic epidermal necrolysis is 1 per 2,000,000 patients per year. We need *at least* 6,000,000 (3 × 2,000,000) patients to see 1 case at 95% CI.
        During clinical development, drug reactions of <0.1% cannot be ruled out if fewer than 3000 patients are exposed to the drug.
        - However, only a minimum of 1500 patients are required to be exposed to an investigational drug before approval.
        Due to the rarity of malignancies, long-term follow-up data with adequate sample sizes are critical to informing malignancy risk in dermatologic therapies.

Issues with Interpreting Malignancy Risk Data in Dermatology Patients

- Patients with inflammatory skin diseases such as psoriasis and atopic dermatitis have higher rates of malignancies than the general population [1, 2].
    - Psoriasis is associated with a 20% higher risk of all cancers when compared to the general population (114 events/10,000 PY vs. 95 events/10,000 PY; SIR 1.2; 95% CI, 1.13–1.26) (Fig. 6.1) [1]. Specific cancers include:
        NMSC—65% higher risk than the general population (129 events/10,000 PY vs. 78 events/10,000 PY)
        - Psoriasis severity is proportional to the risk of developing NMSC.
        Lymphoma—50% higher risk than the general population (9 events/10,000 PY vs. 6 events/10,000 PY)
    - Atopic dermatitis is associated with a 13% higher risk of developing any cancer (SIR 1.13; 95% CI, 1.01–1.25) [2]. Specific cancers include:
        Esophagus cancer (SIR, 3.5; 95% CI, 1.3–7.7)
        Lung cancer (SIR, 2.0; 95% CI, 1.3–2.8)
        Lymphoma (SIR, 2.0; 95% CI, 1.4–2.9)

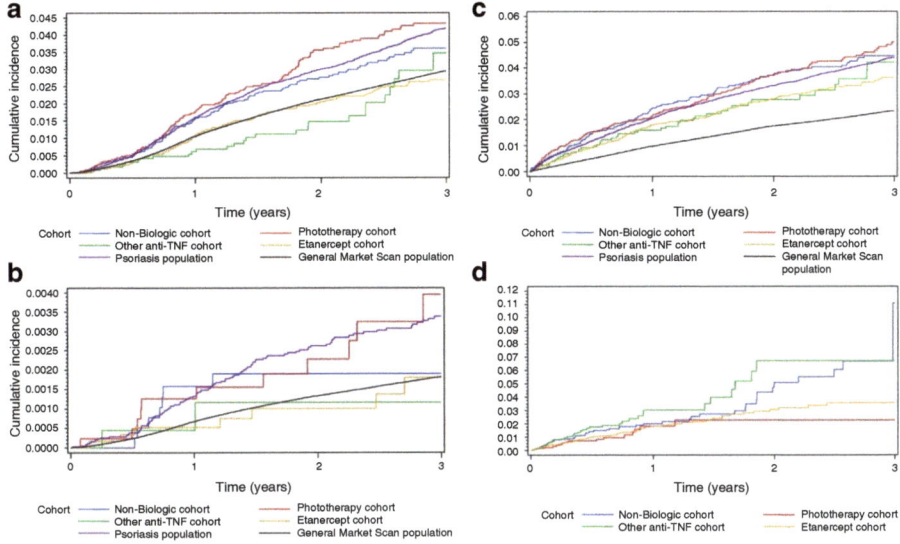

Fig. 6.1 (**a**) Cumulative incidence for all malignancies (excluding non-melanoma skin cancer) during person-time of observation. (**b**) Cumulative incidence for lymphomas during person-time of observation. (**c**) Cumulative incidence for non-melanoma skin cancer (NMSC) during person-time of observation. (**d**) Cumulative incidence for hospitalized infectious events (HIEs) during person-time of exposure. (From: Kimball, A.B., et al., *Incidence rates of malignancies and hospitalized infectious events in patients with psoriasis with or without treatment and a general population in the USA: 2005–2009.* Br J Dermatol, 2014. **170**(2): p. 366–73)

- Malignancy risk of dermatologic therapies may be confounded by the higher risk of malignancies due to the diseases themselves. Therefore, the most informative studies utilize placebo-controlled patients with the same diagnoses as the comparison group.

Importance of Long-Term Follow-Up and Continual Monitoring

- Typically, the placebo period is within 4 months. Therefore, it is difficult to determine the significance of the safety data beyond a placebo-controlled period.
- Malignancies may take years to develop. Thus, shorter-term trials may be less informative regarding malignancy risks.
- Long-term safety surveillance studies with large populations are essential to understanding and evaluating the long-term risks of dermatologic treatments.
- Registries that collect long-term safety data from different drugs will be highly informative of the long-term risks of newer therapies.

Malignancy Risk of Traditional Systemic Agents (Table 6.1)

Where available, the dose and duration of the treatments in each study are included. However, not all studies provided this information.

Table 6.1 Summary table of the malignancy risks of dermatologic therapies (excluding biologics)

Medication	Class	Dermatologic indications	Internal malignancy risk	Skin malignancy risk	Overall malignancy risk
Methotrexate (various brand products available)	Folic acid analog	Off-label for psoriasis and AD	Increased risk of lymphoma and lung cancers in RA patients No increased risk in psoriasis patients	Increased risk of melanoma and NMSC in RA patients No increased risk in psoriasis patients	Increased risk in RA patients
Cyclosporine (various brand products available)	Calcineurin inhibitor	Off-label for psoriasis and AD	Increased risk of lymphoma in psoriasis patients	Increased risk of NMSCs in psoriasis patients (risk may be reduced by using lower doses for limited duration)	Increased risk in psoriasis patients
Azathioprine (various brand products available)	Purine analog	Off-label for psoriasis and AD	Increased risk of NMSC in organ transplant, IBD, and myasthenia gravis patients No malignancy risk data available for dermatologic indications	Increased risk of lymphoma in organ transplant and IBD patients No malignancy risk data available for dermatologic indications	Increased risk in organ transplant and IBD patients No malignancy risk data available for dermatologic indications
Mycophenolate Mofetil (CellCept®)	IMP dehydrogenase inhibitor	Off-label for psoriasis, AD, and Behcet's disease	Data on risk of lymphoma in organ transplant patients is mixed No malignancy risk data available for dermatologic indications	Data on risk of NMSC in organ transplant patients is mixed No malignancy risk data available for dermatologic indications	Data on risk in organ transplant patients is mixed No malignancy risk data available for dermatologic indications

(continued)

Table 6.1 (continued)

Medication	Class	Dermatologic indications	Internal malignancy risk	Skin malignancy risk	Overall malignancy risk
Oral corticosteroids (various brand products available)	Inhibits inflammatory molecules	Used for a variety of inflammatory conditions, including AD, pemphigus vulgaris, contact dermatitis, RA, etc	Higher risk of non-Hodgkin lymphoma and esophageal cancers	Higher risk of NMSC	–
Acitretin (Soriatane®)	Structurally similar to vitamin A; targets retinoic acid receptors	Psoriasis. Used off-label for skin cancer prevention	–	May prevent AKs and NMSCs in special populations	–
Isotretinoin (Accutane®)	Structurally similar to vitamin A; targets retinoic acid receptors	Acne. Used off-label for skin cancer prevention	–	May prevent NMSC in special populations	–
Bexarotene (Targretin®)	Selectively activates retinoid X receptors	CTCL	RCTs did not detect a higher malignancy risk		
Apremilast (Otezla®)	Inhibits PDE-4	Psoriasis	RCTs did not detect a higher malignancy risk		
Vorinostat (Zolinza®)	Inhibits histone deacetylases	CTCL	RCTs did not detect a higher malignancy risk		
Romidepsin (Istodax®)	Inhibits histone deacetylases	CTCL	RCTs did not detect a higher malignancy risk		
Dapsone (Avlosulfon®)	Inhibits dihydropteroate synthase	Lepromatous leprosy and dermatitis herpetiformis	No increased risk of internal malignancies	–	–
PUVA	Psoralen intercalates into the DNA double helix	Psoriasis	–	Higher risk of SCCs and melanoma in a dose-dependent fashion	–
UVB	Absorbed by DNA and urocanic acid, which alters antigen-presenting cell activity	Psoriasis, AD, and vitiligo	–	Patients previously treated with PUVA and subsequently treated with UVB have higher risk of NMSC in a dose-dependent fashion	–

Immunosuppressant Medications

The immune system has a well-established role in preventing cancer development. Therefore, immunosuppressant medications may theoretically promote cancer development.

- **Methotrexate** (various brand products available)
  - *Mechanism of action*: folic acid analog that inhibits dihydrofolate reductase, which blocks the synthesis of DNA, RNA, and purines. MTX also reduces the rate of epidermal proliferation.
  - *Use*: to treat inflammatory diseases such as psoriasis and rheumatoid arthritis. It is also used to treat ectopic pregnancies and malignancies such as lymphomas, leukemias, and choriocarcinomas. It is used off-label for atopic dermatitis.
  - *Common adverse effects*: ulcerative stomatitis, leukopenia, nausea, and abdominal distress.
  - MTX use in patients with RA is linked with melanoma, NMSC, non-Hodgkin's lymphoma, and lung cancer.
      In study of 6841 patients with RA, RA patients treated with ≥12 months of MTX had a 24% higher risk of NMSC than RA patients on other treatments (OR = 1.24; 95% CI, 1.04–1.48) [3].
      In a study comparing dermatology and rheumatology patients on MTX to other dermatology and rheumatology patients, those on MTX had a significantly higher risk of CMM when compared to those not on MTX (0.58% and 0.50%, respectively; $p < 0.001$) [4].
      In a post-market population-based registry study in Melbourne, Australia, there was an estimated 50% higher risk of cancer among patients with RA exposed to methotrexate when compared to the general population [5].
      - Three-fold greater risk of cutaneous malignant melanoma (CMM)
      - Five-fold greater risk of non-Hodgkin's lymphoma
      - Three-fold greater risk of lung cancer
      In a case series study, *Hoshida, et al* found that 11 of 28 cases of MTX-associated lymphoproliferative disorders exhibited spontaneous remission after withdrawal of MTX [6].
      Many of the lymphoma cases associated with MTX were Ebstein-Barr virus-related (EBV) lymphomas [7].
      - It is possible that immunosuppression due to MTX treatment allows EBV-induced polyclonal proliferation of B-cells to occur, leading to B-cell lymphoma.
  - Evidence on malignancy risk in psoriasis patients treated with MTX is mixed. Results from PSOLAR, a long-term follow-up registry study of psoriasis treatments, found that psoriasis patients on MTX did not have higher risk of systemic malignancies with any duration of treatment when compared to psoriasis patients not on methotrexate (Fig. 6.3) [8]:
      - >0 to <3 months; OR = 1.53 (95% CI, 0.36–6.03)
      - ≥3 to <12 months; OR = 1.15 (95% CI, 0.42–3.20)
      - ≥12 months; OR = 0.70 (95% CI, 0.41–1.19)

Stern, et al. found that psoriasis patients treated with PUVA in addition to ≥36 months of MTX had a four-fold higher risk of lymphoma when compared to psoriasis patients on MTX treatment alone (IRR = 4.39; 95% CI, 1.56–12.06) [9]

Suzuki, et al. reported a case of pulmonary lymphoma positive for EBV viral DNA in a psoriasis patient who had been taking 2.5–5 mg of MTX for 15 years [10].

CONCLUSION
Long-term follow-up data comparing psoriasis patients on MTX to other psoriasis patients found no differences in the risk of systemic malignancies.

- **Cyclosporine** (various brand products available)
  - *Mechanism of action:* calcineurin inhibitor leading to suppression of T cells and IL-2 production.
  - *Use:* to prevent organ rejection in organ transplant patients. It is also used off-label to treat inflammatory diseases such as psoriasis, atopic dermatitis, psoriatic arthritis, and rheumatoid arthritis.
  - *Common adverse effects*: renal dysfunction, tremor, hirsutism, hypertension, and gum hyperplasia.
  - CsA is classified as an IARC Group 1 carcinogen due to its links to SCC and non-Hodgkin lymphoma.
  - In a study of 5000 organ transplant patients treated with cyclosporine, CsA use has been linked with skin cancers, lymphoma, and other internal malignancies when compared to the general population [11, 12].
      28-fold higher risk of lymphoma
      Seven-fold higher risk of NMSCs
  - CsA use in patients with psoriasis has been linked with lymphoma and skin cancers [11].
      In a prospective cohort study of 1252 patients with severe psoriasis treated with 3–5 mg/kg/day of cyclosporine, *Paul* et al. found psoriasis patients treated with cyclosporine had a significantly higher malignancy rate when compared to the general population after a mean duration of 4.5 years (SIR = 2.1; 95% CI, 1.6–2.9) [13].
    - There was a six-fold increase in skin malignancies (mostly SCC).
    - Those on cyclosporine for ≥2 years had a 3.3 times higher risk of NMSCs than those on cyclosporine for <2 years.
      In a review of 25 years of malignancy risk data on CsA for treatment of dermatologic conditions, *Muellenhoff* et al. found little evidence supporting an increased risk for skin malignancies when used at the accepted dermatology use safety guidelines of ≤5 mg/kg/day or less [14].
    - Of the reviewed 14 cases of skin cancer in psoriasis patients treated with CsA, the majority of these patients either took doses of CsA exceeding the recommendations of the dermatology safety guidelines or had pre-existing carcinogenic risk factors.

Other cases of BCC, Waldenstrom macroglobulinemia, hairy cell leukemia, and EBV-associated lymphoma, and other lymphoproliferative diseases have also been reported in psoriasis patients treated with cyclosporine [11].

The combined use of cyclosporine and PUVA is thought to increase the risk of skin cancers. Therefore, it is important to screen these patients regularly for skin cancers [11].

CONCLUSION

Psoriasis patients treated with CsA were found to have higher risk of malignancies including lymphoma and skin cancers when compared to the general population. However, there were limited studies comparing psoriasis patients on CsA to other psoriasis patients. Skin cancer risk may be reduced by using lower doses of cyclosporine ($\leq$5 mg/kg/day) for limited durations.

- **Azathioprine** (various brand products available)
  - *Mechanism of Action:* purine analog that inhibits DNA production.
  - *Use:* to prevent organ transplant rejection and treat rheumatoid arthritis. It is also used off-label to treat psoriasis, atopic dermatitis, inflammatory bowel disease, multiple sclerosis, and myasthenia gravis.
  - *Common adverse effects:* leukopenia, infections, and nausea/vomiting.
  - Azathioprine is classified as an IARC Group 1 carcinogen due to its links to lymphoma and skin cancers.
  - Studies show that azathioprine causes accumulations of 6-thioguanine in the DNA, which is believed to trigger cancer when patients are later exposed to UVA light [15].
  - Azathioprine use in organ transplant patients has been linked with lymphomas and skin cancers.

    In a 2016 meta-analysis of 27 studies, *Jiyad, et al* found that organ transplant patients treated with azathioprine had a 56% higher risk of SCC than those not treated with azathioprine. However, there was no association between azathioprine treatment and BCC [16].

    In a population-based cohort study in Australia, *Na, et al* found that organ transplant patients on high doses (>1.30 mg/kg/day) of azathioprine had significantly higher risk of both short-term ($\leq$1 year after transplant) (HR, 2.20; 95% CI, 1.21–4.01) and long-term (>1 year after transplant) (HR, 1.78; 95% CI, 1.12–2.84) non-Hodgkin's lymphoma when compared to organ transplant patients not treated with azathioprine [17].
  - Azathioprine use in IBD has been linked with lymphomas and skin cancers.

    In a 2015 meta-analysis of 18 studies, *Kotlyar, et al* found that treatment of IBD with azathioprine and 6-mercaptopurine is linked with a nearly fivefold increased risk of lymphoma (SIR, 4.92; 95% CI, 3.10–7.78) [18].

    In a study comparing thiopurine-treated patients with IBD and TNF inhibitor-treated patients with IBD, those treated with thiopurines had a significantly higher risk of developing malignancies such as skin cancers and lymphomas (HR, 4.15; 95% CI 1.82–9.44; $p = 0.0007$) [19].

In a systematic review of 18 studies, *Hagen, et al* found that patients with IBD treated with thiopurines had an increased risk of NMSC proportional to the duration of thiopurine therapy [20].
- Azathioprine use in myasthenia gravis patients has been linked with NMSC
  In a case-control study of Danish population-based registries, myasthenia patients treated with azathioprine had significantly higher risk of NMSC (OR, 3.3; 95% CI, 1.5–7.3) [21].
  - This risk became more pronounced with subgroup analysis of patients on long-term (≥5 years) (OR = 4.8; 95% CI, 1.7–13.6) and high-dose (≥150 g) (OR = 4.6; 95% CI, 1.7–2.5) treatment with azathioprine.
- Currently, there are no investigations that studied malignancy risk of azathioprine use in treatment of patients with skin diseases.

CONCLUSION

Although there are no studies studying malignancy risk in azathioprine treatment of patients with skin diseases, it is well-documented that treatment with azathioprine is linked with lymphomas and skin cancers in patients with other diagnoses. This risk appears to be proportional to the duration and dose of treatment.

- **Mycophenolate Mofetil** (CellCept®)
  - *Mechanism of Action:* blocks purine synthesis via inhibition of inosine monophosphate dehydrogenase.
  - *Use:* to prevent organ transplant rejection. It is also used off-label for immune-mediated disorders such as psoriasis, pemphigus vulgaris, atopic dermatitis, SLE, and Behcet's disease.
  - *Common adverse effects:* diarrhea, leukopenia, sepsis, vomiting, and infections.
  - The evidence on malignancy risk with mycophenolate mofetil use is mixed.
    In RCTs of 1483 patients treated with mycophenolate mofetil to prevent renal allograft rejection, malignancy rates among the mycophenolate-treated patients were similar to the malignancy rates for all renal allograft recipients:
    - 0.4–1.0% of patients developed lymphoma within 1 year
    - 1.6–4.2% of patients developed NMSC within 1 year
    In a retrospective population-based study of heart transplant patients, those on mycophenolate mofetil were significantly more likely to be diagnosed with a malignancy such as non-Hodgkin lymphoma, skin cancers, and lung squamous cell carcinoma when compared to heart transplant patients on everolimus (9.91% and 1.80%, respectively; $p = 0.001$) [22].
    However, in a population-based study of organ transplant patients in Taiwan, mycophenolate mofetil use was not associated an increased risk for malignancies (HR, 1.00; 95% CI, 0.58–2.77; $p = 0.9874$) [23].
  - There have been cases of diffuse large B-cell lymphoma in SLE patients treated with mycophenolate.

- Currently, no population-based studies are available that investigated rates of malignancy associated with mycophenolate use in dermatology patients.

CONCLUSION
There are currently no studies analyzing malignancy risk of mycophenolate use in patients with skin diseases. The evidence for malignancy risk in organ transplant patients treated with mycophenolate is mixed, with most studies finding no difference in malignancy risk in organ transplant patients treated with mycophenolate when compared to other organ transplant patients.

- **Oral Corticosteroids** (various brand products available)
  - *Mechanism of Action:* blocks the action of inflammatory mediators and induces anti-inflammatory mediators.
  - *Uses:* for treatment of a variety of inflammatory conditions, such as atopic dermatitis, pemphigus vulgaris, contact dermatitis, rheumatoid arthritis, etc. Not recommended for use in psoriasis because withdrawal of systemic corticosteroids may lead to the development of life-threatening pustular or erythrodermic psoriasis.

    Oral corticosteroids are also used for *treatment* of cancers such as leukemia, lymphoma, and multiple myeloma.
  - *Common adverse effects:* fluid retention, alteration in glucose tolerance, elevated blood pressure, behavioral and mood changes, increased appetite and weight gain.
  - Systemic corticosteroid use may increase the risk of NMSCs and lymphomas in the general population.

    In population-based studies of the Danish Cancer Registry, overall risks for SCC, BCC, non-Hodgkin lymphoma, and esophageal cancer were significantly higher in patients treated with systemic corticosteroids [24, 25].
    - BCC (SIR = 1.52; 95% CI, 1.09–2.07)
    - SCC (SIR = 2.45; 95% CI, 1.37–4.04)
    - Non-Hodgkin lymphoma (SIR = 2.68; 95% CI, 1.16–5.29)
    - Esophageal cancer (SIR = 1.92; 95% CI, 1.34–2.65).

    In another population-based study of the Danish Cancer Registry, treatment with corticosteroids was not associated with a higher risk of breast cancer (aOR = 1.0; 95% CI, 0.96–1.1) [26].
  - Currently, no population-based studies are available that investigated rates of malignancy associated with systemic corticosteroid use in dermatology patients.

CONCLUSION
Population-based studies found that systemic corticosteroid use is associated with BCC, SCC, non-Hodgkin lymphoma, and esophageal cancers in the general population. However, more studies are needed to further evaluate this risk and assess the malignancy risk of systemic corticosteroid treatment for dermatologic conditions.

Retinoids

Retinoids may reduce the risk of cancer due to their anti-proliferative properties.

- **Acitretin (Soriatane®)**
  - *Mechanism of Action:* exact mechanism of action is unknown. Acitretin is structurally similar to vitamin A. Studies have shown that the metabolites of acitretin bind retinoic acid receptors, which alters gene transcription and leads to anti-proliferative and anti-inflammatory effects.

    By normalizing skin proliferation, acitretin may have *anti*-carcinogenic effects.
  - *Use:* for treatment psoriasis. Used off-label for cancer prevention.
  - *Common adverse effects:* dry skin, dry mouth, cheilitis, joint pain, muscle tightness, alopecia, dry eyes, and rise in blood lipids. May cause severe birth defects if used by pregnant women. May also less commonly cause hepatotoxicity and pancreatitis.

    Adverse-effects are dose-dependent.
  - No studies have found that acitretin increases the risk of malignancies.
  - Acitretin may be useful for the prevention of malignancies.

    In a review of three RCTs, *Chen et al* found that immunocompromised organ transplant patients on systemic retinoids had significantly lower risk of SCC, BCC, and actinic keratosis than those in the placebo group after 6–12 month follow-up [27].
    - However, a RCT studying acitretin use in nontransplantation patients at high risk of BCC and SCC found no difference in BCC and SCC risk when compared to the placebo group (OR = 0.41; 95% CI, 0.15–1.13) [28].
    - This study may not have high enough power to detect a significant difference (35 patients in the acitretin group and 35 patients in the placebo group).

    A retrospective study of 32 patients with CTCL treated with acitretin found that response to acitretin as adjuvant or monotherapy is comparable to the response of oral agents currently approved for CTCL [29].

    There are multiple case reports of acitretin used for prevention of skin cancers in solar-damaged skin and in patients with genetic syndromes that predispose to skin cancers [30].

    There are multiple case reports of acitretin used for management of SCCs and BCCs [30].

CONCLUSION

Acitretin may be effective for the prevention of malignancies in certain populations, such as organ transplant patients and patients with genetic syndromes predisposing to skin cancers.

- **Isotretinoin** (Accutane®)
  - *Mechanism of Action:* exact mechanism of action is unknown. Isotretinoin is structurally similar to vitamin A. Studies have shown that isotretinoin induces apoptosis of various cells in the body, such as sebaceous gland cells.

Isotretinoin is also converted to metabolites that bind the retinoic acid receptors, which leads to alterations in gene transcription.
- *Use:* to treat severe acne. Used off-label for cancer prevention and management.
- *Common adverse effects:* dry skin, chapped lips, dry eyes, and dry nose/nosebleeds. May cause severe birth defects if used by pregnant women. May also less commonly cause depression, anxiety, suicidality, pseudotumor cerebri, and alopecia.
  Adverse effects are dose-dependent.
- Like acitretin, isotretinoin may be useful for the prevention of malignancies.

  A cohort study of 135 psoriasis patients found that those treated with isotretinoin in combination to PUVA were 30% less likely to be diagnosed with a SCC than those treated with PUVA alone [31].

  In a RCT of 108 patients with oral leukoplakia, significantly more patients in the isotretinoin treatment group experienced major decreases in the size of their lesions (67% versus 10%; $p = 0.0002$) and reversal of dysplasia (54% versus 10%; $p = 0.01$) when compared to placebo [32].

  In a RCT of 103 patients with personal histories of primary head and neck cancer, patients on isotretinoin had significantly fewer secondary tumors than patients on placebo (2% versus 12%, respectively; $p = 0.005$) [33].
  - However, in another RCT of 176 participants with personal histories of head and neck cancer, there were no significant differences in head and neck cancer recurrence in patients taking isotretinoin compared to patients taking placebo ($p = 0.61$).

  In a RCT of 525 participants with a personal history of NMSC, there were no significant differences in NMSC occurrence after 3-year follow-up in patients taking isotretinoin compared to patients taking placebo [30].

  In a retrospective review of 88 glioblastoma patients, those treated with isotretinoin maintenance therapy experienced significantly longer progression-free survival than those not treated with isotretinoin (25.3 months and 8.3 months, respectively; $p = 0.04$) [34].

  Multiple case reports found that isotretinoin may be effective for skin cancer chemoprevention in patients with xeroderma pigmentosum and nevoid basal cell carcinoma syndrome [30].

  Multiple case reports of isotretinoin use in patients with ultra-high-risk neuroblastoma found that isotretinoin may have contributed to prolonged longevity in these patients [35].

  One case report found that isotretinoin may be successful in treating SCC arising from an epidermoid cyst [36]. This patient achieved complete remission of his SCC after treatment with isotretinoin.

CONCLUSION

Isotretinoin may be useful for the prevention of malignancies in specific groups of patients, such as patients with oral leukoplakia, patients with a history of primary head and neck cancers, psoriasis patients treated with PUVA, and patients with xeroderma pigmentosum.

- **Bexarotene** (Targretin®)
  - *Mechanism of action:* selectively activates retinoid X receptors (RXR), which regulates genes controlling cell differentiation and proliferation. The exact mechanism of action for treatment of CTCL is unknown.
  - *Use:* to treat CTCL.
  - *Common adverse effects:* hyperlipidemia, hypercholesterolemia, headache, hypothyroidism, asthenia, leukopenia, rash, nausea, infection, peripheral edema, abdominal pain, and dry skin.
    Adverse effects are dose-dependent.
  - Bexarotene has anti-carcinogenic properties. No studies have found that bexarotene increases risk of new malignancies.
    In phase II-III RCTs of 58 patients with refractory or persistent early-stage CTCL treated with bexarotene for up to 48 weeks, none of the patients experienced any cases of new malignancies [37].
    In phase II-III RCTs of 94 patients with refractory advanced-stage CTCL treated with bexarotene (56 patients treated with ≤300 mg/m^2/day for median of 19.1 weeks; 38 treated with >300 mg/m^2/day for median of 34.3 weeks), none of the patients experienced any cases of new malignancies [38].
    In a phase III RCT of 46 patients with stage IB-IIA mycosis fungoides treated with bexarotene and PUVA for 16 weeks, none of the patients experienced any cases of new malignancies [39].

> **CONCLUSION**
> Bexarotene is useful for the treatment of CTCL. RCTs did not detect a higher risk of developing new malignancies in CTCL patients treated with bexarotene.

Small Molecule Inhibitors

- **Apremilast** (Otezla®)
  - *Mechanism of action:* selectively inhibits PDE-4 in immune cells, which leads to down-regulation of pro-inflammatory cytokines such as TNF-alpha, IL-17, and IL-23, and up-regulation of anti-inflammatory cytokines such as IL-10.
  - *Uses:* to treat psoriasis and psoriatic arthritis.
  - *Common adverse effects:* diarrhea, nausea, upper respiratory infection, and headache.
  - Pre-clinical studies in mice found that inhibition of PDE-4 in immune cells may increase the risk of lung cancer and diffuse large B-cell lymphoma by upregulating angiogenesis [40, 41].
  - In phase III RCTs of apremilast for treatment of moderate-to-severe plaque psoriasis, no difference in overall malignancy rates was found between the apremilast and placebo treatment groups [42, 43].
  - In a meta-analysis of eight RCTs comprising 3289 patients, overall malignancy rates were not significantly different among the apremilast and placebo treatment groups [44].

- **Vorinostat** (Zolinza®)
  - *Mechanism of action:* inhibits histone deacetylases HDAC1, HDAC2, HDAC3, and HDAC6, which leads to the accumulation of acetylated histones and induces cell cycle arrest and/or apoptosis.
  - *Uses:* to treat CTCL.
  - *Common adverse effects:* diarrhea, fatigue, nausea, thrombocytopenia, anorexia, and dysgeusia.
  - Vorinostat has anti-carcinogenic properties. Phase II RCTs did not detect an increased risk of new malignancies in CTCL patients treated with vorinostat.
    In a phase II RCT of 33 patients with CTCL treated with vorinostat for up to 36 weeks, none of the patients experienced any cases of new malignancies [45].
    In a phase IIb RCT of 74 patients with advanced CTCL treated with vorinostat for 2 years, none of the patients experienced any cases of new malignancies [46].
    In a phase IIb RCT of 74 patients with advanced CTCL treated with vorinostat for 6 months, none of the patients experienced any cases of new malignancies [47].

- **Romidepsin** (Istodax®)
  - *Mechanism of action:* inhibits histone deacetylases, which leads to the accumulation of acetylated histones and induces cell cycle arrest and/or apoptosis.
  - *Uses:* to treat CTCL.
  - *Common adverse effects:* nausea, fatigue, infections, vomiting, anorexia, thrombocytopenia, ECG T-wave changes, neutropenia, and lymphopenia.
  - Romidepsin has anti-carcinogenic properties. Phase II RCTs did not detect an increased risk of new malignancies in CTCL patients treated with romidepsin.
    In a phase II RCT of 71 patients with CTCL treated with romidepsin for a median of about 4 months, none of the patients experienced any cases of new malignancies [48].
    In a phase II pivotal RCT of 96 patients with refractory CTCL treated with romidepsin for up to 6 months, none of the patients experienced any cases of new malignancies [49].

> **CONCLUSION**
> Romidepsin is useful for the treatment of CTCL. RCTs did not detect a higher risk of developing new malignancies in CTCL patients treated with romidepsin.

Antibiotic

- **Dapsone** (Avlosulfon®)
  - *Mechanism of action:* inhibits dihydropteroate synthase by competing with the binding of para-aminobenzoate, which inhibits the bacterial synthesis of dihydrofolic acid.
  - *Uses:* to treat lepromatous leprosy and dermatitis herpetiformis. Also used off-label to treat Behcet's disease, lupus erythematous, and bullous pemphigoid.
  - *Common adverse effects:* nausea, vomiting, loss of appetite, dizziness, blurred vision, tinnitus, headache, insomnia, and photosensitivity.
  - In a follow-up study of 115,407 Vietnam veterans followed for up to 24 years, those exposed to dapsone did not have a higher risk of malignancies when compared to those not exposed to dapsone [50]:
    - Non-Hodgkin's lymphoma (IRR = 1.8; 95% CI, 0.8–3.9)
    - Kidney (IRR = 1.3; 95% CI, 0.4–4.1)
    - Hodgkin's disease (IRR = 1.2; 95% CI, 0.3–5.5)
    - Bladder (IRR = 1.0; 95% CI, 0.5–2.1)
    - Oral (IRR = 0.6; 95% CI, 0.3–1.1)
    - Leukemia (IRR = 0.5; 95% CI, 0.2–1.2)
  - No studies specifically evaluated the malignancy risk of dapsone for treatment of leprosy or dermatitis herpetiformis.

> **CONCLUSION**
> Long-term follow-up data comparing Vietnam veterans treated with dapsone to those not treated with dapsone found no differences in malignancy risk between the two groups.

Malignancy Risk of Phototherapy

Theoretically, the cumulative UV exposure from phototherapy may increase the risk of skin cancers.

- **Psoralen and Ultraviolet A (PUVA)**
  - *Mechanism of action:* intercalates into the DNA double helix and generates DNA-distorting photoproducts, which halts the dividing epidermis.
  - *Uses:* to treat psoriasis.
  - *Common adverse effects:* nausea, itching, and redness of the skin.

- Malignancy risk with PUVA treatment has been most extensively studied with the PUVA Follow-Up Study evaluating 1380 patients for nearly three decades [9, 51, 52].

    Patients treated with at least 337 PUVA treatments had a 100-fold higher risk of SCC when compared to those of the general population.

    There was no increase in BCC risk with PUVA treatment.

    Risk of genital SCC persisted after cessation of PUVA treatment.

    Malignant melanoma risk was significantly higher in patients treated with PUVA in those with at least 250 treatments.

    - Odds of melanoma were three-fold higher in patients with Fitzpatrick skin types I and II when compared to types III and IV.
- In a large-scale study of 4799 Swedish patients who received PUVA treatment, male patients who received more than 200 treatments had a 30-fold increase in SCC when compared to the general population [53].
- Combination therapy of PUVA with other psoriasis treatments may influence malignancy risks [11]:

    Methotrexate treatment with PUVA further increases risk of SCC and lymphoma.

    Cyclosporine treatment with PUVA further increases risk of SCC.

    Retinoid treatment with PUVA reduces risk of SCC when compared to PUVA alone.

CONCLUSION

PUVA treatment increases the risk of SCC and melanoma in a dose-dependent fashion. Combining other psoriasis treatments with PUVA may influence malignancy risk.

- **Ultraviolet B**
    - *Mechanism of action:* UVB is absorbed by DNA and urocanic acid in the skin, which alters antigen-presenting cell activity.
    - UVB therapy can be divided into two modalities: BB-UVB and NB-UVB.

        BB-UVB: involves wider spectrum of UVB wavelengths.

        NB-UVB: emits monochromatic radiation at 311 nm.

        - Found to be superior to BB-UVB for a number of dermatoses including atopic dermatitis, psoriasis, and vitiligo [11].
    - *Uses:* for treatment of psoriasis, atopic dermatitis, and vitiligo.
    - *Common adverse effects:* erythema, dry skin, and hyperplasia.

        Adverse-effects are less pronounced with NB-UVB therapy.
    - UVB is accepted as a safer therapy than PUVA in terms of malignancy potential.

        In the PUVA Follow-Up Study with 1380 adult participants, long-term exposure to BB-UVB with a mean total of 409 treatments was not found to be associated with a higher risk of SCC [51].

        A study of 3876 patients treated with NB-UVB found no association between NB-UVB and skin cancers [52].

– Patients previously exposed to PUVA and subsequently treated with >300 treatments of UVB had a significantly higher risk of SCC (IRR = 1.37; 95% CI, 1.03–1.83) and BCC (IRR = 1.45; 95% CI, 1.07–1.96) when compared to patients treated with UVB without a history of PUVA exposure [52].

CONCLUSION
Neither NB-UVB nor BB-UVB appears to increase malignancy risk. However, patients previously treated with PUVA and subsequently treated with >300 UVB treatments appear to have higher risks of SCCs and BCCs.

Malignancy Risk of Biologics (Table 6.2)

It is important to understand certain limitations when assessing malignancy risks in patients treated with biologics:

Table 6.2 Summary table of the malignancy risks of biologic therapies

Medication	Class	Dermatologic indications	Internal malignancy risk	Skin malignancy risk	Overall malignancy risk
Etanercept (Enbrel®)	TNF inhibitor	Psoriasis	No increased risk of lymphoma	No increased risk of skin cancers	No increased risk
Adalimumab (Humira®)	TNF inhibitor	Psoriasis and hidradenitis suppurativa	No increased risk of lymphomas	Higher risk of NMSC. Risk of melanoma requires further validation	No increased risk
Infliximab (Remicade®)	TNF inhibitor	Psoriasis	Cases of childhood hepatic T-cell lymphomas have been reported	–	No increased risk
All TNF inhibitors (Grouped)	TNF inhibitors	Psoriasis	No increased risk of internal malignancies	Data on NMSC risk are mixed	Data on risk of overall malignancies in RA patients are mixed
Ustekinumab (Stelara®)	IL-12/23 inhibitor	Psoriasis	No increased risk	Lower risk of BCC	No increased risk
Secukinumab (Cosentyx®)	IL-17 inhibitor	Psoriasis	RCTs have not detected a higher malignancy risk. More long-term follow-up data are needed		

(continued)

Table 6.2 (continued)

Medication	Class	Dermatologic indications	Internal malignancy risk	Skin malignancy risk	Overall malignancy risk
Ixekizumab (Taltz®)	IL-17 inhibitor	Psoriasis	RCTs have not detected a higher malignancy risk. More long-term follow-up data are needed		
Brodalumab (Siliq®)	IL-17 inhibitor	Psoriasis	RCTs have not detected a higher malignancy risk. More long-term follow-up data are needed		
Guselkumab (Tremfya®)	IL-23 inhibitor	Psoriasis	RCTs have not detected a higher malignancy risk. More long-term follow-up data are needed		
Tildrakizumab (Ilumya®)	IL-23 inhibitor	Psoriasis	RCTs have not detected a higher malignancy risk. More long-term follow-up data are needed		
Risankizumab (Skyrizi®)	IL-23A inhibitor	Psoriasis	RCTs have not detected a higher malignancy risk. More long-term follow-up data are needed		
Dupilumab (Dupixent®)	IL-4/13 inhibitor	Atopic dermatitis	RCTs have not detected a higher malignancy risk. More long-term follow-up data are needed		
Rituximuab (Rituxan®)	CD20 inhibitor	Used off-label for autoimmune bullous disorders	–	–	No increased risk
Omalizumab (Xolair®)	Anti-IgE	Chronic spontaneous urticaria	–	–	No increased risk
IVIG (various brand products available)	Binds to host antibodies stimulating their removal	Off-label for dermatomyositis and autoimmune bullous diseases	There is currently no evidence that IVIG increases risk of malignancies		

1. While the placebo-controlled periods of RCTs are highly informative, these periods are generally too short to assess true malignancy risks. This is because if malignancies were to be associated with treatment, these malignancies would develop at a later time beyond that of the placebo-controlled period.
2. Sometimes, studies used malignancy rates of the general population from other databases for comparison; however, these comparisons should be interpreted with caution because the controls were not drawn from the same study population.
3. Malignancy data found on package inserts typically do not indicate a comparison group. For example, if the package insert reports that 1/1000 patients will develop a cancer over 5 years, this result is extremely difficult to interpret because one will not know whether the patient would have developed the cancer without the drug.
4. A drug may be indicated for multiple different diseases. On the package insert, malignancy data is presented in an *aggregate* form. For example, adalimumab is approved for both RA and psoriasis. The RA population typically requires multiple concomitant immunosuppressive medications, which may lead to higher malignancy

rates. Because malignancy rates are grouped together, at first reading, it would seem that the reported malignancy rate is attributable to all patients treated with the drug when, in fact, the malignancy rate may only be attributable to the RA population.

5. Some studies group all medications of the same class together and report malignancy rates of the class rather than analyzing individual medications in the class. For example, some studies may report malignancies of TNF inhibitors as a whole. One of the reasons this may be done is to achieve a larger sample size for evaluation. However, this method of handling data has substantial limitations. Although different biologics may be in the same class, mechanisms of actions differ slightly with each medication. Results from studies grouping biologics based on class may not be representative of the malignancy risk of each individual agent.

6. While meta-analyses are considered the highest level of evidence, they need to be examined for the following factors before one can consider the validity in the findings:
   (a) Heterogeneity in collecting outcomes (example: some studies investigating SCC risk may exclude SCC in situ)
   (b) Heterogeneity in the patient population (example: one study may only study Asian patients, while another study may only study Danish patients)
   (c) Heterogeneity in lost to follow-up rates and length to follow-up
   (d) Heterogeneity in study instrumentation (example: one study may utilize physician-verified diagnoses while another study may utilize self-reported diagnoses)
   (e) Publication bias and heterogeneity in statistical approach

Whenever available, we organized each section by *overall*, *internal*, and *skin* malignancies. We also indicated the specific disease groups studied in each study.

TNF Inhibitors

- **Etanercept** (Enbrel®)
  - *Mechanism of action:* a dimeric fusion protein that interferes with TNF-alpha by functioning as a TNF-alpha decoy receptor, reducing the freely available TNF molecules.
  - *Uses:* to treat RA, psoriasis, psoriatic arthritis, ankylosing spondylitis, and polyarticular juvenile idiopathic arthritis.
  - *Common adverse effects:* infections and injection site reactions. Patients with latent TB and hepatitis B infections may experience reactivation of their infections.
  - Overall malignancies:
    In OBSERVE-5, a 5-year surveillance registry of etanercept use in psoriasis, the rate of overall malignancies not including NMSC was not higher than expected malignancy rates in the general psoriasis population from MarketScan® data (SIR = 0.89; 95% CI, 0.68–1.15) [54].
    An integrated study of both RCTs and long-term uncontrolled trials on the malignancy risk of etanercept as monotherapy for multiple indications did not find an increased incidence of overall malignancies in the etanercept treatment group when compared to placebo (SIR = 1.2; 95% CI, 0.8–1.6) [55].

- Internal malignancies:

  In OBSERVE-5, the rate of lymphoma was not higher than expected lymphoma rates in the general psoriasis population taken from MarketScan® data (SIR = 0.50; 95% CI, 0.06–1.80) [54].

  In a meta-analysis of 49 RCTs of etanercept for multiple indications, psoriasis patients treated with etanercept did not have a significantly higher rate of lymphoma when compared to the expected rates from NIH-SEER data of the general psoriasis population (SIR = 2.01; 95% CI, 0.24–7.27) [56].

  Cases of Hodgkin and non-Hodgkin lymphoma have been reported in children and adolescents treated with etanercept [57].

- Skin malignancies:

  In OBSERVE-5, the rate of NMSC in psoriasis patients treated with etanercept was significantly lower than the expected rates from MarketScan® data (SIR = 0.58; 95% CI, 0.45–0.74) [54].

  In a meta-analysis of 49 RCTs of etanercept for multiple indications, psoriasis patients treated with etanercept did not have significantly higher rates of NMSC when compared to the expected rates from NIH-SEER data of the general psoriasis population (SIR = 2.77; 95% CI, 0.59–25.97) [56].

  - However, when subgroup analysis was completed, they observed higher rates of SCC in psoriasis patients treated with etanercept with a history of both low (SIR = 2.09; 95% CI, 1.27–3.22) and high (SIR = 4.96; 95% CI, 3.03–7.66) sun exposures.

CONCLUSION

Treatment of psoriasis with etanercept does not seem to increase overall or internal malignancy risks. Data on the NMSC risk of psoriasis patients treated with etanercept are mixed.

- **Adalimumab** (Humira®)
  - *Mechanism of action:* a human monoclonal antibody that binds directly to TNF-alpha molecules, reducing the inflammatory response.
  - *Uses:* to treat RA, psoriasis, psoriatic arthritis, juvenile idiopathic arthritis, IBD, hidradenitis suppurativa, ankylosing spondylitis, and uveitis.
  - *Common adverse effects:* injection site reactions, infections, rash, and headache. Patients with latent TB and hepatitis B infections may experience reactivation of their infections.
  - Overall malignancies

    The initial 7-year results from ESPRIT, a surveillance registry of adalimumab treatment for psoriasis, found that the overall malignancy rate of patients taking adalimumab (1.0 events/100PY) was within range of the overall malignancy rate of psoriasis patients not receiving biologic therapies (0.5–2.0 events/100PY) [58].

    A meta-analysis of 18 RCTs of adalimumab for psoriasis found that psoriasis patients treated with adalimumab did not have a higher risk of overall

malignancies excluding NMSC when compared to the expected malignancy rate from NIH-SEER (SIR = 0.86; 95% CI, 0.58–1.23) [59].
- Skin malignancies
  A meta-analysis of 18 RCTs of adalimumab for psoriasis found that psoriasis patients treated with adalimumab had a significantly higher risk of melanoma (SIR = 3.04; 95% CI, 1.11–6.62) and NMSC (SIR = 1.55; 95% CI, 1.10–2.13) when compared to data on expected malignancy rate from NIH-SEER [59].
  • However, most of the patients diagnosed with melanoma had a prior history of extensive UV/sun exposure, a family history of melanoma, or developed melanoma within <6 months of adalimumab treatment.
  Safety data from RCTs of adalimumab for RA, JIA, AS, psoriasis, psoriatic arthritis, and CD found patients with psoriasis had significantly higher risk of melanoma (SIR = 4.37; 95% CI, 1.89–8.61) and NMSC (SIR 1.76; 95% CI, 1.29–2.39) when compared to the general population (Fig. 6.2) [60].

> **CONCLUSION**
> Long-term follow-up data comparing psoriasis patients treated with adalimumab to other psoriasis patients found no difference in overall malignancy risk between the two groups. Psoriasis patients treated with adalimumab may have higher risk of NMSC. Data on melanoma need further validation.

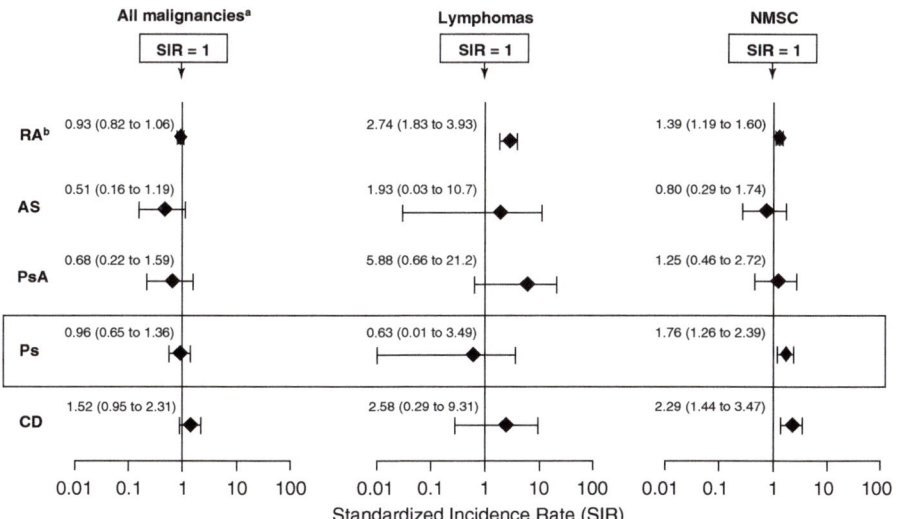

Fig. 6.2 Standardized incidence rates (SIR (95% CI)) for all malignancies excluding NMSC, lymphomas, and NMSC for RA, ankylosing spondylitis, psoriatic arthritis, psoriasis (boxed), and Crohn's Disease when compared to the general population. No malignancies were observed for juvenile idiopathic arthritis. Based on data from 14,160 patients. (From: Burmester, G.R., et al., *Adalimumab: long-term safety in 23,458 patients from global clinical trials in rheumatoid arthritis, juvenile idiopathic arthritis, ankylosing spondylitis, psoriatic arthritis, psoriasis and Crohn's disease*. Ann Rheum Dis, 2013. **72**(4): p. 517–24)

- **Infliximab** (Remicade®)
  - *Mechanism of action:* a chimeric monoclonal antibody that binds directly to TNF-alpha molecules, reducing the inflammatory response.
  - *Uses:* to treat RA, psoriasis, psoriatic arthritis, IBD, and ankylosing spondylitis.
  - *Common adverse effects:* infections, infusion reactions, abdominal pain, and headache. Patients with latent TB and hepatitis B infections may experience reactivation of their infections.
  - Overall malignancies

    In a meta-analysis of 10 IBD trials, Crohn's and UC patients receiving infliximab did not have a higher overall malignancy risk when compared to patients receiving placebo [61].

    In a follow-up study of 764 Japanese psoriasis patients treated with infliximab, the overall malignancy rate after 2 years of follow-up was not significantly higher when compared to the overall malignancy rate among the general Japanese population (SIR = 1.60; 95% CI, 0.89–2.81) [62].
  - Internal malignancies

    Multiple cases of childhood hepatic T-cell lymphoma have been reported with infliximab treatment for IBD [63].

CONCLUSION

Treatment of psoriasis with infliximab does not appear to increase overall malignancy risk. There have been multiple cases of childhood hepatic T-cell lymphoma in children treated with infliximab for IBD, and further studies are needed to evaluate the long-term safety of infliximab in children.

- Grouped analyses of malignancy risk with different TNF inhibitor treatments:
  - Overall malignancies

    A 2006 meta-analysis found a higher risk of overall malignancies in RA patients treated with adalimumab, infliximab, or etanercept when compared to other RA patients (OR = 3.3; 95% CI, 1.20–9.10) [64].

    However, a 2012 updated meta-analysis including 64 RCTs found no increased risk of overall malignancies in RA patients treated with TNF inhibitors when compared to other RA patients (OR = 0.98; 95% CI, 0.51–1.9) [65].
  - Internal malignancies

    Meta-analyses grouping patients with other diseases including psoriasis, psoriatic arthritis, ankylosing spondylitis, and RA did not detect a higher risk of systemic malignancies in those treated with TNF inhibitors when compared to other patients with the same diagnoses [11, 66].

    In a meta-analysis of 20 RCTs, psoriasis patients treated with TNF inhibitors did not have higher risk of systemic malignancies (OR = 1.48; 95% CI, 0.71–3.09) or NMSCs (OR = 1.33; 95% CI, 0.58–3.04) when compared to psoriasis patients on placebo [67].

Results from the PSOLAR registry of 1260 psoriasis patients treated with TNF inhibitors found that psoriasis patients on long-term (≥12 months) TNF inhibitor treatment including adalimumab, etanercept, and infliximab may have higher risk of systemic malignancies when compared to other psoriasis patients (OR = 1.54; 95% CI; 1.10–2.15) (Fig. 6.3) [8].

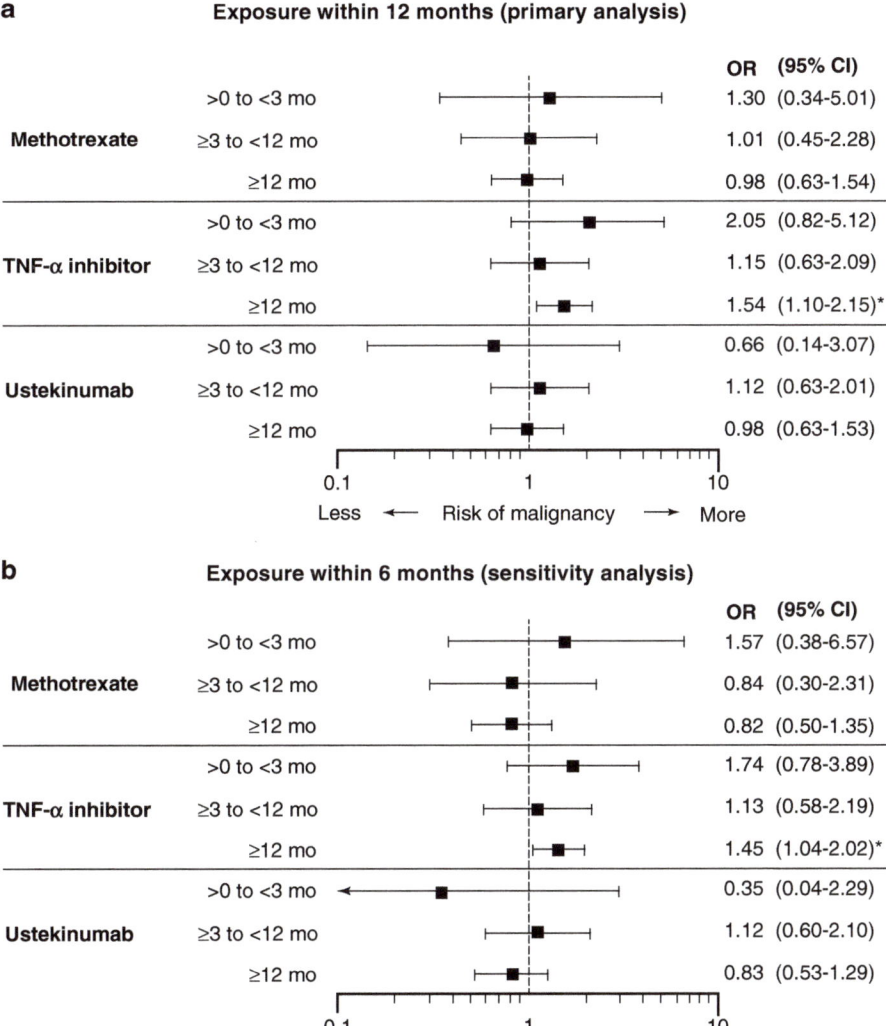

Fig. 6.3 Results from PSOLAR. Adjusted odds ratios (ORs) (95% CI) of overall malignancies in association with duration of therapy with methotrexate, TNF inhibitors, or ustekinumab for treatment of psoriasis within 12 months of exposure (**a**), 6 months of exposure (**b**), and 24 months of exposure (**c**). The *P* value comparing malignancy cases with matched controls was significant for TNF inhibitor treatment within 12-month (*P* = 0.012), 6-month (*P* = 0.028), and 24-month (*P* = 0.12) exposure periods. (From: Fiorentino, D., et al., *Risk of malignancy with systemic psoriasis treatment in the Psoriasis Longitudinal Assessment Registry.* J Am Acad Dermatol, 2017. **77**(5): p. 845–854.e5)

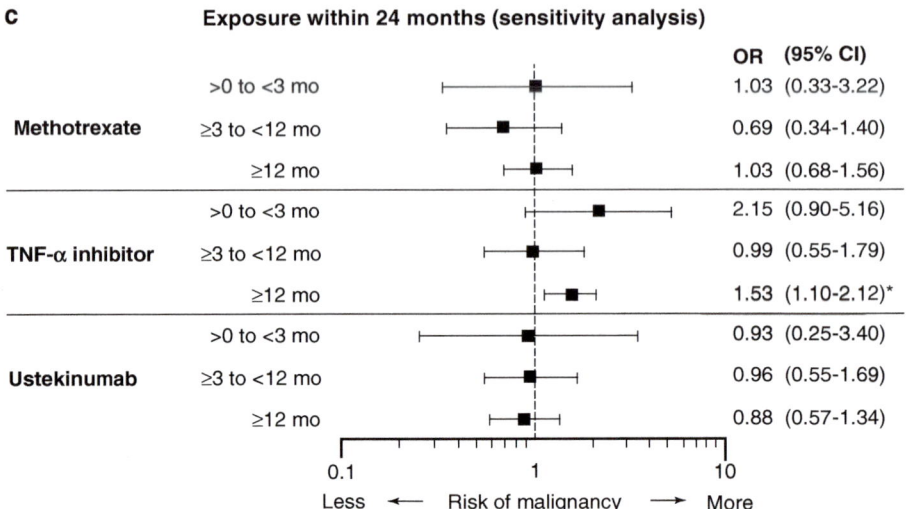

Fig. 6.3 (continued)

- Subgroup analyses of the individual TNF inhibitors found no statistically significant difference in risk.
- The elevated risk from the pooled analysis was mostly driven by etanercept (OR = 1.37; 95% CI, 0.94–2.01) and adalimumab (OR = 1.37; 95% CI, 0.93–2.02) rather than infliximab (OR = 1.01; 95% CI, 0.59–1.74).
- Skin malignancies
  NMSC:
  In a cohort study of 5889 systemically treated psoriasis patients, those on biologics had 42% higher risk of NMSC when compared to nonbiologic users after a mean follow-up time of 5.48PYs (aHR = 1.42; 95% CI, 1.12–1.80) [68]. 97% of patients on biologics were treated with TNF inhibitors.
  - The higher NMSC risk was mostly driven by higher rates of cutaneous SCC (aHR = 1.81; 95% CI, 1.23–2.67).
  In a study of PSOLAR data, psoriasis patients treated with TNF inhibitors did not have a higher risk of NMSC when compared to psoriasis patients treated with MTX (RR = 0.81; 95% CI, 0.37–1.80) [69].
  - Subgroup analyses did not find a higher risk of SCC (RR = 3.49; 95% CI, 0.40–30.1) or BCC (RR = 0.55; 95% CI, 0.22–1.36).
  In a study comparing NMSC incidence of 280 psoriasis patients and 448 RA patients treated with TNF inhibitors, psoriasis patients had significantly higher rates of NMSC with shorter time until first NMSC (HR = 6.0; 95% CI, 1.6–22.4) [70].

Melanoma:
Results from the PSOLAR registry found that psoriasis patients treated with TNF inhibitors did not have a higher risk of melanoma, even after ≥ 12 months of treatment (OR = 0.56; 95% CI, 0.19–1.62) [8].

IL-12/23 Inhibitor

In pre-clinical studies, both IL-12 and IL-23 were found to have anti-tumorigenic properties. Therefore, blockade of these two molecules may theoretically increase malignancy risk.

> **CONCLUSION**
> Based on the results of individual TNF inhibitors, psoriasis patients treated with TNF inhibitors do not appear to have higher rates of overall or internal malignancies. Data from pooled analyses on systemic malignancies need further validation. Psoriasis patients treated with TNF inhibitors do not appear to have higher rates of melanoma. Data on the risk of NMSC are mixed. It is important to note that many of the psoriasis patients in clinical trials have been treated with PUVA in the past, which is known to increase the risk of skin cancers.

- **Ustekinumab** (Stelara®)
  - *Mechanism of Action:* human monoclonal antibody binding the p40 subunit of both IL-12 and IL-23, which suppresses inflammatory response.
  - *Uses:* to treat psoriasis, psoriatic arthritis, and Crohn's disease.
  - *Common adverse effects:* nasopharyngitis, upper respiratory tract infections, headache, and fatigue.
  - Overall malignancies:
    In a 5-year follow-up study of 3117 psoriasis patients treated with ustekinumab, risk of overall malignancies including NMSC was not higher than that of the general population (1.13/100PY; 95% CI, 0.92–1.37) [71].
  - Internal malignancies:
    A safety surveillance study using PSOLAR data did not detect a higher systemic malignancy risk in psoriasis patients treated with ustekinumab when compared to other psoriasis patients regardless of the duration of treatment (Fig. 6.3) [72].
  - Skin malignancies:
    Another study using PSOLAR data found that psoriasis patients treated with ustekinumab had significantly lower rates of NMSC when compared to psoriasis patients treated with MTX (OR = 0.35; 95% CI, 0.14–0.87) [69].

- This finding was mostly driven by a lower risk of BCC (OR = 0.26; 95% CI, 0.09–0.72) rather than SCC (OR = 1.37; 95% CI, 0.14–13.5).

CONCLUSION
Studies did not find a higher risk of overall or internal malignancies in psoriasis patients treated with ustekinumab when compared to other psoriasis patients. Psoriasis patients treated with ustekinumab have a lower risk of NMSC when compared to psoriasis patients treated with MTX; this finding was mostly driven by a lower risk of BCC.

IL-17 Inhibitors

Preclinical models and studies of cancer subjects suggest IL-17 inhibition may have both anti- and pro-tumorigenic effects.

- **Secukinumab** (Cosentyx®)
  - *Mechanism of Action:* human monoclonal antibody binding to the IL-17A cytokine, inhibiting its pro-inflammatory effect. IL-17 inhibition also inhibits stimulation and proliferation of keratinocytes.
  - *Uses:* to treat psoriasis, psoriatic arthritis, and ankylosing spondylitis.
  - *Common adverse effects:* nasopharyngitis, diarrhea, and upper respiratory tract infection.
  - Secukinumab was FDA approved for plaque psoriasis in January of 2015. There is currently limited long-term safety data in patients treated with secukinumab.
  - In a meta-analysis of ten phase II and phase III RCTs of secukinumab for psoriasis, overall malignancy IR in the secukinumab treatment group after 52 weeks (0.48 per 100 PY) was comparable to the expected malignancy IR in the general population (0.45 per 100 PY) [73].

CONCLUSION
Based on the available data, secukinumab is not associated with higher risk of malignancy. More long-term data with a large sample size are needed to evaluate its long-term safety.

- **Ixekizumab** (Taltz®)
  - *Mechanism of Action:* humanized monoclonal antibody binding to the IL-17A cytokine, inhibiting its pro-inflammatory effect. IL-17 inhibition also inhibits the stimulation and proliferation of keratinocytes.
  - *Uses:* to treat psoriasis and psoriatic arthritis.

- *Common adverse effects:* injection site reactions, upper respiratory tract infections, nausea, and tinea infections.
- Ixekizumab was FDA approved for plaque psoriasis in March of 2016. There is currently limited long-term safety data in patients treated with ixekizumab.
- In three phase III RCTs of ixekizumab for treatment of psoriasis, malignancy rates in those in the ixekizumab treatment group were consistent with malignancy rates expected in patients with psoriasis [74].

CONCLUSION
Based on the available data, ixekizumab is not associated with higher risk of malignancy. More long-term data with a large sample size are needed to evaluate its long-term safety.

- **Brodalumab** (Siliq®)
  - *Mechanism of Action:* human monoclonal antibody binding directly to the IL-17 receptor, preventing the IL-17 cytokine from activating the receptor.
  - *Uses:* to treat psoriasis.
  - *Common adverse effects:* arthralgia, headache, fatigue, diarrhea, oropharyngeal pain, nausea, myalgia, injection site reactions, influenza, neutropenia, and tinea infections.
  - Brodalumab was FDA approved for plaque psoriasis in February 2017. There is currently limited long-term safety data in patients treated with brodalumab.
  - A meta-analysis of 6 RCTs of brodalumab for treatment of psoriasis reported an acceptable safety profile with no reports of malignancies [75].

CONCLUSION
Based on the available data, brodalumab is not associated with higher risk of malignancy. More long-term data with a large sample size are needed to evaluate its long-term safety.

IL-23 Inhibitors

- **Guselkumab** (Tremfya®)
  - *Mechanism of action:* human monoclonal antibody targeting the p19 subunit of the cytokine IL-23.
  - *Uses:* to treat psoriasis and psoriatic arthritis.
  - *Common adverse effects:* upper respiratory infections, headache, injection site reactions, arthralgias, diarrhea, gastroenteritis, tinea infections, and herpes simplex infections.
  - Guselkumab was FDA approved for plaque psoriasis in July of 2017. There is currently limited long-term safety data in patients treated with guselkumab.

- In phase III RCTs of guselkumab for treatment of psoriasis, malignancy rates were low in all treatment groups through week 48 with no class effect observed [76].

CONCLUSION
Based on the available data, guselkumab is not associated with higher risk of malignancy. More long-term data with a large sample size are needed to evaluate its long-term safety.

- **Tildrakizumab** (Ilumya®)
  - *Mechanism of action:* humanized monoclonal antibody targeting the cytokine IL-23.
  - *Uses:* to treat psoriasis.
  - *Common adverse effects:* upper respiratory infections, injection site reactions, and diarrhea.
  - Tildrakizumab was FDA approved for plaque psoriasis in March of 2018. There is currently limited long-term safety data in patients treated with tildrakizumab.
  - In phase III RCTs of tildrakizumab for treatment of psoriasis, malignancy rates were low in all groups, and there were no significant differences in malignancy rates between the tildrakizumab, etanercept, and placebo groups [77, 78].

CONCLUSION
Based on the available data, tildrakizumab is not associated with higher risk of malignancy. More long-term data with a large sample size are needed to evaluate its long-term safety.

- **Risankizumab** (Skyrizi®)
  - *Mechanism of action:* humanized monoclonal antibody targeting the cytokine IL-23A.
  - *Uses:* to treat psoriasis.
  - *Common adverse effects:* upper respiratory infections, injection site reactions, and diarrhea.
  - Risankizumab was FDA approved for plaque psoriasis in April of 2019. There is currently limited long-term safety data in patients treated with tildrakizumab.
  - In phase III RCTs of risankizumab for treatment of psoriasis, malignancy rates were low in all groups through week 48 [79].

> **CONCLUSION**
> Based on the available data, tildrakizumab is not associated with higher risk of malignancy. More long-term data with a large sample size are needed to evaluate its long-term safety.

IL-4/13 Inhibitor

- Dupilumab (Dupixent®)
  - *Mechanism of action:* human monoclonal antibody targeting the alpha subunit of the IL-4 receptor, which modulates signaling of both IL-4 and IL-13 pathways.
  - *Uses:* to treat atopic dermatitis.
  - *Common adverse effects:* injection site reactions, conjunctivitis, blepharitis, oral herpes, keratitis, eye pruritus, other herpes simplex infection, and dry eye.
  - Dupilumab was FDA approved for atopic dermatitis in March 2017. There is currently limited long-term safety data in patients treated with dupilumab.
  - Phase III studies of dupilumab for atopic dermatitis reported no cases of malignancies at 52 weeks [80, 81].

> **CONCLUSION**
> Based on the available data, dupilumab is not associated with higher risk of malignancy. More long-term data with a large sample size are needed to evaluate its long-term safety.

CD20 Inhibitor

- **Rituximab** (Rituxan®)
  - *Mechanism of action:* chimeric antibody targeting CD20 protein, which is found on the surface of immune system B cells.
  - *Uses:* to treat leukemias, lymphomas, RA, myasthenia gravis, granulomatosis with polyangiitis, and microscopic polyangiitis. Used off-label for chronic graft-versus-host-disease and autoimmune bullous disorders (e.g. pemphigus vulgaris, bullous pemphigoid, epidermolysis bullosa acquisita, mucous membrane pemphigoid).
  - *Common adverse effects:* upper respiratory tract infections, nasopharyngitis, urinary tract infections, and bronchitis.
  - Overall malignancies:
    - In a 5-year follow-up study of 989 RA patients in the SUNSTONE registry, those treated with rituximab had a lower overall malignancy rate (0.36 events/100 PY; 95% CI, 0.21–0.57) than the expected malignancy rate in the general RA population (1.0 events/100 PY) [82].

In a meta-analysis of nine trials involving 4621 non-Hodgkin lymphoma patients, those randomized to rituximab had no difference in risk of secondary primary malignancies when compared to those on placebo (OR = 0.88; 95% CI, 0.66–1.19) [83].

In a meta-analysis evaluating the long-term safety of rituximab in 3595 RA patients over 11 years, there was no difference in overall malignancy risk in the rituximab treated population (0.74 events/100 PY; 95% CI, 0.60–0.88) when compared to the placebo population (0.81 events/100 PY) [84].

In an RCT of 280 adult renal transplant patients randomized to either rituximab or placebo during transplant surgery, those randomized to rituximab did not have a higher risk of overall malignancy after 24 months (OR = 1.03; 95% CI, 0.40–2.66) [85].

- Skin malignancies:
  There have been 16 case reports of melanoma in patients treated with rituximab for various indications [86].
- In several small observational studies of rituximab for the treatment of pemphigus, there were no reports of malignancies in those treated with rituximab [87–92].

CONCLUSION
Based on the currently available small observational studies of rituximab for use in pemphigus, rituximab does not appear to be associated with a higher malignancy risk. Long-term safety data of rituximab for treatment of RA and non-Hodgkin lymphoma did not find a higher overall malignancy risk in patients treated with rituximab when compared to other patients of the same diagnosis.

Anti-IgE

- **Omalizumab** (Xolair®)
  - *Mechanism of action:* humanized monoclonal antibody targeting free human IgE, which inhibits the binding of IgE to mast cells and basophils.
  - *Uses:* to treat allergic asthma and chronic spontaneous urticaria.
  - *Common adverse effects:* nausea, nasopharyngitis, sinusitis, upper respiratory tract infection, viral upper respiratory tract infection, arthralgia, headache, and cough.
  - In RCTs of omalizumab for treatment of asthma, malignancies were observed in 20/4127 (0.5%) of patients treated with omalizumab and 5/2236 (0.2%) of patients treated with placebo [93].
    Observed malignancies included skin (11 cases of NMSC and 2 cases of melanoma in a total of 6 patients), breast (5 cases), prostate (2 cases), non-Hodgkin lymphoma (1 case), thyroid (1 case), metastatic adenocystic parotid (1 case), bladder (1 case), parotid (1 case), pancreas (1 case), and colorectal (1 case).

Patients were treated with omalizumab for no more than 1 year.
- In a pooled analysis of phase I, II, and III RCTs of omalizumab for all indications, the incidence of malignant neoplasms in the omalizumab treatment group (25/5015 patients; 0.50%) was slightly higher than in the control group (5/2854 patients; 0.18%) [94].

  However, based upon comparisons with data from the NIH SEER database, incidence of malignancy in the omalizumab treatment group was similar to the expected incidence in the general population.
- In a prospective observational cohort study of 7857 asthma patients either treated with or without omalizumab, malignancy rates between the omalizumab and nonomalizumab cohorts were similar for all malignancies (RR = 0.84; 95% CI, 0.62–1.13) and all malignancies excluding NMSC (RR = 0.98; 95% CI, 0.71–1.36) [95].
- In a 40-week RCT of omalizumab treatment for chronic idiopathic urticaria, none of the 319 patients experienced any events of malignancy [96].

CONCLUSION
Meta-analyses found no differences in malignancy risk in asthma and chronic idiopathic urticaria patients treated with omalizumab when compared to controls.

IVIG

- **IVIG** (various brand products available)
  - *Mechanism of action:* IVIG contains a mixture of human antibodies. The exact mechanism of IVIG is not well understood. It has been proposed that IVIG may directly bind to abnormal host antibodies, stimulating their removal via the host's complement system. IVIG may also interact with the receptors on the surface of various host immune cells, modulating their immune response.
  - *Uses:* to treat primary immunodeficiency, idiopathic thrombocytopenic purpura, Kawasaki disease, Guillain-Barre syndrome, and certain cases of HIV/AIDS infections. May also be used off-label for treatment of dermatomyositis and autoimmune bullous diseases [97].
  - *Common adverse effects:* headache, fever, chills, hypertension, rash, and nausea.
  - There have been no reports of IVIG increasing the risk of malignancies.
  - There have been numerous cases reporting cancer regression after IVIG administration, and IVIG has been touted as a possible cancer treatment [98].

CONCLUSION
There is currently no evidence that IVIG increases risk of malignancies. IVIG may be effective as a possible cancer treatment, but more studies are needed to validate the safety and efficacy of IVIG for the treatment of cancers.

Malignancy Risk Analysis in Immunosuppressed Patients

Transplant Patients

- Malignancy, cardiovascular disease, and infection are the three major sources of morbidity and mortality in organ transplant recipients [99].
- Overall malignancy risk is two- to three-fold higher in transplant patients when compared to the general population [100].
  - When stratified by age, younger transplant recipients experience a greater increase in malignancy risk when compared to older recipients (15 to 30-fold higher risk in children and two-fold higher risk in those >65 years old).
  - The following cancers are much higher in organ transplant patients when compared to the general population:
       NMSC (SIR = 13.85; 95% CI, 11.92–16.0)
       Lip cancer (SIR = 16.78; 95% CI, 14.02–19.92)
       Kaposi sarcoma (SIR = 61.46; 95% CI, 50.95–73.49)
       Non-Hodgkin lymphoma (SIR = 7.54; 95% CI, 7.17–7.93)
       Liver cancer (SIR = 11.56; 95% CI, 10.83–12.33)
       Kidney cancer (SIR = 4.65; 95% CI, 4.32–4.99)
  - The following cancers are slightly, but significantly, higher in organ transplant patients when compared to the general population:
       Colorectal (SIR = 1.24; 95% CI 1.15–1.34)
       Lung cancer (SIR = 1.97; 95% CI, 1.86–2.08)
- Transplant patients are vulnerable to viral infections due to immunosuppression. Viral infections have been implicated in carcinogenesis among organ transplant patients treated with immunosuppressive therapies [100]. These infections include:
  - EBV leading to Hodgkin's lymphoma, non-Hodgkin's lymphoma, and naso-pharyngeal carcinomas.
  - HHV-8 leading to Kaposi's sarcoma, multiple myeloma, and primary effusion lymphoma.
  - HPV leading to SCC, anogenital cancers, and head and neck cancers.
  - Merkel cell polyomavirus leading to Merkel cell carcinoma.
  - Hepatitis B and hepatitis C leading to hepatocellular carcinoma.
- Immunosuppressive treatment regimens to prevent organ transplant rejection may increase malignancy risk by reducing the immune system's ability detect and destroy cancer cells.
  - Cyclosporine and azathioprine use in organ transplant patients is linked to a higher risk of lymphoma and SCC [14, 16, 17].
  - There is some evidence that oral corticosteroid use may increase the risk of NMSC, non-Hodgkin lymphoma, and esophageal cancers in the general population [24, 25]. However, more studies are needed to further evaluate this risk.
- With the known malignancy risk in organ transplant patients, clinicians should be vigilant with regular cancer screenings in this population.
  - In a systematic review of clinical practice guidelines for cancer screenings in organ transplant patients, the authors found [101]:

Ten out of ten clinical practice guidelines recommend bi-annual to annual skin exams for skin cancer prevention, with more frequent exams in patients with other dermatologic concerns.

Recommendations with respect to other cancer screenings for organ transplant patients vary based on individual guidelines:

- Breast cancer: Four guidelines recommended breast cancer screening to follow general population guidelines, two guidelines recommend annual mammograms, and one guideline provided no specific recommendations for breast cancer screening.
- Prostate cancer: Four guidelines recommended prostate cancer screening to follow general population guidelines, and three guidelines recommend annual prostate-specific antigen testing and digital rectal examinations.
- Kidney and urothelial cancer: Two guidelines recommended against kidney and urothelial cancer screening, and one guideline recommended annual renal ultrasound screenings in kidney transplant patients.
- Colorectal cancer: Five guidelines recommended following general population guidelines, one guideline recommended annual colonoscopies, and one guideline recommended annual fecal occult blood tests.

HIV Patients

- HIV infection may predispose patients to certain types of viral infections; these types of viral infections have been linked to cancers.
  - Any diagnoses of the following cancers are AIDS-defining [102]:
      Kaposi sarcoma (500 times more likely in HIV patients)
      - Related to risk of HHV-8 infection
      Non-Hodgkin lymphoma (12 times more likely in HIV patients)
      - Related to the risk of EBV infection
      Cervical cancer (3 times more likely in HIV patients)
      - Related to the risk of HPV infection
  - Other non-AIDS-defining malignancies include the following [103]:
      Liver, anus, oral cavity/pharynx, lung, and Hodgkin lymphoma
      - Higher risk of liver cancer appears to be related to the higher risk of hepatitis viral infections in HIV patients.
      - Higher risk of oral cavity/pharynx cancers appears to be related to the higher risk of HPV infection in HIV patients.
- Any HIV patient, regardless of CD4+ count or AIDS-defining illness, should be treated with HAART [104].
  - HAART reduces the risk of developing HIV-associated cancers
- Preventative measures should be taken to lower the risk of malignancies [105]. These measures include:
  - HPV vaccination
  - Pap smears
  - Smoking cessation
  - HAART

Review Questions and Answers

1. The incidence of hepatocellular carcinoma in a new biologic for treatment of psoriasis is 1 case per 300 patients per year. How many patients need to be exposed to the biologic for 1 year in order to adequately assess the risk of hepatocellular carcinoma?
   (a) 300
   (b) 600
   (c) 900
   (d) 1200
   (e) 1500

   Answer: (c) 900
   For rare, adverse drug-related events, the rule of 3's may be utilized to calculate the number of patients needed to see 1 case of the event at 95% confidence interval. To adequately assess the risk of hepatocellular carcinoma that occurs in 1 case per 300 patients treated per year, we would need at least $3n$ trial patients, which equates to $3(300) = 900$ patients.

2. A 76-year-old male presents for care for his psoriasis. He has a history of a kidney transplant 10 years ago due to end stage renal disease. He had also worked as an almond farmer for 30 years and has had multiple skin cancers removed in the past. On exam, his psoriasis involves 10% body surface area and is on his trunk and bilateral upper and lower extremities. The patient does not want injections due to fear of needles and prefers oral medications. Which of the following medications is the most appropriate treatment for this patient?
   (a) Apremilast
   (b) Oral corticosteroids
   (c) Acitretin
   (d) Isotretinoin
   (e) Methotrexate

   Answer: (c) Acitretin
   Acitretin (c) is helpful because this patient has several risk factors for skin cancer, including a history of multiple skin cancers and organ transplant. Acitretin has been associated with prevention of skin malignancies in these high-risk populations.
   Although apremilast (a) and methotrexate (e) are also effective in the treatment of psoriasis, they have not been associated with the same chemopreventive properties seen with acitretin. While oral corticosteroids (b) may have an immediate therapeutic effect for psoriasis, withdrawal from corticosteroids has been associated with aggressive pustular or erythrodermic psoriasis. Isotretinoin (d) is used primarily for acne; evidence is lacking for its use in psoriasis.

3. Which of the following statements is accurate regarding malignancy risk of IL-17 inhibitors (secukinumab, ixekizumab, and brodalumab) among psoriasis patients?
   (a) Phase III data of IL-17 inhibitors (secukinumab, ixekizumab, and brodalumab) did not show increased risk of internal malignancies.
   (b) Phase III data of IL-17 inhibitors (secukinumab, ixekizumab, and brodalumab) showed increased risk of NMSC.
   (c) Of the IL-17 inhibitors, only treatment with secukinumab is associated with a higher risk of lymphomas.
   (d) Of the IL-17 inhibitors, only treatment with brodalumab is associated with a higher risk of lymphomas.
   (e) Of the IL-17 inhibitors, only treatment with ixekizumab is associated with a higher risk of lymphomas.

Answer: (a) Phase III data of IL-17 inhibitors (secukinumab, ixekizumab, and brodalumab) did not show increased risk of internal malignancies.
All three IL-17 inhibitors including secukinumab, ixekizumab, and brodalumab were recently FDA-approved for psoriasis within the past few years. To date, RCT data did not detect higher internal malignancy rates in patients treated with these IL-17 inhibitors. However, long-term data with large populations of patients are needed to continue to evaluate their safety profile. Registries that collect long-term safety data from different drugs will be highly informative of the long-term risks of newer therapies.

4. A 59-year-old female with a history of atopic dermatitis presents for care. She recently underwent liver transplant for cirrhosis secondary to HCV. The operation was successful without complications. On physical exam, the patient's skin is completely clear, and she has a well-healed scar from her surgery. She is currently being treated with tacrolimus, mycophenolate, and prednisone to prevent organ rejection. Which of the following is the most appropriate recommendation for this patient at this time?
   (a) Bi-annual skin examinations
   (b) Annual colonoscopy
   (c) Annual endoscopy
   (d) Annual fecal occult blood test
   (e) Annual renal ultrasound

Answer: (a) Bi-annual skin checks
Organ transplant patients are at a two- to three-fold higher risk of malignancies when compared to the general population. Clinical practice guidelines recommend bi-annual (twice a year) to annual (once a year) skin cancer screenings, with more frequent exams in patients with other dermatologic concerns. There is currently no consensus among guidelines whether annual colonoscopies (c), annual endoscopies (b), annual fecal occult blood tests (d), or annual renal ultrasounds (e) should be done in this population.

References

1. Kimball AB, et al. Incidence rates of malignancies and hospitalized infectious events in patients with psoriasis with or without treatment and a general population in the U.S.A.: 2005-09. Br J Dermatol. 2014;170(2):366–73.
2. Hagstromer L, et al. Incidence of cancer among patients with atopic dermatitis. Arch Dermatol. 2005;141(9):1123–7.
3. Scott FI, et al. Risk of nonmelanoma skin cancer associated with the use of immunosuppressant and biologic agents in patients with a history of autoimmune disease and nonmelanoma skin cancer. JAMA Dermatol. 2016;152(2):164–72.
4. Polesie S, et al. Methotrexate treatment and risk for cutaneous malignant melanoma: a retrospective comparative registry-based cohort study. Br J Dermatol. 2017;176(6):1492–9.
5. Buchbinder R, et al. Incidence of melanoma and other malignancies among rheumatoid arthritis patients treated with methotrexate. Arthritis Rheum. 2008;59(6):794–9.
6. Hoshida Y, et al. Lymphoproliferative disorders in rheumatoid arthritis: clinicopathological analysis of 76 cases in relation to methotrexate medication. J Rheumatol. 2007;34(2):322–31.
7. Kamel OW, et al. Brief report: reversible lymphomas associated with Epstein-Barr virus occurring during methotrexate therapy for rheumatoid arthritis and dermatomyositis. N Engl J Med. 1993;328(18):1317–21.
8. Fiorentino D, et al. Risk of malignancy with systemic psoriasis treatment in the psoriasis longitudinal assessment registry. J Am Acad Dermatol. 2017;77(5):845–54. e5
9. Stern RS. Lymphoma risk in psoriasis: results of the PUVA follow-up study. Arch Dermatol. 2006;142(9):1132–5.
10. Suzuki M, et al. Pulmonary lymphoma developed during long-term methotrexate therapy for psoriasis. Respirology. 2007;12(5):774–6.
11. Patel RV, et al. Treatments for psoriasis and the risk of malignancy. J Am Acad Dermatol. 2009;60(6):1001–17.
12. Cockburn IT, Krupp P. The risk of neoplasms in patients treated with cyclosporine A. J Autoimmun. 1989;2(5):723–31.
13. Paul CF, et al. Risk of malignancies in psoriasis patients treated with cyclosporine: a 5 y cohort study. J Invest Dermatol. 2003;120(2):211–6.
14. Muellenhoff MW, Koo JY. Cyclosporine and skin cancer: an international dermatologic perspective over 25 years of experience. A comprehensive review and pursuit to define safe use of cyclosporine in dermatology. J Dermatolog Treat. 2012;23(4):290–304.
15. O'Donovan P, et al. Azathioprine and UVA light generate mutagenic oxidative DNA damage. Science. 2005;309(5742):1871–4.
16. Jiyad Z, et al. Azathioprine and risk of skin cancer in organ transplant recipients: systematic review and meta-analysis. Am J Transplant. 2016;16(12):3490–503.
17. Na R, et al. Iatrogenic immunosuppression and risk of non-Hodgkin lymphoma in solid organ transplantation: a population-based cohort study in Australia. Br J Haematol. 2016;174(4):550–62.
18. Kotlyar DS, et al. Risk of lymphoma in patients with inflammatory bowel disease treated with azathioprine and 6-mercaptopurine: a meta-analysis. Clin Gastroenterol Hepatol. 2015;13(5):847–858.e4; quiz e48–e50.
19. Beigel F, et al. Risk of malignancies in patients with inflammatory bowel disease treated with thiopurines or anti-TNF alpha antibodies. Pharmacoepidemiol Drug Saf. 2014;23(7):735–44.
20. Hagen JW, Pugliano-Mauro MA. Nonmelanoma skin cancer risk in patients with inflammatory bowel disease undergoing thiopurine therapy: a systematic review of the literature. Dermatol Surg. 2018;44(4):469–80.
21. Pedersen EG, et al. Risk of non-melanoma skin cancer in myasthenia patients treated with azathioprine. Eur J Neurol. 2014;21(3):454–8.
22. Wang YJ, et al. Malignancy after heart transplantation under everolimus versus mycophenolate mofetil immunosuppression. Transplant Proc. 2016;48(3):969–73.

23. Hsiao FY, Hsu WW. Epidemiology of post-transplant malignancy in Asian renal transplant recipients: a population-based study. Int Urol Nephrol. 2014;46(4):833–8.
24. Sorensen HT, et al. Use of systemic corticosteroids and risk of esophageal cancer. Epidemiology. 2002;13(2):240–1.
25. Sorensen HT, et al. Skin cancers and non-hodgkin lymphoma among users of systemic glucocorticoids: a population-based cohort study. J Natl Cancer Inst. 2004;96(9):709–11.
26. Sorensen GV, et al. Use of glucocorticoids and risk of breast cancer: a Danish population-based case-control study. Breast Cancer Res. 2012;14(1):R21.
27. Chen K, Craig JC, Shumack S. Oral retinoids for the prevention of skin cancers in solid organ transplant recipients: a systematic review of randomized controlled trials. Br J Dermatol. 2005;152(3):518–23.
28. Kadakia KC, et al. Randomized controlled trial of acitretin versus placebo in patients at high-risk for basal cell or squamous cell carcinoma of the skin (North Central Cancer Treatment Group Study 969251). Cancer. 2012;118(8):2128–37.
29. Cheeley J, et al. Acitretin for the treatment of cutaneous T-cell lymphoma. J Am Acad Dermatol. 2013;68(2):247–54.
30. Bettoli V, Zauli S, Virgili A. Retinoids in the chemoprevention of non-melanoma skin cancers: why, when and how. J Dermatolog Treat. 2013;24(3):235–7.
31. Nijsten TE, Stern RS. Oral retinoid use reduces cutaneous squamous cell carcinoma risk in patients with psoriasis treated with psoralen-UVA: a nested cohort study. J Am Acad Dermatol. 2003;49(4):644–50.
32. Hong WK, et al. 13-cis-retinoic acid in the treatment of oral leukoplakia. N Engl J Med. 1986;315(24):1501–5.
33. Bhatia AK, et al. Double-blind, randomized phase 3 trial of low-dose 13-cis retinoic acid in the prevention of second primaries in head and neck cancer: long-term follow-up of a trial of the Eastern Cooperative Oncology Group-ACRIN Cancer Research Group (C0590). Cancer. 2017;123(23):4653–62.
34. Chen SE, et al. Isotretinoin maintenance therapy for glioblastoma: a retrospective review. J Oncol Pharm Pract. 2014;20(2):112–9.
35. Cash T, et al. Prolonged isotretinoin in ultra high-risk neuroblastoma. J Pediatr Hematol Oncol. 2017;39(1):e33–5.
36. Skroza N, et al. Isotretinoin for the treatment of squamous cell carcinoma arising on an epidermoid cyst. Dermatol Ther. 2014;27(2):94–6.
37. Duvic M, et al. Phase 2 and 3 clinical trial of oral bexarotene (Targretin capsules) for the treatment of refractory or persistent early-stage cutaneous T-cell lymphoma. Arch Dermatol. 2001;137(5):581–93.
38. Duvic M, et al. Bexarotene is effective and safe for treatment of refractory advanced-stage cutaneous T-cell lymphoma: multinational phase II-III trial results. J Clin Oncol. 2001;19(9):2456–71.
39. Whittaker S, et al. Efficacy and safety of bexarotene combined with psoralen-ultraviolet A (PUVA) compared with PUVA treatment alone in stage IB-IIA mycosis fungoides: final results from the EORTC cutaneous lymphoma task force phase III randomized clinical trial (NCT00056056). Br J Dermatol. 2012;167(3):678–87.
40. Pullamsetti SS, et al. Phosphodiesterase-4 promotes proliferation and angiogenesis of lung cancer by crosstalk with HIF. Oncogene. 2013;32(9):1121–34.
41. Suhasini AN, et al. A phosphodiesterase 4B-dependent interplay between tumor cells and the microenvironment regulates angiogenesis in B-cell lymphoma. Leukemia. 2016;30(3):617–26.
42. Kavanaugh A, et al. Longterm (52-week) results of a phase III randomized, controlled trial of apremilast in patients with psoriatic arthritis. J Rheumatol. 2015;42(3):479–88.
43. Reich K, et al. The efficacy and safety of apremilast, etanercept and placebo in patients with moderate-to-severe plaque psoriasis: 52-week results from a phase IIIb, randomized, placebo-controlled trial (LIBERATE). J Eur Acad Dermatol Venereol. 2017;31(3):507–17.

44. Song GG, Lee YH. Relative efficacy and safety of apremilast, secukinumab, and ustekinumab for the treatment of psoriatic arthritis. Z Rheumatol. 2018;77(7):613–20.
45. Duvic M, et al. Phase 2 trial of oral vorinostat (suberoylanilide hydroxamic acid, SAHA) for refractory cutaneous T-cell lymphoma (CTCL). Blood. 2007;109(1):31–9.
46. Duvic M, et al. Evaluation of the long-term tolerability and clinical benefit of vorinostat in patients with advanced cutaneous T-cell lymphoma. Clin Lymphoma Myeloma. 2009;9(6):412–6.
47. Olsen EA, et al. Phase IIb multicenter trial of vorinostat in patients with persistent, progressive, or treatment refractory cutaneous T-cell lymphoma. J Clin Oncol. 2007;25(21):3109–15.
48. Piekarz RL, et al. Phase II multi-institutional trial of the histone deacetylase inhibitor romidepsin as monotherapy for patients with cutaneous T-cell lymphoma. J Clin Oncol. 2009;27(32):5410–7.
49. Whittaker SJ, et al. Final results from a multicenter, international, pivotal study of romidepsin in refractory cutaneous T-cell lymphoma. J Clin Oncol. 2010;28(29):4485–91.
50. Study of cancer incidence in relation to dapsone use—executive summary. [cited 2019 May 7]. http://amvif.com/government/Veterans/Dapsone/Dapsone.htm.
51. Stern RS, Liebman EJ, Vakeva L. Oral psoralen and ultraviolet-A light (PUVA) treatment of psoriasis and persistent risk of nonmelanoma skin cancer. PUVA follow-up study. J Natl Cancer Inst. 1998;90(17):1278–84.
52. Studniberg HM, Weller P. PUVA, UVB, psoriasis, and nonmelanoma skin cancer. J Am Acad Dermatol. 1993;29(6):1013–22.
53. Lindelof B, et al. PUVA and cancer: a large-scale epidemiological study. Lancet. 1991;338(8759):91–3.
54. Kimball AB, et al. OBSERVE-5: observational postmarketing safety surveillance registry of etanercept for the treatment of psoriasis final 5-year results. J Am Acad Dermatol. 2015;72(1):115–22.
55. Mariette X, et al. Malignancies associated with tumour necrosis factor inhibitors in registries and prospective observational studies: a systematic review and meta-analysis. Ann Rheum Dis. 2011;70(11):1895–904.
56. Gottlieb AB, et al. Clinical trial safety and mortality analyses in patients receiving etanercept across approved indications. J Drugs Dermatol. 2011;10(3):289–300.
57. Hooper M, et al. Malignancies in children and young adults on etanercept: summary of cases from clinical trials and post marketing reports. Pediatr Rheumatol Online J. 2013;11(1):35.
58. Menter A, et al. Long-term safety and effectiveness of Adalimumab for moderate to severe psoriasis: results from 7-year interim analysis of the ESPRIT registry. Dermatol Ther (Heidelb). 2017;7(3):365–81.
59. Leonardi C, et al. Comprehensive long-term safety of adalimumab from 18 clinical trials in adult patients with moderate-to-severe plaque psoriasis. Br J Dermatol. 2019;180(1):76–85.
60. Burmester GR, et al. Adalimumab: long-term safety in 23 458 patients from global clinical trials in rheumatoid arthritis, juvenile idiopathic arthritis, ankylosing spondylitis, psoriatic arthritis, psoriasis and Crohn's disease. Ann Rheum Dis. 2013;72(4):517–24.
61. Lichtenstein GR, et al. A pooled analysis of infections, malignancy, and mortality in infliximab- and immunomodulator-treated adult patients with inflammatory bowel disease. Am J Gastroenterol. 2012;107(7):1051–63.
62. Torii H, et al. Safety profiles and efficacy of infliximab therapy in Japanese patients with plaque psoriasis with or without psoriatic arthritis, pustular psoriasis or psoriatic erythroderma: results from the prospective post-marketing surveillance. J Dermatol. 2016;43(7):767–78.
63. Mackey AC, et al. Hepatosplenic T cell lymphoma associated with infliximab use in young patients treated for inflammatory bowel disease. J Pediatr Gastroenterol Nutr. 2007;44(2):265–7.
64. Bongartz T, et al. Anti-TNF antibody therapy in rheumatoid arthritis and the risk of serious infections and malignancies: systematic review and meta-analysis of rare harmful effects in randomized controlled trials. JAMA. 2006;295(19):2275–85.

65. Solomon DH, Mercer E, Kavanaugh A. Observational studies on the risk of cancer associated with tumor necrosis factor inhibitors in rheumatoid arthritis: a review of their methodologies and results. Arthritis Rheum. 2012;64(1):21–32.
66. Geller S, et al. Malignancy risk and recurrence with psoriasis and its treatments: a concise update. Am J Clin Dermatol. 2018;19(3):363–75.
67. Dommasch ED, et al. The risk of infection and malignancy with tumor necrosis factor antagonists in adults with psoriatic disease: a systematic review and meta-analysis of randomized controlled trials. J Am Acad Dermatol. 2011;64(6):1035–50.
68. Asgari MM, et al. Malignancy rates in a large cohort of patients with systemically treated psoriasis in a managed care population. J Am Acad Dermatol. 2017;76(4):632–8.
69. deShazo R, Soltani-Arabshahi R, Krishnasamy S, et al. The effect of biologic therapy on non-melanoma skin cancer incidence among patients in the Psoriasis Longitudinal Assessment and Registry (PSOLAR). Eur Acad Dermatol Venerol. 2017.
70. van Lumig PP, et al. An increased risk of non-melanoma skin cancer during TNF-inhibitor treatment in psoriasis patients compared to rheumatoid arthritis patients probably relates to disease-related factors. J Eur Acad Dermatol Venereol. 2015;29(4):752–60.
71. Papp KA, et al. Long-term safety of ustekinumab in patients with moderate-to-severe psoriasis: final results from 5 years of follow-up. Br J Dermatol. 2013;168(4):844–54.
72. Papp K, et al. Safety surveillance for ustekinumab and other psoriasis treatments from the psoriasis longitudinal assessment and registry (PSOLAR). J Drugs Dermatol. 2015;14(7):706–14.
73. van de Kerkhof PC, et al. Secukinumab long-term safety experience: a pooled analysis of 10 phase II and III clinical studies in patients with moderate to severe plaque psoriasis. J Am Acad Dermatol. 2016;75(1):83–98. e4
74. Strober B, et al. Short- and long-term safety outcomes with ixekizumab from 7 clinical trials in psoriasis: etanercept comparisons and integrated data. J Am Acad Dermatol. 2017;76:432–440.e17. https://doi.org/10.1016/j.jaad.2016.09.026.
75. Attia A, et al. Safety and efficacy of brodalumab for moderate-to-severe plaque psoriasis: a systematic review and meta-analysis. Clin Drug Investig. 2017;37(5):439–51.
76. Reich K, et al. Efficacy and safety of guselkumab, an anti-interleukin-23 monoclonal antibody, compared with adalimumab for the treatment of patients with moderate to severe psoriasis with randomized withdrawal and retreatment: results from the phase III, double-blind, placebo- and active comparator-controlled VOYAGE 2 trial. J Am Acad Dermatol. 2017;76(3):418–31.
77. Amin M, et al. Review of phase III trial data on IL-23 inhibitors tildrakizumab and guselkumab for psoriasis. J Eur Acad Dermatol Venereol. 2017;31(10):1627–32.
78. Reich K, et al. Tildrakizumab versus placebo or etanercept for chronic plaque psoriasis (reSURFACE 1 and reSURFACE 2): results from two randomised controlled, phase 3 trials. Lancet. 2017;390(10091):276–88.
79. Papp KA, et al. Risankizumab versus ustekinumab for moderate-to-severe plaque psoriasis. N Engl J Med. 2017;376(16):1551–60.
80. Blauvelt A, et al. Long-term management of moderate-to-severe atopic dermatitis with dupilumab and concomitant topical corticosteroids (LIBERTY AD CHRONOS): a 1-year, randomised, double-blinded, placebo-controlled, phase 3 trial. Lancet. 2017;389(10086):2287–303.
81. Simpson EL, Akinlade B, Ardeleanu M. Two phase 3 trials of Dupilumab versus placebo in atopic dermatitis. N Engl J Med. 2017;376(11):1090–1.
82. Winthrop KL, et al. Long-term safety of rituximab in rheumatoid arthritis: analysis from the SUNSTONE registry. Arthritis Care Res (Hoboken). 2018;71(8):993–1003.
83. Fleury I, et al. Rituximab and risk of second primary malignancies in patients with non-Hodgkin lymphoma: a systematic review and meta-analysis. Ann Oncol. 2016;27(3):390–7.
84. van Vollenhoven RF, et al. Longterm safety of rituximab: final report of the rheumatoid arthritis global clinical trial program over 11 years. J Rheumatol. 2015;42(10):1761–6.
85. Cheungpasitporn W, et al. The effectiveness and safety of rituximab as induction therapy in ABO-compatible non-sensitized renal transplantation: a systematic review and meta-analysis of randomized controlled trials. Ren Fail. 2015;37(9):1522–6.

86. Peuvrel L, et al. Melanoma and rituximab: an incidental association? Dermatology. 2013;226(3):274–8.
87. Uzun S, et al. Efficacy and safety of rituximab therapy in patients with pemphigus vulgaris: first report from Turkey. Int J Dermatol. 2016;55(12):1362–8.
88. Gupta J, et al. Low-dose rituximab as an adjuvant therapy in pemphigus. Indian J Dermatol Venereol Leprol. 2017;83(3):317–25.
89. Kasper S. Editorial. Int J Psychiatry Clin Pract. 2009;13(4):243–4.
90. Joly P, et al. First-line rituximab combined with short-term prednisone versus prednisone alone for the treatment of pemphigus (Ritux 3): a prospective, multicentre, parallel-group, open-label randomised trial. Lancet. 2017;389(10083):2031–40.
91. Wang HH, et al. Efficacy of rituximab for pemphigus: a systematic review and meta-analysis of different regimens. Acta Derm Venereol. 2015;95(8):928–32.
92. Allen KJ, Wolverton SE. The efficacy and safety of rituximab in refractory pemphigus: a review of case reports. J Drugs Dermatol. 2007;6(9):883–9.
93. Corren J, et al. Safety and tolerability of omalizumab. Clin Exp Allergy. 2009;39(6):788–97.
94. Belliveau PP. Omalizumab: a monoclonal anti-IgE antibody. MedGenMed. 2005;7(1):27.
95. Long A, et al. Incidence of malignancy in patients with moderate-to-severe asthma treated with or without omalizumab. J Allergy Clin Immunol. 2014;134(3):560–7. e4
96. Saini SS, et al. Efficacy and safety of omalizumab in patients with chronic idiopathic/spontaneous urticaria who remain symptomatic on H1 antihistamines: a randomized, placebo-controlled study. J Invest Dermatol. 2015;135(3):925.
97. Dhar S. Intravenous immunoglobulin in dermatology. Indian J Dermatol. 2009;54(1):77–9.
98. Sapir T, Shoenfeld Y. Uncovering the hidden potential of intravenous immunoglobulin as an anticancer therapy. Clin Rev Allergy Immunol. 2005;29(3):307–10.
99. Chapman JR, Webster AC, Wong G. Cancer in the transplant recipient. Cold Spring Harb Perspect Med. 2013;3(7):a015677.
100. Engels EA, et al. Spectrum of cancer risk among US solid organ transplant recipients. JAMA. 2011;306(17):1891–901.
101. Acuna SA, et al. Cancer screening recommendations for solid organ transplant recipients: a systematic review of clinical practice guidelines. Am J Transplant. 2017;17(1):103–14.
102. Hernandez-Ramirez RU, et al. Cancer risk in HIV-infected people in the USA from 1996 to 2012: a population-based, registry-linkage study. Lancet HIV. 2017;4(11):e495–504.
103. Shiels MS, et al. A meta-analysis of the incidence of non-AIDS cancers in HIV-infected individuals. J Acquir Immune Defic Syndr. 2009;52(5):611–22.
104. WHO. Guideline on when to start antiretroviral therapy and on pre-exposure prophylaxis for HIV. 2015.
105. Goncalves PH, et al. Cancer prevention in HIV-infected populations. Semin Oncol. 2016;43(1):173–88.

Infection, Skin, and Systemic Malignancy

Jina Chung and Karolyn A. Wanat

Abbreviations

CMV	Cytomegalovirus
DLBCL	Diffuse large B-cell lymphoma
EBV	Epstein-Barr virus
EGFR	Epidermal growth factor receptor
HHV8	Human herpes virus 8
HPV	Human papillomavirus
HSV	Herpes simplex virus
I&D	Incision and drainage
MCC	Merkel cell carcinoma
NKTL	Natural killer/T-cell lymphoma]
PTLD	Post-transplant lymphoproliferative disease
RZV	Recombinant glycoprotein E vaccine
SCC	Squamous cell carcinoma
SSTI	Skin and soft tissue infections
Tx	Treatment
VZV	Varicella zoster virus
ZVL	Zoster vaccine live

J. Chung (✉)
Oregon Health & Science University, Department of Dermatology, Portland, OR, USA

K. A. Wanat
Medical College of Wisconsin, Department of Dermatology, Milwaukee, WI, USA

© Springer Nature Switzerland AG 2021
V. Liu (ed.), *Dermato-Oncology Study Guide*,
https://doi.org/10.1007/978-3-030-53437-0_7

Learning Objectives

1. To recognize the cutaneous presentations of infections associated with systemic malignancy.
2. To detect and manage cutaneous infections associated with treatment of malignancy.
3. To identify infections which predispose to malignancy with cutaneous manifestations.

Introduction

The overlap between infections and systemic malignancy can be sorted into several categories:

1. Cutaneous infections related to the malignancy itself, its treatment (chemotherapy and/or radiation), or invasive procedures used for diagnosis or treatment
2. Cutaneous infections related to specific therapies for malignancy
3. Infections which predispose to cutaneous malignancy

Cutaneous Infections in the Setting of Malignancies

Cancer patients are at increased risk of skin and soft tissue infections (SSTIs) due to impaired immune function as a result of hematologic malignancy or chemotherapy related effects.

Risk Factors for SSTIs in Patients with Systemic Malignancy [1]

- Neutropenia
- Reduced humoral immunity due to malignancy itself
- Recent or ongoing chemotherapy
- Recent or ongoing radiotherapy
- History of invasive procedures
- Presence of a central venous catheter

Specific Risk Populations

- Neutropenia (e.g., following treatment for hematologic malignancy): ↑ risk of opportunistic fungal infections
- Reduced humoral immunity (e.g., leukemia, multiple myeloma): ↑ risk of bacterial infections (streptococcal and staphylococcal)
- Reduced cellular immunity (e.g., immunosuppression after solid organ transplant): ↑ risk of viral and mycobacterial infections

Bacterial Infections

Cellulitis/Erysipelas
- Most common SSTI in the cancer population [1]
- Ill-defined area of painful erythema ± fever
- Most common agents: Group A streptococcus and *Staphylococcus aureus*
- Immunosuppressed patients with systemic malignancy may also have Gram-negative organisms as the culprit pathogen (e.g., *Pseudomonas aeruginosa*, *Escherichia coli*)
- **Tx:** Empiric treatment in a neutropenic patient should cover for Gram-positive organisms (e.g., vancomycin) and *Pseudomonas* spp. (e.g., carbapenems, piperacillin/tazobactam, third-generation cephalosporins) [2]

Bacterial Folliculitis
- Folliculitis: superficial bacterial infection of the hair follicles
- Most common agent: *S. aureus*
- **Tx:** chlorhexidine wash, topical antibiotics (mupirocin/clindamycin), oral antibiotics if widespread/suppurative

Furuncle, Carbuncle, Abscess
- Furuncle: suppurative inflammatory nodule involving a hair follicle
- Carbuncle: coalescing of multiple furuncles
- Abscess: walled-off collection of pus within dermis and/or subcutis (Fig. 7.1)
- Most common agent: *S. aureus*
- **Tx:** incision and drainage (I&D)
- Oral antibiotics for most likely pathogen if: + fever, >5 cm of surrounding cellulitis, immunocompromised host [3]

Fig. 7.1 Abscess

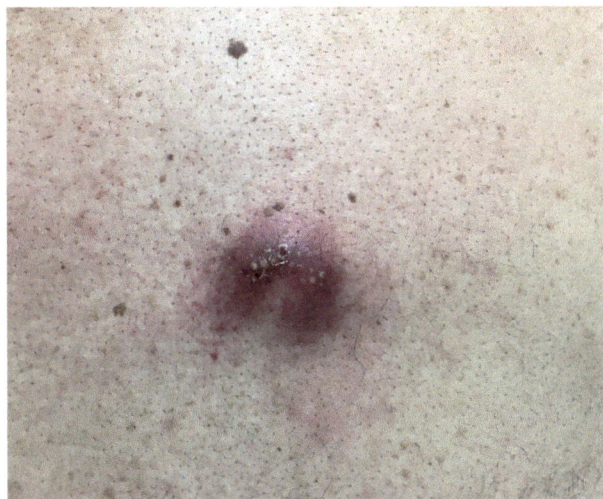

Ecthyma Gangrenosum

- Painless red macule → hemorrhagic bulla → necrotic ulcer with overlying eschar and erythematous rim (Fig. 7.2) [4]
- Most commonly associated with *Pseudomonas* bacteremia (high mortality rate in these cases), but can also be associated with other bacterial or fungal organisms [5]
- Can occur as a primary skin lesion in the absence of bacteremia (better prognosis)
- Diagnosis: skin biopsy, tissue cultures, blood cultures
- **Tx:** antipseudomonal penicillins (piperacillin/tazobactam), aminoglycosides, fluoroquinolones, third-generation cephalosporins

Atypical Mycobacterial Infections

- Can present as primary infection or cutaneous dissemination from systemic disease
- Localized disease: asymptomatic, slowly growing papules/nodules/pustules that may ulcerate (Fig. 7.3); multiple lesions may be in sporotrichoid distribution
- Rapidly-growing mycobacteria (*M. abscessus*, *M. fortuitum*, and *M. chelonae*): most commonly associated with SSTIs in cancer patients, especially surgical site and catheter-associated infections [6, 7]
- *Mycobacterium avium complex*: most common cause of systemic atypical mycobacterial infection in immunocompromised patients, and can rarely present with disseminated cutaneous disease
- **Tx:** regimen dependent upon specific infecting organism and severity; often includes various combination of minocycline, rifampicin, ehtambutol, isoniazid, ciprofloxacin, clarithromycin, azithromycin, cotrimoxazole

Fig. 7.2 Ecthyma gangrenosum

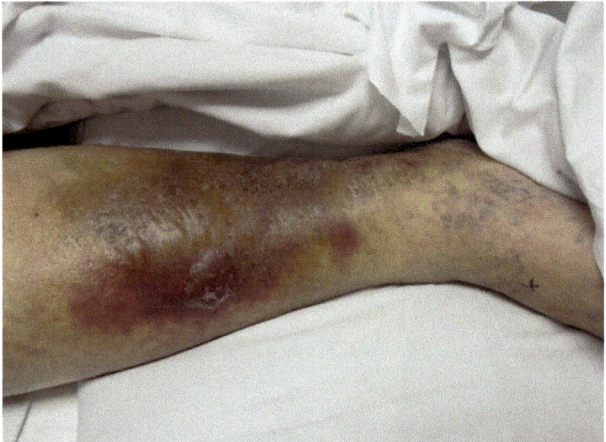

Fig. 7.3 Atypical mycobacterial infection

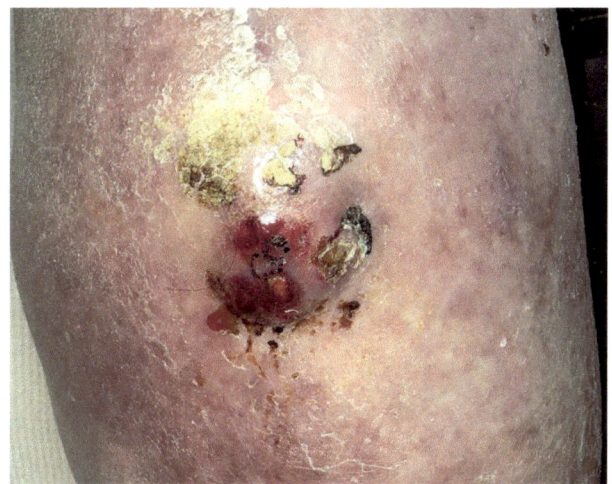

Viral Infections

Herpes Simplex Virus 1 and 2 (HSV1/HSV2)
- Painful vesicles, erosions or ulcers with scalloped borders (Fig. 7.4a)
- Oral lesions may mimic stomatitis secondary to chemotherapy
- Can present as exophytic, verrucous, tumor-like lesions resembling verrucous carcinoma in immunocompromised hosts (herpes vegetans)
- Diagnosis: PCR (most sensitive), Tzanck smear (the presence of multinucleated giant cells indicates infection with either HSV or VZV) (Fig. 7.4b), viral culture (can test for acyclovir resistance in refractory lesions)
- **Tx:** valacyclovir, famciclovir, acyclovir + foscarnet or cidofovir for resistant HSV

Varicella Zoster Virus (VZV)
- Prodrome of localized pain or pruritus followed by grouped vesicles in a dermatomal distribution (Fig. 7.4c)
- Disseminated zoster: >20 vesicles outside area of the primary and adjacent dermatomes
- In immunocompromised patients, antiviral therapy should be initiated even if time after onset >72 h
- Potential complications: ocular (herpes zoster ophthalmicus, acute retinal necrosis), otic (Ramsay Hunt syndrome), neurologic (meningitis, encephalitis), hepatitis, pneumonitis
- Ramsay Hunt syndrome: ipsilateral facial paralysis, ear pain, vesicles in auditory canal
- **Tx:** valacyclovir, famciclovir, acyclovir; IV acyclovir if disseminated/ocular involvement; + prednisone if ocular/optic involvement; + gabapentin/pregabalin for postherpetic neuralgia [8]

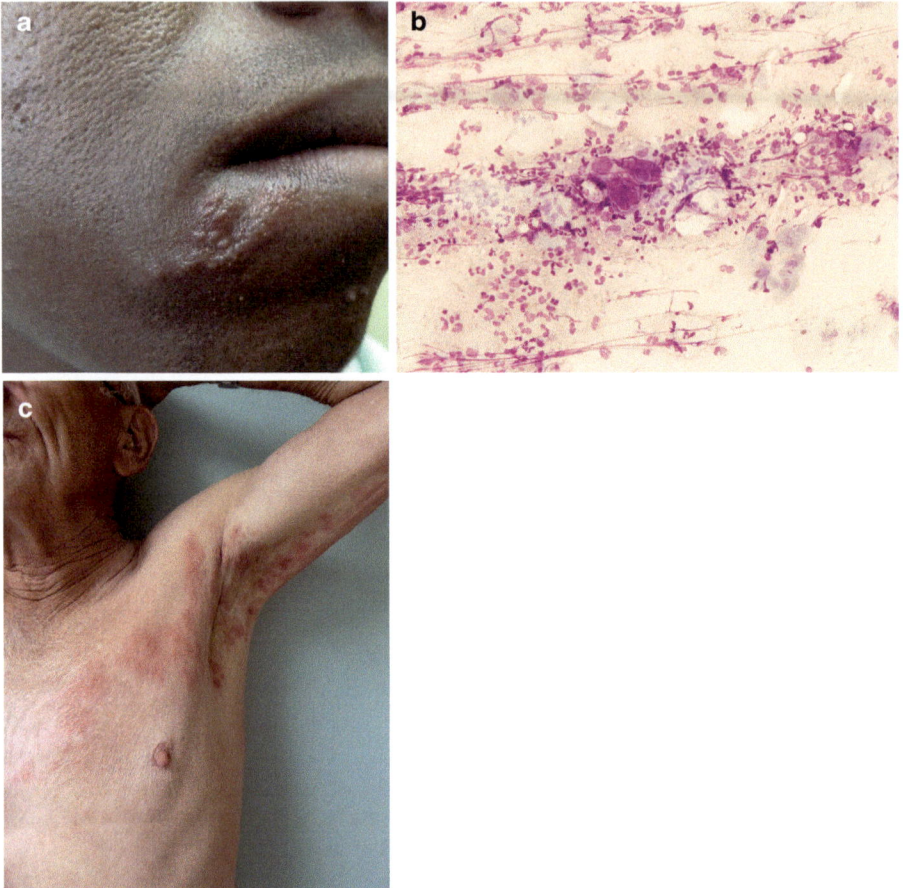

Fig. 7.4 (**a**) HSV infection- clinical, (**b**) Positive Tzanck prep, (**c**) VZV infection- clinical

- **Prevention:** consider vaccination in patients ≥50 years of age, ≥4 weeks prior to starting chemotherapy or immunosuppressive therapy [9]
- Zoster vaccine live (ZVL) should be avoided during chemotherapy
- Non-live recombinant glycoprotein E vaccine (RZV) is preferred

Verruca Vulgaris
- Caused by the human papillomavirus (HPV)
- Can be recalcitrant in immunosuppressed patients
- **Tx:** salicylic acid, topical 5-fluorouracil, podophyllin, cryotherapy, intralesional candida, intralesional chemotherapy (5-fluorouracil, bleomycin), laser treatment, topical cidofovir

Cytomegalovirus (CMV)
- Anogenital ulcerations in immunocompromised patients (due to hematologic malignancy or transplant)

- Biopsy may show characteristic "owl's eye" eosinophilic inclusions surrounded by a halo [10]
- **Tx:** valganciclovir

Oral Hairy Leukoplakia
- Benign, EBV-associated disease in immunocompromised patients; presents as white/gray plaques on the tongue (lateral/dorsolateral aspect)
- Only the superficial layers can be removed by scraping, unlike oral candidiasis [11]
- **Tx:** no tx required; can consider gentian violet (2%), topical retinoids, podophyllin, or topical acyclovir if symptomatic [12]

Trichodysplasia Spinulosa (Viral Associated Trichodyplasia)
- Causative agent: Trichodysplasia spinulosa-associated polyomavirus (TSPyV)
- Skin-colored, spine-like keratotic papules on the nose and forehead in immunocompromised patients, resulting in eyebrow/eyelash alopecia [13]
- **Tx:** oral valganciclovir, topical cidofovir, acitretin [14–16]

Fungal Infections

- Patients with severe neutropenia (absolute neutrophil count <500/mm) are at highest risk

Dermatophyte Infections
- Most common: tinea pedis, onychomycosis
- Most common causative organisms: *Microsporum*, *Epidermophyton*, and *Trichophyton* spp. (Fig. 7.5)
- Immunocompromised patients may have disseminated tinea corporis with large (>20 cm) lesions
- **Tx:** topicals- 1% terbinafine, azoles (ketoconazole, clotrimazole, miconazole, econazole, imidazole, etc); systemic therapies- terbinafine, itraconazole, griseofulvin, fluconazole (if diffuse or involving hair follicles)

Fig. 7.5 KOH positive for dermatophytes

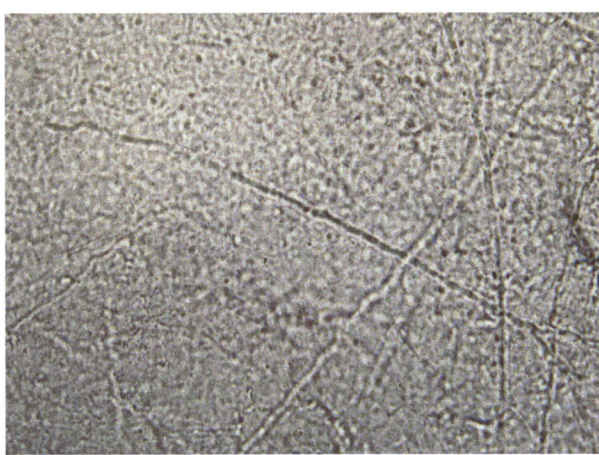

Candidiasis
- Immunocompromised patients are at higher risk for oral candidiasis
- Patients with frequent episodes may need thrush prophylaxis (fluconazole, nystatin, etc.)
- May also present as intertrigo, vaginitis, balanitis, paronychia [17]
- Disseminated candidiasis also can occur in setting of immunodeficiency and invasive candidiasis is the leading cause of mycosis-associated mortality in the United States [18]

Opportunistic Fungal Infections
- Most common in neutropenic patients with hematologic malignancies, patients with history of bone marrow transplantation (BMT)

Cryptococcosis (*Cryptococcus neoformans*)
- Found in pigeon droppings; transmission through inhalation
- Disseminated disease from primary pulmonary infection in immunocompromised hosts
- Cutaneous disseminated disease: subcutaneous nodules, ulcerated plaques, umbilicated papules (molluscum contagiosum-like lesions)
- Mucoid capsule stains red with mucicarmine; India ink stains yeast in CSF
- **Tx:** amphotericin B ± flucytosine

Fungal Sinusitis with Cutaneous Extension
- Risk factors: poorly controlled diabetes mellitus, neutropenia (especially s/p BMT) [19]
- Most common causative organisms: *Aspergillus*, *Rhizopus*, mucormycetes
- Symptoms: fever, cough, headache, nasal congestion; cutaneous extension may present as red edematous plaques on cheeks, nasal sidewall, or eyelids (± periorbital edema, proptosis); may be misdiagnosed as facial cellulitis [17]
- Work-up: CT or MRI, skin biopsy and tissue cultures
- *Aspergillus* spp.: septate hyphae with acute-angle (<90°) branching
- *Mucor* spp.: non-septate hyphae with wide-angle (>90°) branching
- **Tx:**
  - Aspergillosis (*Aspergillus* spp.): voriconazole, amphotericin B, caspofungin, itraconazole
  - Mucormycosis (*Rhizopus* spp., *Mucor* spp.): amphotericin B, surgical debridement

Cutaneous Disseminated Fungal Disease
- Most common causative organisms: *Candida*; *Aspergillus*, *Fusarium*, *Rhizopus*, *Mucor* spp.
- Presentation: red macules/plaques with central purpura → may develop central eschar

- Angioinvasive fungi will involve blood vessels and cause thrombosis, necrosis
- 75% of patients with *Fusarium* fungemia may manifest skin lesions [17]
- Diagnosis: biopsy and tissue cultures
- Histopathology:
  - *Aspergillus and Fusarium* spp.: septate hyphae with acute-angle (<90°) branching
  - Mucormycetes: broad hyphae without septations, with wide-angle (>90°) branching
- Additional work-up if disseminated fungal infection is suspected: blood fungal cultures, high-resolution chest CT, bronchoscopy, serum markers (galactomannan, beta-D-glucan)
- **Tx:**
  - Aspergillosis (*Aspergillus* spp.), Mucormycosis (*Rhizopus* spp., *Mucor* spp.): same as above
  - Fusariosis (*Fusarium* spp.): voriconazole

Cutaneous Infections Associated with Specific Cancer Therapeutics

Taxanes (Docetaxel, Paclitaxel), Antimetabolites (Capecitabine), Topoisomerase II Inhibitors (Etoposide, Teniposide, Amsacrine)

- Acute exudative paronychia with onycholysis and onychomadesis, that may progress to subungual abscesses [20, 21]
  - **Tx:** topical/oral antibiotic therapy, I&D if indicated

Epidermal Growth Factor Receptor (EGFR) Inhibitors

- Monoclonal antibodies to EGFR (cetuximab, panitumumab), small molecule tyrosine kinase inhibitors (erlotinib, gefitinib, osimertinib), dual kinase inhibitors of EGFR and HER2 (lapatinib, neratinib, afatinib)
- EGFR inhibits IL-1 dependent inflammation of the hair follicle [22]
  - Papulopustular eruption: folliculocentric, pruritic papules and pustules involving the head/neck, trunk, and proximal upper extremities which are initially aseptic but can become superinfected with bacteria (most commonly *S.aureus*)
    - **Tx:** low potency topical steroids, topical clindamycin, oral tetracyclines, low-dose isotretinoin (20–30 mg/day)
  - Paronychia
    - Nail fold inflammation, may also lead to pyogenic granuloma-like lesions

○ Can be superinfected with bacteria or fungi
○ **Tx:** Antiseptic soaks, topical steroids, topical or systemic antibiotics/anti-fungals, silver nitrate [22]

MEK Inhibitors, mTOR Inhibitors

• MEK inhibitors: trametinib, cobimetinib, binimetinib
• mTOR inhibitors: rapamycin, sirolimus, temsirolimus, everolimus
• Papulopustular eruption, paronychia (side effect profile similar to EGFR inhibitors)
• **Tx:** same as above for EGFR inhibitors

Infections Associated with Cutaneous Malignancies

• 15–20% of cancers are associated with viral infections [23]

Squamous Cell Carcinoma (SCC)

• Higher incidence of cutaneous squamous cell carcinoma in patients with Sezary syndrome, chronic lymphocytic leukemia [24, 25]
• Hematopoietic cell and solid organ transplantation patients: ↑ risk of SCCs [26, 27]
• High-risk HPV subtypes (HPV 16, HPV 18) are associated with oropharyngeal and anogenital SCCs [28]

Epstein-Barr Virus (EBV)

• Associated with lymphoproliferative disorders including endemic Burkitt's lymphoma, Hodgkin's lymphoma, natural killer/T-cell lymphoma (NKTL), nasopharyngeal carcinoma, post-transplant lymphoproliferative disease (PTLD), diffuse large B-cell lymphoma (DLBCL) of the elderly, lymphomatoid granulomatosis [23]
• Cutaneous lesions of PTLD: violaceous papules or plaques that resemble leukemia cutis
• NKTL: aggressive malignancy that most commonly involves the nose; primarily occurs in East Asia and Latin America
• DLBCL of the elderly
  – Occurs in patients >50 years of age
  – EBV+ in 10–15% of cases in Asia and South America, in 5% of cases in Western countries [29]
  – Skin is the most common extranodal site involved
• Lymphomatoid granulomatosis

- EBV-driven lymphoproliferative disorder, typically in patients with immunodeficiency (iatrogenic immunosuppression, HIV, primary immunodeficiency) [30]
- Cutaneous lesions can present as plum-colored nodules with ulceration, often disseminated
- Pulmonary involvement (>90%), CNS involvement

Human Herpes Virus 8 (HHV8)

- Causative agent of Kaposi sarcoma
  - Four types of KS: Classic (extremities of elderly men in Mediterranean countries), AIDS-related, iatrogenic (related to immunosuppression), endemic (young patients in sub-Saharan Africa)
  - Clinical presentation can vary from purpuric patches/plaques to violaceous nodules (Fig. 7.6)
  - Frequently involves head/neck, mucosa, gastrointestinal tract
- Associated with Castleman's disease, primary effusion lymphoma

Polyomavirus

- Merkel cell polyomavirus
  - Associated with Merkel cell carcinoma (MCC); present in 69–85% of MCC tumors [31]
  - MCC: rapidly growing red to violaceous nodule on sun-exposed areas
  - ↑ risk in immunosuppressed patients (s/p organ transplant, chronic lymphocytic leukemia)

Fig. 7.6 Kaposi sarcoma

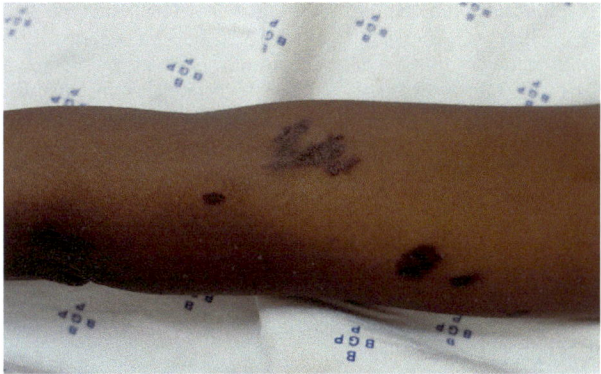

Review Questions and Answers

1. What is the most sensitive method of diagnosing HSV infection?
   (a) Polymerase chain reaction (PCR)
   (b) Tzanck smear
   (c) Viral culture
   (d) Viral serologies

   Answer: (a) Polymerase chain reaction (PCR)

2. Which of these is the initial treatment of choice for disseminated cryptococcosis in an immunocompromised patient?
   (a) Amphotericin B
   (b) Fluconazole
   (c) Itraconazole
   (d) Voriconazole

   Answer: (a) Amphotericin B

3. Which HPV subtype is most frequently associated with oropharyngeal SCCs?
   (a) HPV 1
   (b) HPV 6
   (c) HPV 11
   (d) HPV 16

   Answer: (d) HPV 16

References

1. Kofteridis DP, Valachis A, Koutsounaki E, Maraki S, Mavrogeni E, Economidou FN, et al. Skin and soft tissue infections in patients with solid tumours. Sci World J. 2012;2012:804518.
2. Freifeld AG, Bow EJ, Sepkowitz KA, Boeckh MJ, Ito JI, Mullen CA, et al. Clinical practice guideline for the use of antimicrobial agents in neutropenic patients with cancer: 2010 update by the Infectious Diseases Society of America. Clin Infect Dis. 2011;52(4):427–31.
3. Stevens DL, Bisno AL, Chambers HF, Dellinger EP, Goldstein EJ, Gorbach SL, et al. Practice guidelines for the diagnosis and management of skin and soft tissue infections: 2014 update by the infectious diseases society of America. Clin Infect Dis. 2014;59(2):147–59.
4. Grunwald MR, McDonnell MH, Induru R, Gerber JM. Cutaneous manifestations in leukemia patients. Semin Oncol. 2016;43(3):359–65.
5. Vaiman M, Lazarovitch T, Heller L, Lotan G. Ecthyma gangrenosum and ecthyma-like lesions: review article. Eur J Clin Microbiol Infect Dis. 2015;34(4):633–9.
6. Lai CC, Tan CK, Cheng A, Chung KP, Chen CY, Liao CH, et al. Nontuberculous mycobacterial infections in cancer patients in a medical center in Taiwan, 2005-2008. Diagn Microbiol Infect Dis. 2012;72(2):161–5.
7. Henkle E, Winthrop KL. Nontuberculous mycobacteria infections in immunosuppressed hosts. Clin Chest Med. 2015;36(1):91–9.
8. Lau CH, Missotten T, Salzmann J, Lightman SL. Acute retinal necrosis features, management, and outcomes. Ophthalmology. 2007;114(4):756–62.

9. Rubin LG, Levin MJ, Ljungman P, Davies EG, Avery R, Tomblyn M, et al. 2013 IDSA clinical practice guideline for vaccination of the immunocompromised host. Clin Infect Dis. 2014;58(3):e44–100.

10. Fasanya AA, Pedersen FT, Alhassan S, Adjapong O, Thirumala R. Cytomegalovirus Cutaneous Infection in an Immunocompromised Patient. Cureus. 2016;8(5):e598.

11. Hall LD, Eminger LA, Hesterman KS, Heymann WR. Epstein–Barr virus: Dermatologic associations and implications: Part I. Mucocutaneous manifestations of Epstein–Barr virus and nonmalignant disorders. J Am Acad Dermatol. 2015;72(1):1–19.

12. Brasileiro CB, Abreu M, Mesquita RA. Critical review of topical management of oral hairy leukoplakia. World J Clin Cases. 2014;2(7):253–6.

13. Rouanet J, Aubin F, Gaboriaud P, Berthon P, Feltkamp MC, Bessenay L, et al. Trichodysplasia spinulosa: a polyomavirus infection specifically targeting follicular keratinocytes in immunocompromised patients. Br J Dermatol. 2016;174(3):629–32.

14. Wanat KA, Holler PD, Dentchev T, Simbiri K, Robertson E, Seykora JT, et al. Viral-associated trichodysplasia: characterization of a novel polyomavirus infection with therapeutic insights. Arch Dermatol. 2012;148(2):219–23.

15. Santesteban R, Feito M, Mayor A, Beato M, Ramos E, de Lucas R. Trichodysplasia Spinulosa in a 20-month-old girl with a good response to topical cidofovir 1%. Pediatrics. 2015;136(6):e1646–9.

16. Aleissa M, Konstantinou MP, Samimi M, Lamant L, Gaboriaud P, Touze A, et al. Trichodysplasia spinulosa associated with HIV infection: clinical response to acitretin and valganciclovir. Clin Exp Dermatol. 2018;43(2):231–3.

17. Mays SR, Bogle MA, Bodey GP. Cutaneous fungal infections in the oncology patient: recognition and management. Am J Clin Dermatol. 2006;7(1):31–43.

18. Shields BE, Rosenbach M, Brown-Joel Z, Berger A, Ford BA, Wanat KA. Angioinvasive fungal infections impacting the skin: background, epidemiology, and clinical presentation (part 1). J Am Acad Dermatol. 2019;80(4):869–880.e5.

19. Iwen PC, Rupp ME, Hinrichs SH. Invasive mold sinusitis: 17 cases in immunocompromised patients and review of the literature. Clin Infect Dis. 1997;24(6):1178–84.

20. Payne AS, James WD, Weiss RB. Dermatologic toxicity of chemotherapeutic agents. Semin Oncol. 2006;33(1):86–97.

21. Reyes-Habito CM, Roh EK. Cutaneous reactions to chemotherapeutic drugs and targeted therapies for cancer: part I. Conventional chemotherapeutic drugs. J Am Acad Dermatol. 2014;71(2):203.e1–203.e12. quiz 15–6.

22. Macdonald JB, Macdonald B, Golitz LE, LoRusso P, Sekulic A. Cutaneous adverse effects of targeted therapies: part I: inhibitors of the cellular membrane. J Am Acad Dermatol. 2015;72(2):203–18. quiz 19–20.

23. McLaughlin-Drubin ME, Munger K. Viruses associated with human cancer. Biochimica et Biophysica Acta. 2008;1782(3):127–50.

24. Scarisbrick JJ, Child FJ, Evans AV, Fraser-Andrews EA, Spittle M, Russell-Jones R. Secondary malignant neoplasms in 71 patients with sézary syndrome. Arch Dermatol. 1999;135(11):1381–5.

25. Kaplan AL, Cook JL. Cutaneous squamous cell carcinoma in patients with chronic lymphocytic leukemia. Skinmed. 2005;4(5):300–4.

26. DePry JL, Vyas R, Lazarus HM, Caimi PF, Gerstenblith MR, Bordeaux JS. Cutaneous malignant neoplasms in hematopoietic cell transplant recipients: a systematic review. JAMA Dermatol. 2015;151(7):775–82.

27. Cheng JY, Li FY, Ko CJ, Colegio OR. Cutaneous squamous cell carcinomas in solid organ transplant recipients compared with immunocompetent patients. JAMA Dermatol. 2018;154(1):60–6.

28. Castellsagué X, Alemany L, Quer M, Halec G, Quirós B, Tous Belmonte S, et al. HPV involvement in head and neck cancers: comprehensive assessment of biomarkers in 3680 patients. J Natl Cancer Inst. 2016;108(6):djv403.

29. Castillo JJ, Beltran BE, Miranda RN, Young KH, Chavez JC, Sotomayor EM. EBV-positive diffuse large B-cell lymphoma, not otherwise specified: 2018 update on diagnosis, risk-stratification and management. Am J Hematol. 2018;93(7):953–62.
30. Hristov AC. Primary cutaneous diffuse large B-cell lymphoma, leg type: diagnostic considerations. Arch Pathol Lab Med. 2012;136(8):876–81.
31. Houben R, Shuda M, Weinkam R, Schrama D, Feng H, Chang Y, et al. Merkel cell polyomavirus-infected Merkel cell carcinoma cells require expression of viral T antigens. J Virol. 2010;84(14):7064–72.

Dermatologic Adverse Effects of Anticancer Therapy I: General Principles

8

Timothy Dang, Vincent Liu, and Bernice Kwong

Abbreviations

ACTH	Adrenocorticotropic hormone
AFND	Acute febrile neutrophilic dermatosis (Sweet syndrome)
AGEP	Acute generalized exanthematous pustulosis
AIDS	Acquired immunodeficiency syndrome
ALK	Anaplastic lymphoma kinase
ALL	Acute lymphoblastic leukemia
AML	Acute myeloid leukemia
APML	Acute promyelocytic leukemia
ATG	Antithymocyte globulin
ATRA	All-*trans* retinoid acid
BTK	Bruton tyrosine kinase
CLL	Chronic lymphocytic leukemia
CML	Chronic myelogenous leukemia
CMV	Cytomegalovirus
CNS	Central nervous system
CTCL	Cutaneous T-cell lymphoma
DAE	Dermatologic adverse effect
DMSO	Dimethyl sulfoxide

T. Dang (✉)
Santa Clara Valley Medical Center, San Jose, CA, USA
e-mail: timdang@stanford.edu

V. Liu
Department of Dermatology, University of Iowa Health Care, Iowa, IA, USA
e-mail: vincent-liu@uiowa.edu

B. Kwong
Stanford Health Care; Stanford Medicine, Palo Alto, CA, USA
e-mail: bernicek@stanford.edu

© Springer Nature Switzerland AG 2021
V. Liu (ed.), *Dermato-Oncology Study Guide*,
https://doi.org/10.1007/978-3-030-53437-0_8

DRESS   Drug reaction with eosinophilia and systemic symptoms
FGFR   Fibroblast growth factor receptor
GM-CSF Granulocyte macrophage colony-stimulating factor
GVHD   Graft versus host disease
HBV   Hepatitis B virus
HHV-8   Human herpesvirus-8
HIF   Hypoxia-inducible factor
HIV   Human immunodeficiency virus
HPV   Human papillomavirus
HSV   Herpes simplex virus
IDH-2   Isocitrate dehydrogenase-2
IL   Interleukin
ILP   Isolated limb perfusion
KA   Keratoacanthoma
mTOR   Mechanistic target of rapamycin
nbUVB   Narrow band ultraviolet B
NEH   Neutrophilic eccrine hidradenitis
NSAID   Nonsteroidal anti-inflammatory drug
PABA   Para-aminobenzoic acid
PARP   Poly (ADP-ribose) polymerase
PCR   Polymerase chain reaction
PDGFR   Platelet-derived growth factor receptor
PDT   Photodynamic therapy
PGF   Placental growth factor
PI3K   Phosphatidylinositol 3-kinase
PIH   Postinflammatory hyperpigmentation
PKC   Protein kinase C
RANKL   Receptor activator of nuclear factor kappa-B ligand
SCC   Squamous cell carcinoma
SCF   Stem cell factor
SCLE   Subacute cutaneous lupus erythematosus
SDRIFE Symmetrical drug-related intertriginous and flexural exanthema
SJS   Stevens-Johnson syndrome
SLAMF7 Signaling lymphocytic activation molecule family member 7
SLE   Systemic lupus erythematosus
SLL   Small lymphocytic lymphoma
SNRI   Serotonin-norepinephrine reuptake inhibitor
SSRI   Selective serotonin reuptake inhibitor
TEN   Toxic Epidermal Necrolysis
TLGLL   T-cell large granular lymphocytic leukemia
T-PLL   T-cell prolymphocytic leukemia
UPF   Ultraviolet protection factor
UVR   Ultraviolet radiation
VEGF   Vascular endothelial growth factor
VEGFR   Vascular endothelial growth factor receptor
VZV   Varicella zoster virus

Learning Objectives

1. To describe the common dermatologic symptoms frequently experienced by oncology patients.
2. To outline an approach to management of common dermatologic conditions encountered by cancer patients.
3. To detail the risks and benefits of both skin-directed and systemic therapies in management of frequent skin, hair, and nail complications of cancer therapy.

Introduction

More people with a history of cancer are living today than ever before as a result of an increased number of cancer diagnoses in combination with increased survival rates. This population will continue to increase in size, especially as our anticancer armamentarium continues to evolve. These individuals face a diverse range of dermatologic challenges that includes paraneoplastic phenomena, cutaneous metastases, and adverse effects of anticancer therapy and hematologic transplant. The goal of supportive oncodermatology is to provide early and effective intervention to improve patient quality of life while maintaining optimal anticancer therapy [1].

General Principles of Management

General Principles for Steroid Therapy

Introduction
- Topical and systemic steroids are frequently used to treat dermatologic conditions, including the adverse effects of anticancer therapies. Erythematous rashes (e.g., maculopapular rashes, injection site reactions, fixed drug eruptions) are often associated with underlying inflammation in the skin and may respond to the anti-inflammatory effects of steroid therapy. Though short-term steroid therapy is typically well-tolerated, judicious consideration of steroid vehicle, potency, and treatment duration is necessary to avoid adverse effects.

Topical Steroids
- **Vehicle**
  - Topical steroid vehicles differ in their relative composition of powders, oils, and liquids. The properties of different vehicles should be considered to optimize efficacy, tolerability, and adherence to treatment. For patients who require multiple topical agents, the selection of similar vehicle agents facilitates the ease of applying medications. Application under occlusive dressings further increases the effectiveness of topical medications by increasing percutaneous drug absorption (Table 8.1).

- **Potency**
  - Topical steroids are subdivided into different potency groups that vary from super-high potency (group 1) to least potent (group 7). Factors that should be considered when selecting different topical steroids include region of the body and severity of the condition treated. As a general rule, regions of the body with thick skin (e.g., the palms and soles) warrant higher potency topical steroids, while regions of the body with thin skin (e.g., the face) merit lower potency topical steroids to optimize efficacy and avoidance of adverse effects.
- **Duration**
  - Topical steroids should be used for the minimum duration necessary to achieve effective results. Super-high potency steroids should not be used for >3 weeks of treatment if possible. Medium to high potency steroids should not be used for >8 weeks of treatment if possible. Low potency steroids do not typically cause adverse effects but can when used persistently on normal to near-normal skin.
  - Rebound flares of rashes can occur with discontinuation of topical steroid therapy; these rebound flares can be reduced with tapering of potency and frequency of use. For dermatologic conditions that necessitate long-term use of topical anti-inflammatory therapy, consider use of topical calcineurin inhibitors (e.g., tacrolimus, pimecrolimus), either with or without concurrent topical steroid therapy (Table 8.1).

Table 8.1 Vehicle comparison of topical steroids

Vehicle	Characteristics	Advantages	Disadvantages
Ointment	• ~80% oil, ~20% water	• Effective skin emollient • Occlusive effect enhances medication absorption • Ideal for hair-free areas and areas of thick skin, including the palms and soles	• Greasiness can be uncomfortable and/or cosmetically intolerable • Difficult to use in hair-bearing areas
Cream	• ~50% oil, ~50% water	• More` comfortable and cosmetically appealing than ointments and oils	• Less potent than ointments and oils due to minimal occlusive effect • Contains preservatives that can cause contact dermatitis
Oil	• Predominantly oil formulation; can contain small amounts of alcohol solution	• Easy to apply in hair-bearing areas • Occlusive effect enhances medication absorption	• Greasiness can be uncomfortable and/or cosmetically intolerable
Solution	• ~25% alcohol solution	• Easy to apply in hair-bearing areas	• Alcohol base can be irritating

Systemic Corticosteroids

- Systemic steroids (e.g., oral and/or intravenous formulations) are typically reserved for severe inflammatory disorders, including dermatologic conditions that affect a large body surface area.
- Oral steroids are typically dosed once daily in the morning to minimize suppression of the hypothalamic–pituitary–adrenal axis. Split-dosing typically increases both efficacy and the risk of developing adverse effects. Low-dose therapies (e.g., <20 mg prednisone) and short-term (e.g., <3 weeks) therapies do not typically require drug tapering to facilitate adrenal recovery.
- The decision to drug taper should balance the risk of rebound flares of the rash, adrenal recovery, and adverse effects of long-term systemic steroid therapy. At doses > 60 mg per day, prednisone can typically be tapered in 20 mg increments. At doses 30–60 mg per day, prednisone can typically be tapered in 10 mg increments. At doses < 30 mg per day, tapering by 5 mg per day (or even more gradually) may be necessary. Alternating between doses every other day can further facilitate successful steroid taper (Table 8.2).

Table 8.2 Common adverse effects of chronic corticosteroid therapy

Dermatologic	• Skin atrophy • Ecchymoses • Cushingoid appearance (e.g., truncal obesity, round face) • Acne (typically monomorphic and lacks comedones) • Hirsutism • Facial erythema • Striae
Cardiovascular	• Fluid retention • Hypertension • Arrhythmias
Gastrointestinal	• Gastritis • Ulcer formation • Gastrointestinal bleeding
Musculoskeletal	• Osteoporosis
Metabolic	• Hyperglycemia • Adrenal insufficiency
Immunologic	• Immunosuppression
Hematologic	• Leukocytosis
Ophthalmologic	• Cataracts • Increased intraocular pressure
Neuropsychiatric	• Mood disorders (e.g., emotional lability, hypomania, mania) • Delirium • Memory disturbances • Sleep disturbances

Management of Alopecia

Incidence: up to 80%

Natural Course: hair loss within 2–3 weeks of first cycle; regrowth typically 3–6 months after last cycle

- **Clinical Features**:
  - Diffuse, patchy, or total alopecia at any hair-bearing area (Fig. 8.1)
  - Not typically painful, can have associated tingling or tenderness
  - Individuals hairs can thin
- **Additional Considerations**:
  - Persistent texture or color change (>33% incidence)
  - Persistent diffuse, non-scarring alopecia (>2% incidence)—counseling, rule-out other etiologies

Management
- Prior to chemotherapy:
  - Cut hair short (can reduce scalp irritation and psychosocial stress)
- During chemotherapy:
  - Scalp cooling: 50% reduction in maximal hair loss in 50–75% of cases
- After chemotherapy:
  - Topical minoxidil 5% (for scalp) and bimatoprost (for eyelashes): can accelerate rate of regrowth
  - Evaluate for other causes of alopecia (e.g., telogen effluvium, alopecia areata, androgenetic alopecia) [2]

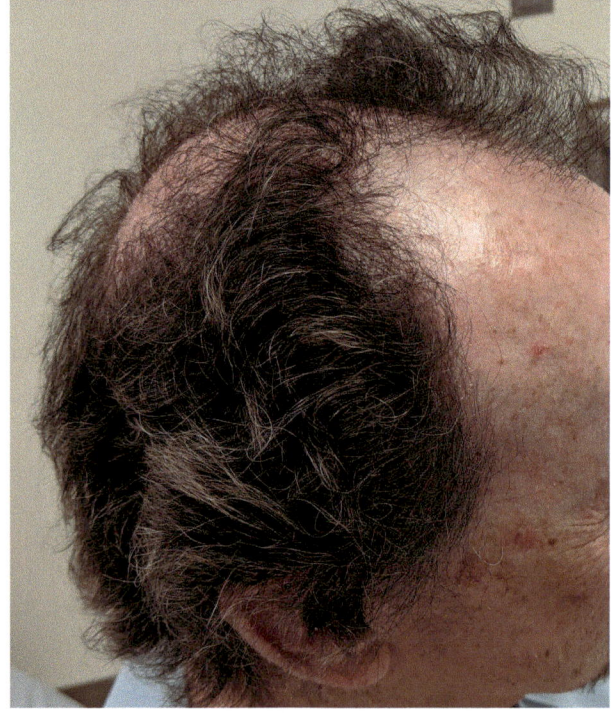

Fig. 8.1 Cancer-related alopecia

Management of Xerosis

Introduction
- Several anticancer therapies exacerbate xerosis. In addition, maintenance of a healthy skin barrier with frequent emolliation may reduce the risk of developing dermatologic adverse effects of anticancer therapies. Accordingly, it is prudent to practice diligent dry skin care measures. These strategies should balance effectiveness and practicality to optimize quality of life.

Emollients
- All skin should be moisturized at least once per day and more frequently if possible. Topical emollients should be applied immediately after water exposure, which tends to lead to net water loss in the skin if not followed by diligent moisturization. Hot water leads to more water loss than tepid or cool water and should be avoided if possible or followed with aggressive moisturization.
- Creams are more effective emollients than lotions, which contain more water content. However, consistent use of lotions may be more tolerable and thus overall more effective than inconsistent use of creams. In general, formulations with fewer ingredients (e.g., that lack fragrances) are better tolerated and have a lower risk of causing adverse effects such as contact dermatitis.

Additional Strategies
- The use of gloves when performing wet work can minimize water loss in the skin. Gloves can also provide occlusion that may enhance the penetration of moisturizers into the skin. Humidifiers can increase the humidity of indoor air to reduce xerosis. Mild cleansers (e.g., synthetic detergent cleansers) that have a low pH may be less irritating than other cleansers and exacerbate xerosis.

Fissure Care
- Cyanoacrylate adhesives can bind wound edges to repair fissures caused by persistent xerosis (e.g., in individuals who receive EGFR inhibitor therapy). In particular, wounds that are clean, under low tension, and approximated readily are more suitable for cyanoacrylate closure. Advantages of cyanoacrylate closure include wide availability (over-the-counter) and low cost of this intervention, reduction of pain associated with fissures, and the creation of a barrier to water and pathogens [3].

Management of Pruritus

- Nonpharmacologic management
  - Dry skin care—see section "Management of xerosis"
  - Cool environment—lightweight clothing, use of warm (not hot) water, air-conditioned environments
  - Physical intervention—keep nails short, occlude pruritic areas

- Local therapy
  - Topical corticosteroids—for inflamed areas, can use under occlusion
  - Topical capsaicin
  - Topical calcineurin inhibitors (e.g., tacrolimus, pimecrolimus)
  - Topical anesthetics—pramoxine cream, lidocaine and prilocaine cream, menthol cream
  - Topical amitriptyline-ketamine
  - Topical doxepin
- Systemic therapy
  - Antihistaminic therapy
  - Opioid receptor antagonists/agonists (naltrexone)
  - Oral doxepin
  - Gabapentin/pregabalin
  - Aprepitant (neurokinin receptor 1 antagonist)
  - Dronabinol
  - Mirtazapine
  - Duloxetine
  - Selective serotonin and norepinephrine reuptake inhibitor antidepressant (SSNRI)
  - Phototherapy

Management of Edema

- Nonpharmacologic management
  - Cool compresses
  - Compression therapy—compression stockings for legs
- Local therapy
  - Topical corticosteroids—for inflamed areas, can use under occlusion
  - Topical calcineurin inhibitors (e.g., tacrolimus, pimecrolimus)
- Systemic therapy
  - Diuretics as dictated by cardiovascular and renal status
  - Dietary salt restrictions as indicated

Management of Mucositis

- **Natural Course**: typically develops within days of initiating therapy and begins with erythema and a burning sensation, followed by desquamation; mucosal lesions typically heal 10–14 days after chemotherapy administration
- **Clinical Features**: can be mild (e.g., erythema, shallow ulceration) to severe (e.g., large, painful ulcerations that result in dysphagia)
- **Management**:
  - optimization of dentition prior to chemotherapy
  - oral cryotherapy (e.g., swishing of ice chips), dietary modification (e.g., avoidance of alcohol, acidic foods, and foods that can traumatize the oral mucosa),
  - topical agents (e.g., anesthetics, anti-inflammatory agents, analgesics, mucosal coating agents),

- treatment of superinfection (e.g., oral candidiasis, HSV infection);
- palifermin (a keratinocyte growth factor)
- low-level laser therapy may have some benefit, but their use is limited by cost and convenience
- **Additional Considerations**: "magic" or "miracle" mouthwashes are commonly prescribed; formulations differ between institutions and can contain topical anesthetic, antacid suspension, diphenhydramine, nystatin, steroids, and antibiotics; data are insufficient to recommend any specific mouthwash

Management of Photosensitivity

Introduction
- Several anticancer therapies and drugs used to manage complications of anticancer therapy are photosensitizing and/or cause dermatologic adverse effects that are exacerbated by sun exposure. Accordingly, it is prudent to practice diligent photoprotective measures. These strategies should balance effectiveness and practicality to optimize quality of life.

Management
- Avoidance of sun during peak hours (between 10 a.m. and 4 p.m.)
- Photoprotective clothing
  - Long-sleeved shirts and pants, broad-brim hats
  - Tightly woven and thick fabrics, dark colors
  - Commercially available products with UPF
- Sunscreen (see Table 8.3)
  - Inorganic (physical) sunscreens tend to be more tolerable for patients with inflamed skin.

Table 8.3 Characteristics of sunscreens

Type	Organic (chemical) sunscreens	Inorganic (physical) sunscreens
Components	• Avobenzone • Octocrylene • Salicylates (e.g., octisalate, homosalate) • Benzophenones (e.g., oxybenzone, sulisobenzone) • PABA derivatives • Cinnamates (e.g., octinoxate, cinoxate)	• Zinc oxide • Titanium dioxide
Mechanism	• Absorbs UVR	• Reflects and scatters UVR and visible radiation
Advantages	• More cosmetically appealing • Requires less frequent reapplication	• Less irritating and sensitizing
Disadvantages	• Can cause an irritated or burning sensation on sensitive skin • Can rarely cause allergic contact dermatitis • Recent question of absorption and impact on environment	• Opacity can be cosmetically unappealing for patients • Micronized formulations are more cosmetically appealing but less capable of scattering radiation • Requires more frequent reapplication

- Sunscreen-containing cosmetic products can increase overall use of sunscreens.
- The amount of UVB radiation absorbed by SPF 15, 30, and 50 sunscreens is 93%, 97%, and 98%, respectively.
    Consistent use of a lower SPF sunscreen (which may be more tolerable) may increase overall photoprotection compared to a higher SPF sunscreen that is used less frequently.
- Proper use:
    1 teaspoon to the face and neck, 1 teaspoon to the frontal torso, 1 teaspoon to the back torso, 1 teaspoon to each upper extremity, 2 teaspoons to each lower extremity
    Application 15–30 min before sun exposure to allow formation of protective film on the skin
    Wait 10–20 min after application before dressing
    Reapplication at least every 2 h and after water exposure
- Window films that block UVA, which can penetrate window glass (indoors, cars)

Management of Nail Changes

Introduction

- Several anticancer therapies cause nail changes, which range from asymptomatic concerns to dose-limiting toxicities that severely impact quality of life. Strategies that protect the nail may reduce the risk of developing dermatologic adverse effects of anticancer therapies. Accordingly, it is prudent to practice diligent general nail care. These strategies should balance effectiveness and practicality to optimize quality of life (Table 8.4).

Natural Course of Nail Changes

- Nail changes that develop as a result of anticancer therapy typically manifest within weeks to months of initiating treatment. This delay in clinical manifestations is likely due to the slow growth of the nail plate. For example, cytotoxic chemotherapies can arrest nail growth and result in Beau lines (Fig. 8.2). However, the proximal end of the nail plate lies deep to the proximal nail fold and is not visible. Thus, nail changes will not be apparent until nail growth returns and the involved portion of the nail grows distal to the cuticle where it becomes visible.
- Unless there is scarring to the nail matrix (e.g., from severe infection), nails typically regrow normally. On average, fingernail regrowth takes approximately 6 months; toenail growth takes 12–18 months. Older age, vascular disease, and systemic illness can slow the rate of nail regrowth.

Management

- General nail care measures include avoidance of mechanical trauma and wet work as much as possible. Unnecessary manipulation of the nail and cuticle (e.g.,

Table 8.4 Types of nail changes

Onychodystrophy	Any abnormality of nail morphology
Onychorrhexis	Distal, longitudinal splitting or ridging of the nail
Onychoschizia	Horizontal splitting of the nail
Onychomadesis	Detachment of the nail plate from the proximal nail fold
Onycholysis	Separation of the nail plate from the hyponychium, nail bed, and/or lateral nail folds
Paronychia	Inflammation of the lateral and/or proximal nail folds
Trachyonychia	Roughness, brittleness, longitudinal ridging, and pitting of the nail; can have associated thickening of the cuticle
Koilonychia	Upward curvature of the distal nail
Leukonychia	White opacity of the nail; can manifest over the entirety of the nail, in transverse striae (Mees lines), or with punctate lesions
Melanonychia	Pigmentation of the nail from melanin
Beau lines	Transverse grooves in the nail caused by temporary arrest of nail growth
Splinter hemorrhages	Small, thin, longitudinal, red to black lines under the nail from rupture of nail bed capillaries
Subungual hemorrhage	Bleeding underneath the nail; can cause pain and pressure that results in onycholysis
Nail pitting	Punctate pits in the nail

Fig. 8.2 Beau lines

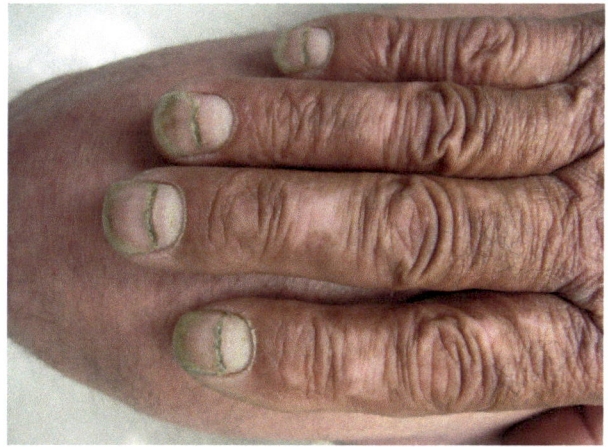

from manicuring) should be avoided. Gloves can be worn during wet work to minimize trauma to the nails [4].

- Biotin supplementation (5 mg daily) can possibly strengthen and thicken nails. Of note, biotin is used in many laboratory tests because it can bond with specific proteins (e.g., hormones, troponin); individuals who take regular biotin supplementation should be cautioned that biotin supplementation may result in false test results [5].
- Protective nail lacquers can be applied to further strengthen the nail and protect from trauma. Though dark coats may protect nails from phototoxic

adverse effects, individuals should be cautioned that this may reduce the ability to visually detect early, asymptomatic nail changes. Nail polish removal can exacerbate nail brittleness and irritate nearby skin. Use of nail polish removal should be avoided; new coats of nail lacquer can be applied over older layers.

- Compromise of the nail plate creates a method of entry for pathogens. Antiseptic soaks may reduce the risk of developing infection. Common recommendations include dilute bleach soaks and dilute vinegar soaks [6]. Prophylactic antimicrobial therapy (e.g., mupirocin ointment, gentamicin solution, ciclopirox solution) may also be useful. However, evidence that supports the use of antiseptic soaks and antimicrobial therapy is lacking. The decision to start therapy should be contextualized with the potential risk of breeding antimicrobial-resistant bacteria. All of these therapies can be drying to the skin and should be followed with frequent moisturization. If there is evidence of infection (e.g., exudate, purulence), a wound culture should be obtained, and empiric antimicrobial therapy should be begun. In severe cases of nail pathology, partial or total nail avulsion can relieve pain (e.g., from pressure underneath the nail, ingrown nail) and facilitate topical treatment.

Management of Extravasation Reactions [7]

Introduction

- Extravasation of chemotherapy agents during infusions, occurring in approximately 0.1–6% of patients receiving chemotherapy [8] may induce a spectrum of irritant to vesicant reactions, depending upon the nature and degree of toxicity and necrosis triggered by the extravasated agent. Irritant reactions reflect local inflammation, manifesting pain, erythema, swelling, and phlebitis. Vesicant reactions produce blistering. Predisposition to injurious extravasation may be attributed to the nature of the agent (e.g. vasodilating, etc.), degree of perfusion, and other host factors, such as diabetes, etc.
- Nonpharmacologic management
  - Stop infusion immediately
  - Aspirate drug via cannula
  - Enhance circulation (remove constricting bands)
  - Immobilize limb and raise above heart level
- Local therapy
  - Keep limb warm
  - Caution with cold packs- need to avoid with vinca alkaloids as can cause ulceration
  - Local antidotes (e.g. sodium thiosulfate to counteract mechlorethamine cisplatin; desrazoxane; topical DMSO, etc.)
  - Surgical considerations (debridement)

Review Questions and Answers

1. What of the following are measures to treat an extravasation reaction?
   (a) Antidotes, as indicated by the nature of the extravasated agent
   (b) Surgical debridement
   (c) Aspiration via cannula following discovery of extravasation
   (d) Avoidance of ice packs with vinca alkaloid extravasation
   (e) All of the above

   Answer: (e) All of the above

2. The following is not an option for management of pruritus in cancer patients
   (a) Phototherapy
   (b) Gabapentin
   (c) Morphine
   (d) Mirtazapine
   (e) Doxepin

   Answer: (c) Morphine

3. The following are components for management of mucositis in cancer patients
   (a) Evaluation and treatment of Candida, if presen
   (b) Evaluation and treatment of herpesvirus infection, if present
   (c) Topical corticosteroids
   (d) Topical analgesics
   (e) All of the above

   Answer: (c) Topical corticosteroids

References

1. Lacouture ME. Dermatologic principles and practice. In: Oncology: conditions of the skin, hair, and nails in cancer patients. 1st ed. Hoboken: Wiley; 2014. isbn:978-0-470-62188-2.
2. Freites-Martinez A, Shapiro J, Chan D, et al. Endocrine therapy–induced alopecia in patients with breast cancer. JAMA Dermatol. 2018;154(6):670–5. https://doi.org/10.1001/jamadermatol.2018.0454. PMID: 29641806.
3. Vlahovic TC, Hinton EA, Chakravarthy D, Fleck CA. A review of cyanoacrylate liquid skin protectant and its efficacy on pedal fissures. J Am Col Certif Wound Spec. 2011;2(4):79–85. https://doi.org/10.1016/j.jcws.2011.02.003. PMID: 24527155.
4. Dimitris R, Ralph D. Management of simple brittle nails. Dermatol Ther. 2012;25(6):569–73. https://doi.org/10.1111/j.1529-8019.2012.01518.x. PMID: 23210755.
5. Willeman T, Casez O, Faure P, Gauchez AS. Evaluation of biotin interference on immunoassays: new data for troponin I, digoxin, NT-Pro-BNP, and progesterone. Clin Chem Lab Med. 2017;55(10):e226–9. https://doi.org/10.1515/cclm-2016-0980. PMID: 28222017.

6. Lacouture ME, Anadkat MJ, Bensadoun RJ, Bryce J, Chan A, Epstein JB, Eaby-Sandy B, Murphy BA, MASCC Skin Toxicity Study Group. Clinical practice guidelines for the prevention and treatment of EGFR inhibitor-associated dermatologic toxicities. Support Care Cancer. 2011;19(8):1079–95. https://doi.org/10.1007/s00520-011-1197-6. PMID: 21630130.
7. Boulanger J, Ducharme A, Dufour A, Fortier S, Almanric K, Comité de l'évolution de la pratique des soinspharmaceutiques (CEPSP); Comité de l'évolution des pratiquesenoncologie (CEPO). Management of the extravasation of anti-neoplastic agents. Support Care Cancer. 2015;23(5):1459–71. https://doi.org/10.1007/s00520-015-2635-7. PMID: 25711653.
8. Kreidieh FY, Moukadem HA, El Saghir NS. Overview, prevention and management of chemotherapy extravasation. World J Clin Oncol. 2016;7(1):87–97. https://doi.org/10.5306/wjco.v7.i1.87. PMID: 26862492.

Dermatologic Adverse Effects of Anticancer Therapy II: Cytotoxic Agents

9

Timothy Dang, Hannah Thompson, Vincent Liu, and Bernice Kwong

Learning Objectives

- To list the clinical indications (approved and off-label) for each anticancer therapy class.
- To recognize dermatologic adverse effects for each anticancer therapy class.
- To describe the incidence, natural course, clinical features, and management of DAEs from each therapy class summarized in the chapter.

Cytotoxic Therapies

Platinum Agents (Table 9.1)

Dermatologic Adverse Effects
- Maculopapular rash (Fig. 9.1)
- Alopecia [1]
- Xerosis
- Pruritus

T. Dang (✉)
Santa Clara Valley Medical Center, San Jose, CA, USA
e-mail: timdang@stanford.edu

H. Thompson
University of Iowa Health Care, Iowa City, IA, USA
e-mail: hannah-j-thompson@uiowa.edu

V. Liu
Department of Dermatology, University of Iowa Health Care, Iowa, IA, USA
e-mail: Vincent-liu@uiowa.edu

B. Kwong
Stanford Health Care, Stanford Medicine, Palo Alto, CA, USA
e-mail: bernicek@stanford.edu

© Springer Nature Switzerland AG 2021
V. Liu (ed.), *Dermato-Oncology Study Guide*,
https://doi.org/10.1007/978-3-030-53437-0_9

Table 9.1 Platinum cytotoxic agents

Agents	Indications
Cisplatin	• Approved
	– Bladder cancer
	– Ovarian cancer
	– Testicular cancer
	• Off-label
	– Anal carcinoma
	– Blood cancers (e.g., Hodgkin lymphoma, non-Hodgkin lymphoma, multiple myeloma)
	– Breast cancer
	– CNS tumors
	– Germ cell tumors
	– Esophageal cancer
	– Gastric cancer
	– Gestational trophoblastic neoplasia
	– Gynecologic cancers (e.g., cervical, endometrial)
	– Head and neck cancer
	– Hepatobiliary cancer
	– Lung cancers (e.g., non-small cell lung cancer, small cell lung cancer)
	– Melanoma
	– Mesothelioma
	– Medulloblastoma
	– Neuroblastoma
	– Neuroendocrine tumors
	– Osteosarcoma
	– Pancreatic cancer
	– Penile cancer
	– Prostate cancer
	– Thymoma and thymic malignancies
Carboplatin	• Approved
	– Ovarian cancer
	• Off-label
	– Anal cancer
	– Bladder cancer
	– Blood cancers (e.g., Hodgkin lymphoma, non-Hodgkin lymphoma)
	– Breast cancer
	– CNS tumors
	– Esophageal cancer
	– Gastric cancer
	– Gynecologic cancers (e.g., cervical, endometrial)
	– Head and neck cancer
	– Lung cancers (e.g., non-small cell lung cancer, small cell lung cancer)
	– Mesothelioma
	– Melanoma
	– Merkel cell carcinoma
	– Neuroendocrine tumors
	– Retinoblastoma
	– Sarcomas (e.g., Ewing sarcoma, osteosarcoma)
	– Testicular cancer
	– Thymic malignancies
	– Thyroid cancer
	– Wilms tumor

Table 9.1   (continued)

Agents	Indications

Oxaliplatin	• Approved
	  – Colorectal cancer
	• Off-label
	  – Biliary adenocarcinoma
	  – Blood cancers (e.g., CLL, non-Hodgkin lymphoma)
	  – Esophageal cancer
	  – Gastric cancer
	  – Neuroblastoma
	  – Neuroendocrine tumors
	  – Ovarian cancer
	  – Pancreatic cancer
	  – Testicular cancer

Fig. 9.1 Representative "maculopapular" morbilliform rash

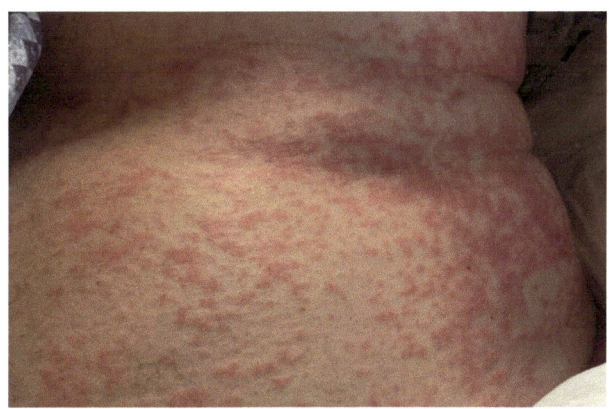

- Mucositis
- Nail changes (e.g., Beau lines, periungual hyperpigmentation [2])
- Radiation recall dermatitis [3]
- Infusion reactions
- Rare: erythroderma [4], extravasation reactions, SJS/TEN [5]

Clinical Summary
- The DAE profiles of different platinum agents (e.g., cisplatin, carboplatin, oxaliplatin) are similar.
- DAEs from platinum agent therapy are uncommon; attribution specifically to platinum agents is challenging because of frequent administration with other agents (e.g., taxanes).
- DAEs include maculopapular rash, alopecia, xerosis, pruritus, mucositis, and nail changes, which can be managed with conventional treatment (see "General principles of management").
- Infusion reactions and extravasation reactions from platinum agents have been reported.

Platinum Agent Infusion Reactions

- **Incidence**: 5–25%; incidence increases with repeated drug exposure.
- **Natural Course**: symptoms can be mild (e.g., localized pruritus and erythema within several hours of receiving treatment) to severe (e.g., immediate, life-threatening hypotension), requiring immediate treatment.
- **Clinical Features**: flushing, urticaria, pruritus, facial swelling, wheezing, chest tightness, dyspnea, abdominal cramping, hypotension.
- **Management**: premedication is not routine but can include systemic steroids, H_1 blockade, and H_2 blockade; there is a lack of data that demonstrates that premedication prevents infusion reactions; standard resuscitation (e.g., epinephrine, antihistamines, bronchodilators) if required.
- **Additional Considerations**: >50% of patients who had an infusion reaction will react again if rechallenged; desensitization protocols [6, 7] have been successfully used.

Platinum Agent Extravasation Reactions

- **Incidence**: rare; more commonly reported for cisplatin and oxaliplatin than carboplatin.
- **Natural Course**: symptoms can be mild (e.g., localized pain, erythema, and edema) to severe (e.g., ulceration and necrosis) and typically develop immediately but can be delayed for days to weeks.
- **Clinical Features**: burning, pain, erythema, pruritus, edema at extravasation site; can progress to increased erythema, pain, induration, desquamation, and blistering; severe cases can cause ulceration and necrosis.
- **Management**: local injection of sodium thiosulfate for cisplatin extravasation reactions; high-dose oral dexamethasone (8 mg BID × 14 days) for large volumes of oxaliplatin extravasation; there is a lack of data supporting the use of topical DMSO and dry, cold compresses (which may improve the reaction) and warm compresses (which may worsen the reaction) [8].
- **Additional Considerations**: cisplatin is considered an irritant at lower concentrations and a vesicant at higher concentrations.

Alkylating Agents

Agents

- Nitrogen mustards
  - Mechlorethamine (systemic)
  - Cyclophosphamide (Table 9.2)
  - Ifosfamide (Table 9.2)
  - Melphalan (Table 9.3)
  - Chlorambucil
  - Bendamustine

Table 9.2 Cyclophosphamide and ifosfamide

Agent	Indications

Cyclophosphamide	• Approved
	  – Blood cancers (e.g., ALL, AML, CLL, CML, Hodgkin lymphoma, non-Hodgkin lymphoma, multiple myeloma)
	  – Breast cancer
	  – Mycosis fungoides
	  – Neuroblastoma
	  – Ovarian cancer
	  – Retinoblastoma
	• Off-label
	  – Ewing sarcoma
	  – Gestational trophoblastic tumors
	  – Ovarian germ cell tumors
	  – Pheochromocytoma
	  – Small cell lung cancer
	  – Wilms tumor
	  – Rhabdomyosarcoma
	  – Waldenström macroglobulinemia

Ifosfamide	• Approved
	  – Testicular germ cell tumors
	• Off-label
	  – Bladder cancer
	  – Blood cancers (e.g., Hodgkin lymphoma, non-Hodgkin lymphoma)
	  – Cervical cancer
	  – Ewing sarcoma
	  – Gynecologic cancers (e.g., cervical, ovarian)
	  – Osteosarcoma
	  – Soft tissue sarcomas
	  – Thymoma and thymic malignancies

Table 9.3 Melphalan

Route	Indications	Dermatologic adverse effects

Systemic	• Approved
	  – Multiple myeloma
	  – Ovarian cancer
	• Off-label
	  – Amyloidosis
	  – Hodgkin lymphoma
	  – Non-Hodgkin lymphoma
	  – Conditioning for stem cell transplantation
		• Maculopapular rash
		• Alopecia
		• Pruritus
		• Edema
		• Mucositis
		• Nail changes (e.g., Beau lines, melanonychia)
		• Urticaria
		• Radiation recall dermatitis

Isolated limb perfusion (ILP)	• Approved – Localized tumors (e.g., melanoma, sarcoma)	• Erythema • Edema • Blistering • Nail changes • Arrest of hair growth • Rare: localized scleroderma [19]

- Aziridines and Epoxides
  - Thiotepa
  - Mitomycin C
- Alkyl sulfonates
  - Busulfan
- Nitrosoureas
  - Carmustine (BCNU, bischloroethylnitrosourea)
  - Streptozocin
  - Lomustine

- Hydrazines and Triazine derivatives
  - Procarbazine
  - Dacarbazine
  - Temozolomide
- Ribonucleotide reductase inhibitor
  - Hydroxyurea

Mechlorethamine (Systemic)

Indications
- Approved
  - Blood cancers (e.g., Hodgkin lymphoma, Non-Hodgkin lymphoma)
  - Bronchogenic carcinoma
  - CLL
  - CML
  - Lymphosarcoma
  - Malignant effusion
  - Mycosis fungoides
  - Polycythemia vera

Dermatologic Adverse Effects
- Maculopapular rash
- Alopecia [9]
- Xerosis
- Pruritus
- Rare: angioedema, erythema multiforme [10], extravasation reactions, urticaria

Clinical Summary
- DAEs from systemic mechlorethamine therapy are uncommon and include maculopapular rash, alopecia, xerosis, and pruritus, which can be managed with conventional treatment (see "General principles of management").
- Mechlorethamine extravasation reactions have been reported and should be managed with local injections of sodium thiosulfate followed by dry, cold compresses.

Mechlorethamine Extravasation Reactions

- **Incidence**: rare
- **Natural Course**: mechlorethamine is a potent vesicant; extravasation typically causes severe and prolonged symptoms that develop immediately but can be delayed for days to weeks.
- **Clinical Features**: burning, pain, erythema, pruritus, and edema at extravasation site; usually progresses to induration, desquamation, blistering, ulceration, and necrosis.
- **Management**: local injections of sodium thiosulfate followed by dry, cold compresses.

Dermatologic Adverse Effects

- Maculopapular rash
- Alopecia
- Edema
- Mucositis
- Nail changes
- Pigmentary changes
- Rare: acneiform rash [11], dermatitis herpetiformis [12], dermatofibromas [13], eccrine squamous syringometaplasia [14], extravasation reactions, hypersensitivity reactions, neutrophilic eccrine hidradenitis [15], panniculitis [16], radiation recall dermatitis [16], SJS/TEN [17], vasculitis [18]

Clinical Summary

- The DAE profiles of cyclophosphamide and ifosfamide are similar.
- DAEs from cyclophosphamide and ifosfamide therapy are common, though attribution specifically to these agents is challenging because of frequent administration with other agents (e.g., taxanes).
- DAEs from cyclophosphamide and ifosfamide therapy include maculopapular rash, alopecia, edema, mucositis, and nail changes, which can be managed with conventional treatment (see "General principles of management").
- Several rare DAEs from cyclophosphamide and ifosfamide therapy have been reported, including dermatitis herpetiformis, panniculitis, and SJS/TEN.
- Cyclophosphamide and ifosfamide extravasation reactions have been reported and should be managed with dry, cold compresses.

Cyclophosphamide- and Ifosfamide-induced Alopecia

- **Incidence**: 30–70% of patients (cyclophosphamide); 50–100% of patients (ifosfamide); more common with high-dose, intravenous therapy; ~ 50% of patients develop total alopecia; persistent alopecia has been reported.
- **Natural Course**: typically begins 3–6 weeks after initiating therapy; hair regrowth typically is visible 3–6 months after cessation of therapy; new hair often has color or texture change (65% of patients); most hair regrowth is total but can be partial or even absent.

- **Clinical Features**: patchy or diffuse hair loss at any site; hair thinning; associated tingling, burning, and pain is uncommon.
- **Management**: cutting the hair short before it falls out may alleviate scalp irritation and decrease psychosocial stress; data on the efficacy of scalp cooling is lacking; topical minoxidil (for scalp) and topical bimatoprost (for eyelashes) can decrease the time to maximal hair regrowth but does not decrease the incidence or severity of alopecia.
- **Additional Considerations**: patients should be counseled on the rare but notable risk of developing persistent alopecia and/or hair changes; patients should be evaluated for other potentially reversible causes of hair loss (e.g., thyroid disorders).

Cyclophosphamide and Ifosfamide Induced Pigmentary Changes
- **Incidence**: ~ 1% (likely underreported)
- **Natural Course**: typically develop 5–6 months after initiating therapy and fade 6–12 months after cessation of therapy.
- **Clinical Features**: can be widespread or localized; cyclophosphamide: predilection for flexural surfaces and palmar creases; ifosfamide: predilection for flexural areas, dorsal and plantar surfaces of the feet, extensor surfaces of the fingers and toes, and under occlusive dressings; nail involvement includes diffuse, longitudinal, and transverse hyperpigmentation; oral mucosal hyperpigmentation is common and can be persistent (particularly around the gingival margin); hair color can change.
- **Management**: not typically necessary; there is a lack of data supporting the use of general principles of treatment for pigmentary changes (e.g., topical therapies such as hydroquinone, laser therapy) specifically for cyclophosphamide-induced pigmentary changes, but conservative approaches (e.g., sun protection, cosmetic camouflage) are prudent.

Clinical Summary
- DAEs from systemic melphalan therapy include maculopapular rash, alopecia, pruritus, edema, mucositis, and nail changes, which can be managed with conventional treatment (see "General principles of management").
- DAEs from isolated limb perfusion of melphalan are common and can be managed with conventional, symptomatic treatment (see "General Principles of Management").
- Localized scleroderma from ILP using melphalan has been reported.

Chlorambucil

Indications
- Approved
  - CLL
  - Blood cancers (e.g., Hodgkin lymphoma, non-Hodgkin lymphoma)

- Off-label
  - Waldenström macroglobulinemia

Dermatologic Adverse Effects
- Maculopapular rash
- Mucositis
- Rare: alopecia, DRESS [20] (Fig. 9.2), erythema multiforme [21], exfoliative dermatitis [22], hypersensitivity reactions, radiation recall dermatitis [23], SJS/TEN [24]

Clinical Summary
- DAEs from chlorambucil therapy include maculopapular rash and mucositis, which can be managed with conventional treatment (see "General principles of management").
- Several rare DAEs from chlorambucil therapy have been reported, including erythema multiforme, DRESS, and SJS/TEN.

Bendamustine

Indications
- Approved
  - B-cell Non-Hodgkin lymphoma
  - CLL
- Off-label
  - Other lymphomas (e.g., Hodgkin, follicular, marginal zone, mantle cell)
  - Multiple myeloma
  - Waldenström macroglobulinemia

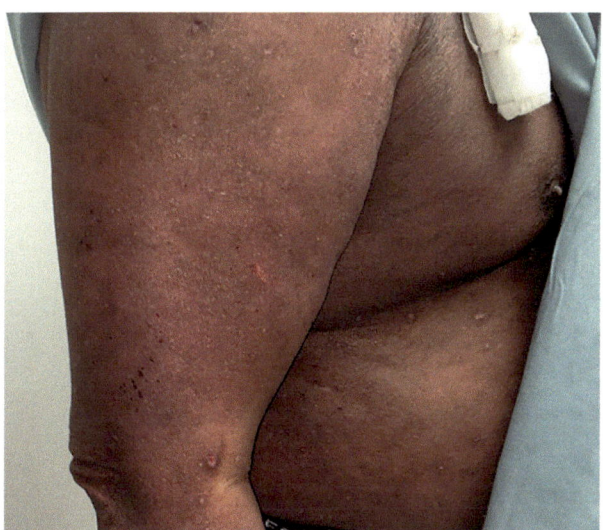

Fig. 9.2 Representative DRESS cutaneous reaction

Dermatologic Adverse Effects
- Maculopapular rash
- Xerosis
- Pruritus
- Mucositis
- Hyperhidrosis
- Rare: AGEP [25] (Fig. 9.3), alopecia, bullous pemphigoid [26], DRESS [26], generalized purpura [27], hypersensitivity reactions, SJS/TEN [27]

Clinical Summary
- DAEs from bendamustine therapy include maculopapular rash, xerosis, pruritus, and mucositis, which can be managed with conventional treatment (see "General principles of management").
- Several rare DAEs from bendamustine therapy have been reported, including AGEP, DRESS, and SJS/TEN.

Thiotepa

Indications
- Approved
  - Conditioning for stem cell transplantation
  - Bladder cancer
  - Breast cancer
  - Malignant effusions
  - Ovarian cancer

Dermatologic Adverse Effects
- Alopecia
- Pruritus
- Mucositis
- Urticaria
- Pattern of erythema, desquamation, and hyperpigmentation [28]
- Rare: angioedema, eccrine squamous syringometaplasia [14], hypersensitivity reactions, leukoderma [29]

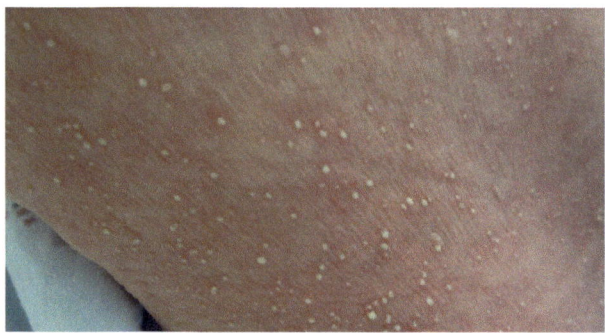

Fig. 9.3 Representative Acute Generalized Exanthematous Pustulosis (AGEP)

Clinical Summary
- Thiotepa therapy in the adult population has decreased in favor of other anticancer therapies; thiotepa therapy is more commonly used in the pediatric population, often as part of a conditioning regimen for stem cell transplantation.
- DAEs from thiotepa therapy include alopecia, pruritus and mucositis, which can be managed with conventional treatment (see "General principles of management").
- Thiotepa therapy commonly causes a self-limited pattern of erythema, desquamation, and hyperpigmentation.
- Other DAEs from thiotepa therapy include alopecia, pruritus, and mucositis, which can be managed with conventional treatment (see "General principles of management").

Thiotepa-induced Erythema, Desquamation, and Hyperpigmentation [28]
- **Incidence**: > 80% of pediatric patients
- **Natural Course**: onset 3–9 days after initiating therapy; progression to diffuse involvement 2–4 days after onset; re-epithelialization of desquamated areas within ~18 days in ~50% of patients.
- **Clinical Features**: mild erythema that becomes generalized and then progresses to bronze or tan hyperpigmentation with superficial desquamation; predilection for intertriginous areas and occluded skin.
- **Management**: liberal emolliation, management of pruritus, barrier creams, avoidance of friction, warmth, and occlusion, which may exacerbate toxicity (evidence suggests that thiotepa is excreted onto the skin through sweat).

Mitomycin C

Indications
- Approved
  - Gastric adenocarcinoma
  - Pancreatic adenocarcinoma
- Off-label
  - Anal cancer
  - Bladder cancer (intravesical therapy)
  - Breast cancer
  - Cervical cancer
  - Esophageal cancer
  - Vulva cancer
  - Ocular surface neoplasms (topical therapy)

Dermatologic Adverse Effects
- Alopecia
- Pruritus
- Edema
- Mucositis
- Diffuse erythema and exfoliative dermatitis

- Hypersensitivity reactions
- Rare: bullous dermatitis [30], erythema multiforme [31, 32], extravasation reactions, generalized pustular folliculitis [33], SDRIFE [34]

Clinical Summary

- DAEs from mitomycin C therapy include alopecia, pruritus, edema, and mucositis, which can be managed with conventional treatment (see "General principles of management").
- Mitomycin C extravasation can result in extensive tissue injury and should be treated with subcutaneous injection of DMSO, followed by topical application of DMSO and dry, cold compresses.
- Intravesical mitomycin C therapy can cause diffuse erythema and exfoliative dermatitis.
- Topical mitomycin C therapy can cause an allergic reaction characterized by localized pruritus, erythema, and edema [35].

Mitomycin C Extravasation Reactions

- **Incidence**: rare
- **Natural Course**: mitomycin C is a potent vesicant; extravasation typically causes severe and prolonged symptoms that develop immediately but can be delayed for days to weeks.
- **Clinical Features**: burning, pain, erythema, pruritus, and edema at extravasation site; usually progresses to induration, desquamation, blistering, ulceration, and necrosis.
- **Management**: subcutaneous injection of DMSO, followed by topical application of DMSO and dry, cold compresses to limit extravasation; surgical excision to remove tissue-bound drug and/or necrosis in severe cases.

Mitomycin C-induced Diffuse Erythema and Exfoliative Dermatitis [36, 37]

- **Incidence**: rare
- **Natural Course**: considered to be a delayed-type contact allergy (type IV hypersensitivity reaction) that can develop after several treatments with intravesical mitomycin C; likely a systemic contact dermatitis due to absorption of mitomycin C; may be exacerbated by skin contact (e.g., of the hands) with urine; worsens after bladder instillations and subsides with time.
- **Clinical Features**: exfoliative dermatitis of the palms and soles, associated with generalized itch; can become vesiculated, more generalized, and leave residual hyperpigmentation.
- **Management**: can identify allergy to mitomycin C with patch testing; topical and/or systemic steroids in combination with antihistamines may control mild toxicity; consider discontinuation of mitomycin C.
- **Additional Considerations**: concurrent palpable purpura on the legs attributed to a type III hypersensitivity reaction has been reported.

Topical Mitomycin C-induced Allergic Reaction [35]
- **Incidence**: ~ 34% of patients
- **Natural Course**: considered to be a delayed hypersensitivity reaction; typically develops after the second or third therapy with topical mitomycin C.
- **Clinical Features**: pruritus with conjunctival and periocular erythema and edema; can be associated with epiphora.
- **Management**: cold compresses and topical lubricants; consider discontinuing mitomycin C until symptoms improve.

Busulfan

Indications
- Approved
  - CML, conditioning for stem cell transplantation
- Off-label
  - Polycythemia vera

Dermatologic Adverse Effects
- Alopecia
- Mucositis
- Hyperpigmentation
- Urticaria
- Rare: eccrine squamous syringometaplasia [14], erythema multiforme [38], extravasation reactions, hypersensitivity reactions, porphyria cutanea tarda [39, 40]

Clinical Summary
- DAEs from busulfan therapy include alopecia and mucositis, which can be managed with conventional treatment (see "General principles of management").
- Hyperpigmentation is a common, benign DAE from busulfan therapy that should be distinguished from primary adrenal insufficiency (Addison disease).
- Busulfan extravasation reactions have been reported and should be managed with local injections of sodium thiosulfate followed by dry, cold compresses.

Hyperpigmentation from Busulfan Therapy ("Busulfan Tan") [41–43]
- **Incidence**: common, up to 46% of patients; more common in patients with darker skin types.
- **Natural Course**: typically develops weeks after initiating therapy; more common with long-term busulfan therapy; thought to be a toxic effect on melanocytes; resolves within months after drug discontinuation in most cases.
- **Clinical Features**: predilection for the neck, upper trunk, nipples, abdomen, flexural areas (including palmar creases); may be associated with burning and pain.

- **Management**: screen for primary adrenal insufficiency (Addison disease) with morning cortisol and plasma ACTH; topical retinoids, hydroquinone, steroids, sun avoidance.

Carmustine (Systemic)

Indications
- Approved
  - Brain tumors (e.g., glioblastoma, medulloblastoma, ependymoma)
  - Hodgkin lymphoma
  - Multiple myeloma
  - Non-Hodgkin lymphoma
- Off-label:
  - Conditioning for stem cell transplantation
  - Mycosis fungoides

Dermatologic Adverse Effects
- Rare: eccrine squamous syringometaplasia [14], extravasation reactions

Clinical Summary
- Systemic carmustine therapy is rarely associated with DAEs.
- Carmustine extravasation reactions have been reported and should be managed with local injections of sodium thiosulfate followed by dry, cold compresses.
- Topical carmustine therapy (see "Topical Anticancer Therapies") is more commonly associated with DAEs.

Streptozocin

Indications
- Approved
  - Pancreatic neuroendocrine tumors (e.g., islet cell carcinoma)
- Off-label
  - Adrenocortical carcinoma
  - Gastrointestinal neuroendocrine tumors

Dermatologic Adverse Effects
- Rare: nonspecific rash, alopecia, pruritus, mucositis, extravasation reactions

Clinical Summary
- DAEs from streptozocin therapy are rare and not well-documented.
- Streptozocin extravasation reactions have been reported and should be managed with local injections of sodium thiosulfate followed by dry, cold compresses.

Lomustine

Indications
- Approved
  - Brain tumors
  - Hodgkin lymphoma

Dermatologic Adverse Effects
- Rare: alopecia, mucositis, neutrophilic eccrine hidradenitis [44], SJS/TEN [45]

Clinical Summary
- DAEs from lomustine therapy are rare and not well-documented.
- Several rare DAEs from lomustine therapy have been reported, including neutrophilic eccrine hidradenitis and SJS/TEN.

Procarbazine

Indications
- Approved
  - Hodgkin lymphoma
- Off-label
  - CNS tumors (e.g., oligodendroglioma, oligoastrocytoma
  - Non-Hodgkin lymphoma

Dermatologic Adverse Effects
- Alopecia
- Pruritus
- Mucositis
- Urticaria
- Flushing (as part of disulfiram-like reaction) [46]
- Hypersensitivity reactions
- Rare: exfoliative dermatitis [47], fixed drug eruption [48], SJS/TEN [45]

Clinical Summary
- DAEs from procarbazine therapy include alopecia, pruritus, and mucositis, which can be managed with conventional treatment (see "General principles of management").
- Hypersensitivity reactions to procarbazine therapy typically manifest with a diffuse, pruritic, maculopapular rash that can be life-threatening.

Procarbazine-induced Hypersensitivity Reaction
- **Incidence**: common; increased risk with concurrent use of anticonvulsants and rechallenge of procarbazine after first reaction.

- **Natural Course**: can develop after first course; can be mild or life-threatening.
- **Clinical Features**: diffuse, pruritic, maculopapular rash that can be associated with fever, cough, and pulmonary infiltrates.
- **Management**: prophylaxis and treatment with steroids; consider discontinuation of procarbazine.

Dacarbazine

Indications
- Approved
  - Hodgkin lymphoma
  - Melanoma
- Off-label
  - Medullary thyroid cancer
  - Pancreatic neuroendocrine tumors
  - Pheochromocytoma
  - Soft tissue sarcomas

Dermatologic Adverse Effects
- Alopecia
- Mucositis
- Photosensitivity
- Hypersensitivity reactions
- Phototoxic reactions [49]
- Rare: extravasation reactions, fixed drug eruption [50], hypopigmentation [51], inflammation of actinic keratoses [52], radiation recall dermatitis [53]

Clinical Summary
- DAEs from dacarbazine therapy include alopecia, mucositis, and photosensitivity, which can be managed with conventional treatment (see "General principles of management").
- Dacarbazine extravasation reactions have been reported and should be managed with local injections of sodium thiosulfate followed by dry, cold compresses.
- Hypersensitivity reactions to dacarbazine therapy are common and should be managed symptomatically with steroids and antihistamines; severe reactions have been reported.
- Dacarbazine causes a phototoxic reaction; patients should practice diligent sun-protective measures.

Dacarbazine Extravasation Reactions
- **Incidence**: rare
- **Natural Course**: dacarbazine is an irritant; extravasation can cause symptoms that develop immediately but can be delayed for days to weeks.
- **Clinical Features**: burning, pain, erythema, pruritus, and edema at extravasation site; cellulitis and phlebitis.

- **Management**: local injection of sodium thiosulfate, followed by dry, cold compresses to limit extravasation; immobilization and avoidance of sunlight.

Dacarbazine-induced Hypersensitivity Reactions
- **Incidence**: common, up to 20% of patients
- **Natural Course**: typically develops after first or second course; not dose-dependent; ranges from mild to severe reaction.
- **Clinical Features**: erythematous and urticarial rashes; may be associated with fever, chills, and myalgia; hypereosinophilia, liver dysfunction, and delayed medullar aplasia have been reported in severe cases.
- **Management**: prophylaxis and treatment with steroids and antihistamines; consider discontinuation of dacarbazine.

Dacarbazine-induced Phototoxic Reactions
- **Incidence**: uncommon
- **Natural Course**: usually develops late (after the third course); resolves after drug discontinuation.
- **Clinical Features**: pruritic, erythematous, maculourticarial rash in sun-exposed areas; can cause blistering in severe cases; typically followed by PIH.
- **Management**: protect infusion bags from sunlight; avoid sun exposure; use sun-protective measures; topical steroids for symptomatic rash; consider dose reduction or switch therapy to temozolomide (which does not cause phototoxicity).

Temozolomide

Indications
- Approved
  - Anaplastic astrocytoma
  - Glioblastoma multiforme
- Off-label:
  - Other brain tumors (e.g., oligodendroglioma, astrocytoma, primary CNS lymphoma)
  - Cutaneous T-cell lymphoma
  - Ewing sarcoma
  - Melanoma
  - Neuroblastoma
  - Neuroendocrine tumors
  - Soft tissue sarcomas

Dermatologic Adverse Effects
- Alopecia
- Xerosis
- Pruritus
- Mucositis
- Hypersensitivity reactions
- Rare: HSV reactivation [54], hand-foot syndrome [55], SJS/TEN [56]

Clinical Summary

- DAEs from dacarbazine therapy include alopecia, xerosis, pruritus, and mucositis, which can be managed with conventional treatment (see "General principles of management").
- Hypersensitivity reactions to temozolomide therapy typically manifest with a diffuse, pruritic, maculopapular rash; severe reactions are rare.
- Several rare DAEs from temozolomide therapy have been reported, including hand-foot syndrome and SJS/TEN.

Hydroxyurea

Indications

- Approved
  - CML
  - Head and neck cancer
- Off-label
  - Meningioma
  - Other myeloproliferative disorders (e.g., AML, essential thrombocytosis, polycythemia vera)

Dermatologic Adverse Effects

- Alopecia
- Xerosis
- Pruritus
- Mucositis
- Hyperpigmentation
- Ulcerations [57]
- Dermatomyositis-like skin changes [58]
- Non-melanoma skin cancers [58]
- Nail changes (e.g., blue lunula [59], leukonychia [60], melanonychia [61])
- Rare: fixed drug eruption [62], radiation recall dermatitis [63], Raynaud phenomenon [64], SDRIFE [65], vasculitis [60]

Clinical Summary

- DAEs from hydroxyurea therapy include alopecia, xerosis, pruritus, and mucositis, which can be managed with conventional treatment (see "General principles of management").
- Hydroxyurea-induced hyperpigmentation is a benign, common DAE that can persist even with drug discontinuation.
- Hydroxyurea-induced cutaneous ulcerations typically develop at sites of trauma, heal slowly, and merit discontinuation of hydroxyurea therapy until the wounds heal.
- Hydroxyurea can induce dermatomyositis-like skin changes, including Gottron-like papules, which do not require therapy and typically resolve upon drug discontinuation.

- Hydroxyurea therapy increases the risk of developing non-melanoma skin cancers; patients should receive regular dermatologic evaluation.

Hydroxyurea-induced Hyperpigmentation
- **Incidence**: common
- **Natural Course**: typically develops after 1–3 years of therapy but can occur within months; usually fades with drug discontinuation but can persist; more common in patients with darker skin types.
- **Clinical Features**: localized or generalized hyperpigmentation (e.g., face, neck, arms, palms); can be exacerbated in pressure-bearing areas; buccal mucosa rare involved; transverse and longitudinal nail discoloration is common.
- **Management**: not typically necessary; there is a lack of data supporting the use of general principles of treatment for pigmentary changes (e.g., topical therapies such as hydroquinone, laser therapy) specifically for cyclophosphamide-induced pigmentary changes, but conservative approaches (e.g., sun protection, cosmetic camouflage) are appropriate.

Hydroxyurea-induced Cutaneous Ulcerations
- **Incidence**: common, up to ~30% of patients
- **Natural Course**: wound healing is delayed (typically 1–9 months), in part due to impaired blood flow in the microcirculation.
- **Clinical Features**: predilection for sites predisposed to trauma (e.g., malleoli, dorsal feet, toes, shins).
- **Management**: discontinue hydroxyurea therapy until wounds heal; wounds may recur with drug reinitiation; apply moist, occlusive wound dressings; several successful wound care strategies (e.g., topical GM-CSF, prostaglandin E1, pentoxyfilline) have been reported) [57].

Hydroxyurea-induced Dermatomyositis-Like Skin Changes
- **Incidence**: rare
- **Natural Course**: develops with long-term hydroxyurea therapy.
- **Clinical Features**: Gottron-like papules and erythema over the dorsal aspects of the fingers; can have associated atrophic and telangiectatic changes; edema and heliotrope rash; not typically associated with muscle tenderness or weakness; lesions can mimic graft-versus-host disease clinically and histopathologically; symptoms are typically mild.
- **Management**: does not typically require therapy; symptoms resolve upon drug discontinuation (range: 10 days–18 months reported).

Hydroxyurea-induced Non-melanoma Skin Cancers
- **Incidence**: common
- **Natural Course**: increases with age and extent of sun exposure; thought to be due to increased mutagenic activity.
- **Clinical Features**: typical features of non-melanoma skin cancers; typically develop at photodistributed sites.

• **Management**: sun-protective measures and regular dermatologic examinations; biopsy and excision per standard management.

Antimetabolites

Agents

• Folate antagonists
  – Methotrexate
  – Pemetrexed
• Pyrimidine analogs
  – Capecitabine
  – Cytarabine
  – Gemcitabine
• Purine analogs
  – Mercaptopurine
  – Thioguanine
  – Fludarabine
  – Cladribine
  – Clofarabine
  – Nelarabine
• Nucleoside analog with thymidine phosphorylase inhibitor
  – Trifluridine/tipiracil

Methotrexate

Indications
• Acute lymphoblastic leukemia (ALL)
• Trophoblastic neoplasms (e.g., choriocarcinoma, hydatidiform mole)
• Breast cancer
• Head and neck cancer
• Lung cancer (e.g., squamous cell cancer, small cell cancer)
• Cutaneous T-cell lymphoma
• Non-Hodgkin lymphomas (NHL)
• Osteosarcoma
• Bladder cancer
• Soft tissue sarcomas

Dermatologic Adverse Effects
• Hand-foot syndrome
• Mucocutaneous ulceration
• Phototoxic reactions
• Alopecia
• Nail changes (e.g., onychodystrophy, discoloration, paronychia)

- Rare: vasculitis, pigmentary changes (e.g., diffuse hyperpigmentation, hair banding), erythema multiforme, SJS/TEN

Clinical Summary
- Common DAEs from methotrexate therapy include mucocutaneous ulceration and radiation recall dermatitis, both of which are typically self-limited and should be managed with topical steroids and local wound care.
- Hand-foot syndrome has been reported as a rare DAE of methotrexate therapy and should be managed with systemic steroids and local wound care.

Hand-Foot Syndrome (Palmar-plantar Erythrodysesthesia)
- **Incidence**: rare
- **Natural Course**: typically develops 1–3 days after initiating therapy but can be delayed.
- **Clinical Features**: painful, erythematous lesions on the palms and plantar surfaces of the feet; predilection for pressure points; can rarely cause bullous lesions and severe disease with epidermal necrosis and exfoliative dermatitis (Fig. 9.4). Distinguish from hand-foot skin reaction of tyrosine kinase inhibitors (Chap. 10).
- **Management**: systemic steroids, local wound care.

Mucocutaneous Ulceration [66, 67]
- **Incidence**: common; mucositis occurs more frequently than cutaneous ulceration.
- **Natural Course**: occurs in the setting of inappropriately high serum concentrations of methotrexate; thus, typically develops within days to weeks of initiating methotrexate therapy, increasing the dose, and/or in the setting of renal damage (e.g., as an adverse effect of NSAID therapy); re-epithelialization occurs within weeks of drug discontinuation or dose reduction.
- **Clinical Features**: often begins with mucosal ulceration and a burning sensation on the skin that progresses to ulceration (Fig. 9.5); predilection for sites of trauma and existing pathology (e.g., psoriatic lesions).
- **Management**: reduce use of NSAIDs and other nephrotoxic medications and monitor methotrexate levels to prevent toxicity; drug discontinuation, dose reduction, or dose splitting; intravenous or intramuscular formulation rather than oral therapy if mucositis is predominant; with local wound care.

Fig. 9.4 Methotrexate-induced hand-foot syndrome

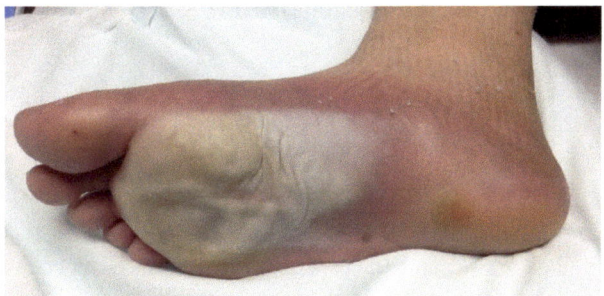

Fig. 9.5 Methotrexate-induced oral ulceration

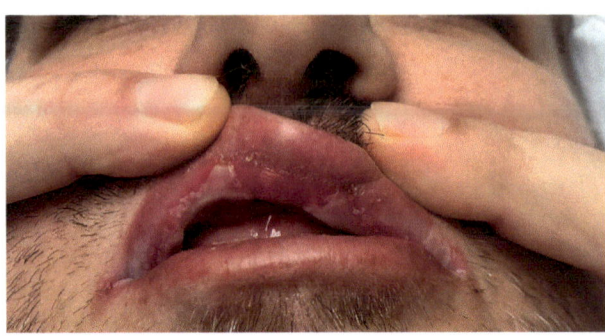

Radiation Recall Dermatitis
- **Incidence**: common
- **Natural Course**: self-limited; typically develops within days after initiating therapy and resolves with time and symptomatic management.
- **Clinical Features**: pruritic, maculopapular rash in areas of previous sun exposure; can vesiculate and desquamate in severe cases.
- **Management**: topical steroids for intact skin; local wound care for desquamated areas.

Pemetrexed

Indications
- Non-small cell lung cancer
- Mesothelioma
- Bladder cancer
- Gynecologic cancers (e.g., ovarian, cervical)
- Thymic malignancies

Dermatologic Adverse Effects
- Maculopapular rash
- Radiation recall dermatitis
- Edema
- Pruritus
- Mucositis
- Rare: urticarial vasculitis, AGEP, SJS/TEN

Clinical Summary
- Common DAEs from pemetrexed therapy include mild maculopapular rashes; patients should receive pretreatment with systemic steroids.
- Radiation recall dermatitis is an uncommon DAE of pemetrexed therapy that can typically be managed symptomatically, though rare, severe cases may require drug discontinuation.
- Periorbital and lower extremity edema have been reported as DAEs of pemetrexed therapy and are typically mild.

Maculopapular Rash

- **Incidence**: common, up to ~66% of patients
- **Natural Course**: self-limited; typically develops within days after initiating therapy and resolves with time and symptomatic management.
- **Clinical Features**: maculopapular rash; occasionally associated with edema and desquamation.
- **Management**: pretreatment with systemic steroids (dexamethasone 5 mg BID × 3 days); topical steroids for intact skin; local wound care for desquamated areas.

Radiation Recall Dermatitis [68, 69]

- **Incidence**: uncommon
- **Natural Course**: self-limited; typically develops ~1 week after initiating therapy and resolves with time and symptomatic management.
- **Clinical Features**: affects areas of previous irradiation; causes mild, blanchable erythema in mild cases; can cause severe soft tissue necrosis in severe cases.
- **Management**: topical steroids for intact skin; systemic steroids in severe cases; local wound care for desquamated areas; consider discontinuation of pemetrexed until wounds heal.

Edema [70]

- **Incidence**: uncommon
- **Natural Course**: self-limited; typically develops within days after initiating therapy and resolves with time and symptomatic management; can recur with additional therapy.
- **Clinical Features**: can affect the periorbital area and lower extremities.
- **Management**: systemic steroids for severe cases.

Capecitabine

Indications

- Breast cancer
- Colorectal cancer
- Anal cancer
- Esophageal cancer
- Gastrointestinal cancers
- Hepatobiliary cancers
- Neuroendocrine tumors
- Gynecologic cancers (e.g., ovarian, fallopian tube)
- Pancreatic cancer

Dermatologic Adverse Effects

- Hand-foot syndrome
- Fingerprint loss
- Inflammation of actinic keratoses

- Hyperpigmentation
- Mucositis
- Alopecia
- Radiation recall dermatitis
- Nail changes (e.g., melanonychia, paronychia, onycholysis)
- Edema
- Rare: SCLE

Clinical Summary
- DAEs from capecitabine therapy are relatively common and heterogeneous.
- Hand-foot syndrome is a very common DAE of capecitabine therapy and should be managed with topical steroids for intact skin, local wound care, emollients, topical keratolytics, and pain control; drug discontinuation should be considered for severe cases refractory to symptomatic treatment.
- Fingerprint loss is a relatively specific DAE of capecitabine therapy that occurs with long-term use but typically resolves upon drug discontinuation.

Hand-Foot Syndrome (HFS)
- **Incidence**: common, up to ~60% of patients
- **Natural Course**: dose-related toxicity; typically develops 1–3 days after initiating therapy but can be delayed; typically resolves 2–4 weeks after drug discontinuation but can cause persistent palmoplantar keratoderma.
- **Clinical Features**: typically begins with tingling sensation of the ventral palms and/or soles and progresses to edematous, erythematous, painful lesions; can rarely cause bullous lesions and severe disease with epidermal necrosis and exfoliative dermatitis (Fig. 9.6).
- **Management**: topical steroids for intact skin; local wound care for desquamated areas; emollients and topical keratolytics for hyperkeratotic areas; pain control; consider discontinuation of capecitabine for severe toxicity; insufficient evidence to suggest use of pyridoxine therapy, local cooling during therapy, and systemic steroids to prevent HFS; oral celecoxib 200–400 mg BID for 12–18 weeks significantly decreases the risk of developing HFS but is associated with cardiovascular and gastrointestinal adverse effects.

Fig. 9.6 Hand-Foot syndrome from doxorubicin

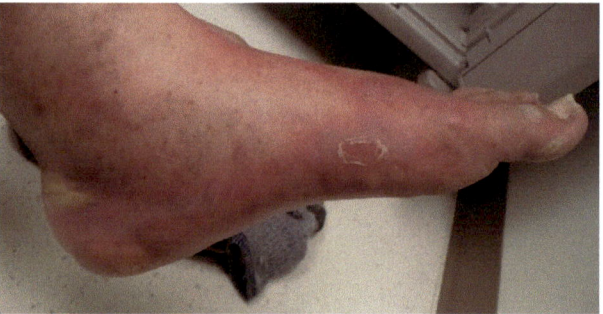

Fingerprint Loss
- **Incidence**: uncommon, ~ 14% of patients
- **Natural Course**: self-limited; typically develops 6–10 weeks after initiating therapy and resolves 2–4 weeks after drug discontinuation.
- **Clinical Features**: loss of fingerprints.
- **Management**: systemic steroids for severe cases.

Cytarabine

Indications
- Blood cancers (e.g., AML, ALL, CML, lymphomas)

Dermatologic Adverse Effects
- Maculopapular rash
- Hand-foot syndrome
- Mucositis
- Alopecia
- Nail changes
- Hyperpigmentation
- Pruritus
- Edema
- Rare: AGEP, NEH, SJS/TEN, mucositis, vasculitis

Clinical Summary
- Common DAEs from cytarabine therapy include maculopapular rashes, mucositis, and alopecia, which can be managed with conventional, symptomatic treatment (see "General Principles of Management").
- Hand-foot syndrome (HFS) is another common DAE from cytarabine therapy and should be managed similarly to HFS from capecitabine therapy.
- Toxic epidermal necrolysis from cytarabine therapy has been reported.
- Neutrophilic eccrine hidradenitis is a rare DAE from cytarabine therapy; though associated with several other anticancer agents, drug-induced NEH was first described in association with cytarabine therapy.

Neutrophilic Eccrine Hidradenitis
- **Incidence**: rare
- **Natural Course**: self-limited; typically develops 1–2 weeks after initiating cytarabine therapy and resolves after 1–2 weeks upon drug discontinuation.
- **Clinical Features**: asymptomatic, erythematous, edematous plaques; can be associated with pruritus and pain; commonly involved sites include the extremities, trunk, and face (Fig. 9.7); histopathologic examination shows neutrophils surrounding eccrine glands, vacuolar interface dermatitis in glands and ducts, necrosis, and epidermal keratinocyte atypia; neutrophils may be absent in patients with neutropenia.

Fig. 9.7 Neutrophilic eccrine hidradenitis secondary to cytarabine

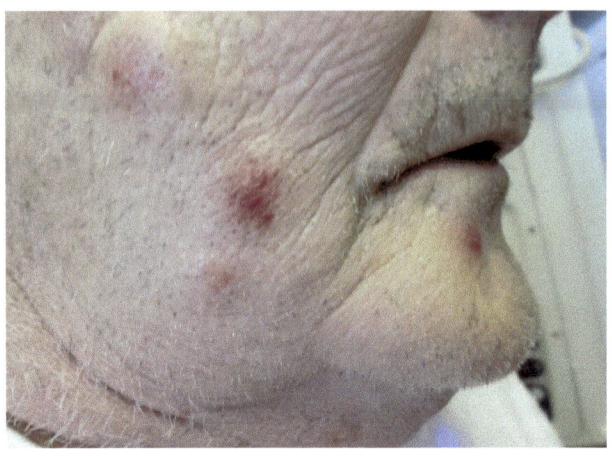

- **Management**: no specific therapy necessary, though benefit from systemic steroids and dapsone has been reported; drug eruption with cytarabine rechallenge is common.

Gemcitabine

Indications
- Breast cancer
- Lung cancers (e.g., non-small cell lung cancer, small cell lung cancer, mesothelioma)
- Gynecologic cancers (e.g., ovarian, cervical, uterine)
- Pancreatic cancer
- Bladder cancer
- Head and neck cancer
- Hepatobiliary cancer
- Osteosarcoma
- Renal cell carcinoma
- Testicular cancer
- Soft tissue sarcomas
- Thymic malignancies
- Lymphomas

Dermatologic Adverse Effects
- Maculopapular rashes
- Mucositis
- Alopecia
- Pruritus
- Edema
- Radiation recall dermatitis

- Hand-foot syndrome
- Rare: Drug-induced linear IgA, SDRIFE, SCLE, scleroderma-like changes, SJS/TEN

Clinical Summary
- Common DAEs from gemcitabine therapy include maculopapular rashes, mucositis, alopecia, and pruritus, which can be managed with conventional, symptomatic treatment (see "General Principles of Management").
- Hand-foot syndrome (HFS) from gemcitabine therapy and should be managed similarly to HFS from capecitabine therapy.
- Rare DAEs from gemcitabine therapy include drug-induced linear IgA and symmetrical drug-related intertriginous and flexural exanthema (SDRIFE).

Drug-induced Linear IgA
- **Incidence**: rare
- **Natural Course**: self-limited; typically develops within days after initiating therapy and resolves with time and symptomatic management.
- **Clinical Features**: vesicular lesions; immunofluorescence studies show linear IgA deposits along the basement membrane.
- **Management**: topical steroids for intact skin; local wound care for desquamated areas; consider systemic steroids and/or drug discontinuation for severe cases.

Symmetrical Drug-Related Intertriginous and Flexural Exanthema (SDRIFE)
- **Incidence**: rare
- **Natural Course**: self-limited; typically develops within days after initiating therapy and resolves within weeks after drug discontinuation with symptomatic management.
- **Clinical Features**: symmetric, erythematous, well-demarcated eruption with predilection for buttocks, genital areas, intertriginous areas, and flexural surfaces; histopathologic examination typically shows superficial perivascular infiltrate of inflammatory cells; can mimic cutaneous lymphoma.
- **Management**: topical steroids for intact skin; local wound care for desquamated areas; consider systemic steroids and/or drug discontinuation for severe cases.

Mercaptopurine

Indications
- Blood cancers (e.g., ALL, APML, lymphomas)

Dermatologic Adverse Effects
- Hand-foot syndrome
- Alopecia
- Hypersensitivity reactions

- Mucositis
- Eruptive nevi
- Nonmelanoma skin cancers
- Rare: photo-onycholysis

Clinical Summary
- Hand-foot syndrome (HFS) from mercaptopurine therapy occurs in up to 10% of patients and should be managed similarly to HFS from capecitabine therapy.
- Common DAEs from mercaptopurine therapy include alopecia, hypersensitivity reactions, and mucositis, which can be managed with conventional, symptomatic treatment (see "General Principles of Management").
- Eruptive nevi and an increased risk of developing nonmelanoma skin cancers has been reported from mercaptopurine therapy; patients should receive regular dermatologic evaluation.

Thioguanine

Indications
- Blood cancers (e.g., AML, ALL)

Dermatologic Adverse Effects
- Maculopapular rashes
- Mucositis
- Pruritus
- Photosensitivity
- Nonmelanoma skin cancers

Clinical Summary
- Common DAEs from thioguanine therapy include maculopapular rashes, mucositis, and pruritus, which can be managed with conventional, symptomatic treatment (see "General Principles of Management").
- Increased photosensitivity and an increased risk of developing nonmelanoma skin cancers has been reported from thioguanine therapy; patients should receive regular dermatologic evaluation.

Fludarabine

Indications
- Blood cancers (e.g., CLL, AML, ALL, lymphomas)

Dermatologic Adverse Effects
- Maculopapular rashes
- Mucositis
- Alopecia
- Edema

- Nonmelanoma skin cancers
- Rare: transfusion-associated graft versus host disease

Clinical Summary
- DAEs from fludarabine therapy are relatively uncommon but include maculo-papular rashes, mucositis, alopecia, and pruritus, which can be managed with conventional, symptomatic treatment (see "General Principles of Management").
- Increased risk of developing aggressive skin cancers have been reported from fludarabine therapy; patients should receive regular dermatologic evaluation.
- Rare, fatal cases of transfusion-associated graft versus host disease have been reported; patients who have received fludarabine therapy should receive irradiated blood products.

Transfusion-Associated Graft Versus Host Disease [71]
- **Incidence**: rare
- **Natural Course**: mediated by immunocompetent allogeneic lymphocytes in blood products; typically develops 4–30 days after transfusion of blood products that have not been irradiated and is usually fatal.
- **Clinical Features**: eruptive maculopapular rash that progresses to erythroderma with severe desquamation, typically associated with fever, abdominal pain, diarrhea, aminotransferase elevations, elevated bilirubin, and pancytopenia; histopathologic examination of the skin is consistent with GVHD (e.g., vacuolization of the stratum basale, histiocytic infiltrate, satellite dyskeratosis).
- **Management**: prevention with irradiation of blood products for patients who have received fludarabine therapy; treatment strategies that have been attempted without success include systemic steroids, azathioprine, ATG, methotrexate, and cyclosporine.

Cladribine

Indications
- Blood cancers (e.g., hairy cell leukemia, AML, Langerhans cell histiocytosis, lymphomas)

Dermatologic Adverse Effects
- Allergic reactions (e.g., injection site reactions, maculopapular rashes, edema, extensive pruritus)
- Erythroderma
- Rare: SJS/TEN, transfusion-associated graft versus host disease

Clinical Summary
- DAEs from cladribine therapy are include maculopapular rashes and pruritus, which can be managed with conventional, symptomatic treatment (see "General Principles of Management").
- Toxic epidermal necrolysis from cytarabine therapy has been reported.

- Rare, fatal cases of transfusion-associated graft versus host disease in patients who have received cladribine therapy have been reported; patients should be managed similarly to patients who have received fludarabine therapy (e.g., irradiated blood products).

Clofarabine

Indications
- Blood cancers (e.g., ALL, AML, Langerhans cell histiocytosis)

Dermatologic Adverse Effects
- Xerosis
- Pruritus
- Edema
- Hand-foot syndrome

Clinical Summary
- Hand-foot syndrome (HFS) from clofarabine therapy is uncommon and should be managed similarly to HFS from capecitabine therapy.
- Common DAEs from mercaptopurine therapy include xerosis and pruritus, which can be managed with conventional, symptomatic treatment (see "General Principles of Management").

Nelarabine

Indications
- Blood cancers (e.g., ALL)

Dermatologic Adverse Effects
- Mucositis
- Edema

Clinical Summary
- DAEs from nelarabine therapy are uncommon and poorly documented.

Trifluridine/Tipiracil

Indications
- Colorectal cancer

Dermatologic Adverse Effects
- Alopecia
- Mucositis

Clinical Summary
- DAEs from trifluridine/tipiracil therapy are uncommon and poorly documented.

Topoisomerase-Interacting Agents

Agents

- Topoisomerase I inhibitors
  - Camptothecin derivatives
       Irinotecan
       Topotecan
- Topoisomerase II inhibitors
  - Anthracyclines (Table 9.4)

Table 9.4 Anthracyclines

Agent	Indications
Doxorubicin	• Breast cancer • Blood cancers (e.g., ALL, AML, lymphomas, multiple myeloma) • Wilms tumor • Neuroblastoma • Soft tissue sarcomas • Bone sarcomas • Gynecologic cancers (e.g., ovarian, uterine) • Bladder cancer • Thyroid cancer • Gastric cancer • Bronchogenic carcinoma • Hepatocellular carcinoma • Renal carcinoma • Salivary gland cancer • Thymic malignancies
Pegylated liposomal doxorubicin	• Breast cancer • Kaposi sarcoma • Multiple myeloma • Hodgkin lymphoma • Gynecologic cancers (e.g., ovarian, uterine) • Cutaneous T-cell lymphoma • Soft tissue sarcomas
Daunorubicin	• ALL, AML
Liposomal daunorubicin	• AML, Kaposi sarcoma
Epirubicin	• Breast cancer • Esophageal cancer • Gastric cancer • Osteosarcoma • Soft tissue sarcomas
Idarubicin	• AML
Valrubicin (intravesical)	• Bladder cancer

Table 9.5 Etoposide and teniposide

Agent	Indications
Etoposide	• Lung cancers (e.g., non-small cell lung cancer, small cell lung cancer) • Testicular cancer • Blood cancers (e.g., AML, lymphoma, multiple myeloma) • CNS tumors • Gestational trophoblastic neoplasia • Thymic malignancies • Breast cancer • Ewing sarcoma • Merkel cell cancer • Neuroendocrine tumors • Osteosarcoma • Prostate cancer • Retinoblastoma • Soft tissue sarcomas • Wilms tumor
Teniposide	• ALL • Non-Hodgkin lymphoma

  Doxorubicin (including pegylated liposomal doxorubicin)
  Liposomal doxorubicin
  Daunorubicin
  Epirubicin
  Idarubicin
  Valrubicin
 – Actinomycin
  Actinomycin D (Dactinomycin)
 – Podophyllotoxin derivatives
  Etoposide (Table 9.5)
  Teniposide (Table 9.5)
 – Anthracenedione
  Mitoxantrone

Irinotecan

Indications
• Colorectal carcinoma
• Gynecologic cancers (e.g., cervical, ovarian)
• CNS tumors
• Esophageal cancer
• Ewing sarcoma
• Gastric cancer
• Lung cancers (e.g., non-small cell lung cancer, small cell lung cancer)
• Pancreatic cancer
• Rhabdomyosarcoma

Dermatologic Adverse Effects
- Alopecia
- Maculopapular rashes
- Pruritus
- Mucositis
- Hyperhidrosis
- Hand-foot syndrome
- Extravasation reactions

Clinical Summary
- Common DAEs from irinotecan therapy include alopecia, maculopapular rashes, pruritus, and mucositis, which can be managed with conventional, symptomatic treatment (see "General Principles of Management").
- Hand-foot syndrome (HFS) from irinotecan therapy occurs in up to 5% of patients and should be managed similarly to HFS from capecitabine therapy.
- Irinotecan is an irritant that can cause extravasation reactions (see "General Principles of Extravasation Reactions").

Topotecan

Indications
- Gynecologic cancers (e.g., cervical, ovarian)
- Small cell lung cancer
- AML
- CNS tumors
- Ewing sarcoma
- Neuroblastoma
- Rhabdomyosarcoma

Dermatologic Adverse Effects
- Alopecia
- Maculopapular rashes
- Pruritus
- Mucositis
- Extravasation reactions
- Nail changes (e.g., melanonychia [72])
- Rare: NEH [73], scleroderma-like changes [74]

Clinical Summary
- Common DAEs from topotecan therapy include alopecia, maculopapular rashes, pruritus, and mucositis, which can be managed with conventional, symptomatic treatment (see "General Principles of Management").
- Topotecan is an irritant that can cause extravasation reactions (see "General Principles of Extravasation Reactions").
- Rare DAEs from topotecan therapy include neutrophilic eccrine hidradenitis (which resolves upon drug discontinuation) and scleroderma-like changes.

Dermatologic Adverse Effects
- Alopecia
- Maculopapular rashes
- Pruritus
- Mucositis
- Xerosis
- Hand-foot syndrome
- Nail changes (e.g., paronychia, onycholysis, melanonychia, leukonychia, Beau lines)
- Photosensitivity
- Radiation recall dermatitis
- Inflammation of actinic keratoses
- Urticaria
- Edema
- Acneiform rash
- Pigmentary changes (Table 9.6)
- Exfoliative dermatitis
- Discoloration of sweat
- Extravasation reactions
- Rare: SCLE, NEH

Clinical Summary
- Common DAEs from anthracycline therapy include alopecia, maculopapular rashes, pruritus, mucositis, xerosis, and photosensitivity, which can be managed with conventional, symptomatic treatment (see "General Principles of Management").
- The adverse effect profiles of different anthracycline agents are similar.
- Hand-foot syndrome (HFS) from anthracycline therapy (particularly liposomal formulations) is common and should be managed similarly to HFS from capecitabine therapy.
- Anthracyclines are potent vesicants; extravasation (particularly of nonliposomal formulations) can result in extensive tissue injury and should be treated with cold compresses and dexrazoxane infusion (or topical DMSO if dexrazoxane is unavailable).

Table 9.6 Specific clinical manifestations of anthracycline hyperpigmentation

Agent	Clinical manifestations
Doxorubicin	Predilection for flexural areas and palmar creases; can affect the tongue and buccal mucosa
Pegylated liposomal doxorubicin	Macular hyperpigmentation on the trunk, extremities, palms, and soles [75]
Daunorubicin	Predilection for sun-exposed areas; can cause annular or polycyclic hyperpigmentation of scalp [76]
Idarubicin	Reticular hyperpigmentation on the trunk and lower extremities; can have associated pruritus [77, 78]

Extravasation Reactions

- **Incidence**: rare
- **Natural Course**: anthracyclines are potent vesicants; symptoms typically develop immediately but can be delayed for days to weeks; often causes severe symptoms.
- **Clinical Features**: burning, pain, erythema, pruritus, edema at extravasation site; usually progresses to increased erythema, pain, induration, desquamation, blistering, ulceration, and necrosis.
- **Management**: cold compresses (removed 15 min prior to and during infusion of dexrazoxane); dexrazoxane infusion in the arm opposite the extravasation immediately (ideally within 6 h): 1000 mg/m^2 (maximum 2000 mg/dose) on days 1 and 2; 500 mg/m^2 (maximum 1000 mg/dose) on day 3; if dexrazoxane is unavailable, then apply topical DMSO at the site of extravasation immediately (ideally within 10 min): 4 drops/10 cm^2 every 6–8 h for 7–14 days; let dry and do not cover with occlusive dressing; lack of evidence supporting the use of subcutaneous steroid injection.
- **Additional Considerations**: liposomal formulations of anthracyclines are not associated with necrotic injury; dexrazoxane therapy is not indicated unless the patient is symptomatic beyond mild inflammation.

Actinomycin D (Dactinomycin)

Indications

- Ewing sarcoma
- Gestational trophoblastic neoplasia
- Rhabdomyosarcoma
- Lung cancers
- Testicular cancer
- Wilms tumor
- Ovarian germ cell tumors

Dermatologic Adverse Effects

- Alopecia
- Pruritus
- Mucositis
- Hyperpigmentation (particularly of previously irradiated skin)
- Acneiform rash
- Inflammation of actinic keratoses
- Radiation recall dermatitis
- Extravasation reactions
- Rare: erythema multiforme, SJS/TEN

Clinical Summary

- Common DAEs from actinomycin D therapy include alopecia, pruritus, and mucositis, which can be managed with conventional, symptomatic treatment (see "General Principles of Management").
- Actinomycin D is a potent vesicant; extravasation can result in extensive tissue injury and should be treated with cold compresses and topical DMSO.

Dermatologic Adverse Effects

- Alopecia
- Pruritus
- Maculopapular rashes
- Mucositis
- Edema
- Hyperpigmentation (Fig. 9.8)
- Radiation recall dermatitis
- Flushing reactions
- Nail changes (e.g., paronychia, onycholysis, Beau lines)
- Extravasation reactions
- Rare: erythema multiforme, SJS/TEN

Clinical Summary

- Common DAEs from etoposide and teniposide therapy include alopecia, pruritus, maculopapular rashes, and mucositis, which can be managed with conventional, symptomatic treatment (see "General Principles of Management").
- Etoposide and teniposide are irritants; extravasation can result in tissue damage and should be treated with warm compresses; cold compresses and topical steroids may worsen ulceration [79].

Mitoxantrone

Indications

- Acute nonlymphocytic leukemias (eg., AML, APML, lymphomas)
- Prostate cancer

Fig. 9.8 Etoposide-associated hyperpigmentation

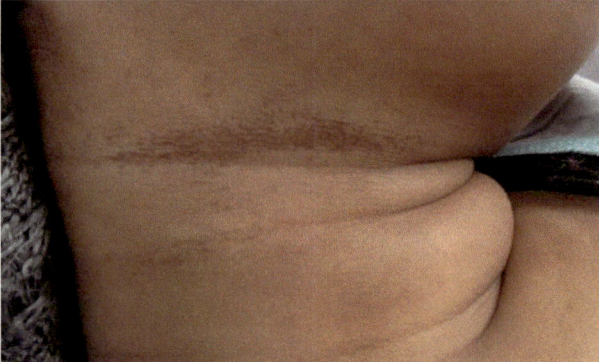

- Breast cancer
- Hepatocellular carcinoma

Dermatologic Adverse Effects
- Alopecia
- Nail changes (e.g., onycholysis, melanonychia)
- Mucositis
- Hyperpigmentation (face, dorsal hands)
- Hypersensitivity reactions (e.g., anaphylaxis, urticaria, angioedema)
- Extravasation reactions [80]

Clinical Summary
- Common DAEs from mitoxantrone therapy include alopecia and mucositis, which can be managed with conventional, symptomatic treatment (see "General Principles of Management").
- Mitoxantrone is an irritant; extravasation can result in tissue damage and should be treated with cold compresses.

Antimicrotubule Agents

Agents

- Taxanes (Table 9.7)
  - Paclitaxel
  - Nanoparticle albumin-bound paclitaxel (nab-paclitaxel)
  - Docetaxel
  - Cabazitaxel
- Vinca Alkaloids (Table 9.8)
  - Vincristine (also vincristine sulfate liposome)
  - Vinblastine
  - Vinorelbine
- Estradiol derivatives
  - Estramustine
- Other agents
  - Eribulin mesylate
  - Ixabepilone

Dermatologic Adverse Effects
- Maculopapular rash
- Alopecia
- Hyperpigmentation
- Edema
- Pruritus
- Xerosis

Table 9.7 Taxanes

Agent	Indications
Paclitaxel	• Breast cancer • Kaposi sarcoma • Lung cancers (e.g., non-small cell lung cancer, small cell lung cancer) • Gastric cancer • Head and neck cancers • Bladder cancer • Esophageal cancer • Ewing sarcoma • Osteosarcoma • Gynecologic cancers (e.g., ovarian, cervical, endometrial) • Penile cancer • Testicular germ cell tumors • Thymoma and thymic carcinoma • Thyroid cancer • Melanoma • Soft tissue sarcomas
Nanoparticle albumin-bound paclitaxel (nab-paclitaxel)	• Breast cancer • Non-small cell lung cancer • Pancreatic cancer • Bladder cancer • Gynecologic cancers (e.g., ovarian, cervical, fallopian tube, primary peritoneal) • Melanoma
Docetaxel	• Breast cancer • Lung cancers (e.g., non-small cell lung cancer, small cell lung cancer) • Prostate cancer • Gastric cancer • Head and neck cancer • Bladder cancer • Esophageal cancer • Ewing sarcoma • Osteosarcoma • Ovarian cancer • Soft tissue sarcomas
Cabazitaxel	• Prostate cancer

- Mucositis
- Photosensitivity
- Flushing reactions
- Inflammation of actinic keratoses [81, 82]
- Periarticular thenar erythema and onycholysis (PATEO)
- Nail changes (e.g., Beau lines, melanonychia, onycholysis)
- Radiation recall dermatitis
- Fixed drug eruption
- Hypersensitivity reactions

Table 9.8 Vinca Alkaloids

Agent	Indications	Dermatologic adverse effects
Vincristine	• ALL • Hodgkin lymphoma • Non-Hodgkin lymphoma • Neuroblastoma • Rhabdomyosarcoma • Wilms tumor • CNS tumors (e.g., gliomas, medulloblastoma, primary CNS lymphoma) • CLL/SLL • Ewing sarcoma • Gestational trophoblastic tumors • Multiple myeloma • Ovarian germ cells tumors • Retinoblastoma • Small cell lung cancer • Thymoma	• Alopecia • Maculopapular rash • Extravasation reactions • Nail changes (e.g., Beau lines, leukonychia) • Pruritus • Hyperpigmentation • Raynaud phenomenon [90]
Vinblastine	• Hodgkin lymphoma • Non-Hodgkin lymphomas • Kaposi sarcoma • Langerhans cell histiocytosis • Testicular cancer • Choriocarcinoma • Bladder cancer • Melanoma • Non-small cell lung cancer • Soft tissue sarcomas	• Alopecia • Maculopapular rash • Photosensitivity • Extravasation reactions • Radiation recall dermatitis • Rare: erythema multiforme [91]
Vinorelbine	• Lung cancers (e.g., non-small cell lung cancer, small cell lung cancer) • Breast cancer • Gynecologic cancers (e.g., cervical, ovarian) • Hodgkin lymphoma • Mesothelioma • Salivary gland cancer • Soft tissue sarcomas	• Alopecia • Extravasation reactions • Pruritus • Radiation recall dermatitis • Hyperpigmentation

- Extravasation reactions
- Rare: erythema multiforme, AGEP, scleroderma-like changes, SCLE, SJS/TEN [83–85]

Clinical Summary
- Common DAEs from taxane therapy include maculopapular rash, alopecia, edema, pruritus, xerosis, mucositis, and photosensitivity, which can be managed with conventional, symptomatic treatment (see "General Principles of Management").
- Hypersensitivity reactions to taxane therapy are common, even with premedication; severe reactions have been reported.

- Taxanes are considered irritants at lower concentrations and vesicants at higher concentrations; extravasation reactions from taxanes are typically mild but can be severe (see "General Principles of Management").
- Periarticular thenar erythema and onycholysis (PATEO) is a clinically distinct form of hand-foot syndrome that, in contrast to the classical form, typically manifests on the dorsal aspects of the hands and feet, can involve the Achilles tendon region, and may be associated with onycholysis and/or other nail changes.
- Cold therapy during infusions can potentially prevent PATEO and nail changes but can be challenging for patients.

Taxane-induced Hypersensitivity Reactions

- **Incidence**: relatively common, even with standard premedication; minor reactions develop in up to 10% of patients; life-threatening reactions develop in up to 3% of patients.
- **Natural Course**: can develop after the first or second infusion; likely caused by solvents used to solubilize taxanes; more common with paclitaxel (solubilized in Cremophor EL®) than docetaxel (solubilized in polysorbate 80); nab-paclitaxel is solvent free.
- **Clinical Features**: typically first manifests with dyspnea, followed by flushing reaction, maculopapular rash, and urticaria; often accompanied by hypotension, pruritus, edema, wheezing, chest tightness, and abdominal cramping.
- **Management**: premedication (i.e., systemic steroids, H_1 blockade, H_2 blockade); close monitoring at the beginning of infusions; standard resuscitation (e.g., epinephrine, antihistamines, bronchodilators) if required; consider changing taxane agents.
- **Additional Considerations**: it is not well-documented if prolonged infusion times decrease the risk of developing hypersensitivity reactions but this is a reasonable strategy to pursue; successful use of desensitization protocols have been reported [86, 87].

Taxane Extravasation Reactions

- **Incidence**: rare
- **Natural Course**: symptoms can be mild (e.g., localized pain, erythema, and edema) to severe (e.g., ulceration and necrosis) and typically develop immediately but can be delayed for days to weeks.
- **Clinical Features**: burning, pain, erythema, pruritus, edema at extravasation site; can progress to increased erythema, pain, induration, desquamation, and blistering; severe cases can cause ulceration and necrosis; typically followed by PIH.
- **Management**: see "General Principles of Management"; no specific antidote; there is a lack of data supporting the use of subcutaneous hyaluronidase; warm compresses (which further disperse the drug) may worsen extravasation; cold compresses (which limit drug diffusion) may be beneficial.

- **Additional Considerations**: taxanes are considered an irritant at lower concentrations and a vesicant at higher concentrations.

Periarticular Thenar Erythema and Onycholysis (PATEO)
- **Incidence**: uncommon; more common with paclitaxel (up to 10% of patients) than docetaxel (up to 5% of patients).
- **Natural Course**: dose-dependent; typically develops weeks after initiating taxane therapy and worsens with continued infusions.
- **Clinical Features**: clinically distinct form of hand-foot syndrome; prodromal symptoms include tingling, burning, pain, and numbness; progresses to erythematous, desquamating lesions predominantly on the dorsal hands (especially overlying the joints) and thenar eminences; can affect dorsal feet and skin over malleoli and Achilles tendon regions; associated with nail changes (see below).
- **Management**: avoidance of wet work and mechanical trauma; frequent emolliation; cold therapy during infusions; topical antiseptics; topical steroids; pain control; directed antimicrobial infection and local wound care; photoprotection to reduce the risk of developing PIH [88, 89].

Taxane-induced Nail Changes
- **Incidence**: common; up to 90% of patients
- **Natural Course**: typically develop within weeks after initiating therapy; changes in the nail plate typically resolve within months as the nail plate grows out; pathology under the nail plate can impede regrowth.
- **Clinical Features**: nail changes include onychorrhexis, Beau lines, splinter hemorrhages, subungual hemorrhages, melanonychia and leukonychia, onychoschizia, onychomadesis, paronychia, onycholysis; secondary infection is common, particularly in patients with onycholysis.(Fig. 9.9).
- **Management**: avoidance of wet work and mechanical trauma; regular gentle file to keep nails short to avoid trauma; frequent emolliation; protective varnish on the nail plate (and reapplication without removal of prior coat); biotin supplementation; cold therapy during infusions; topical antiseptics; pain control; directed antimicrobial infection and local wound care [88, 89].
- **Additional Considerations**: cold therapy (e.g., intermittent submerging of gloved hands in ice water, use of frozen gel packs) can potentially prevent taxane-induced hand-foot syndrome and nail changes but can be logistically challenging and uncomfortable for patients.

Clinical Summary
- Common DAEs from vinca alkaloid therapy include maculopapular rash, alopecia, pruritus, and photosensitivity, which can be managed with conventional, symptomatic treatment (see "General Principles of Management").
- Vinca alkaloids are potent vesicants; extravasation reactions from vinca alkaloids can be severe and should be managed with warm compresses and subcutaneous hyaluronidase.

Fig. 9.9 Taxane-induced
onychodystrophy

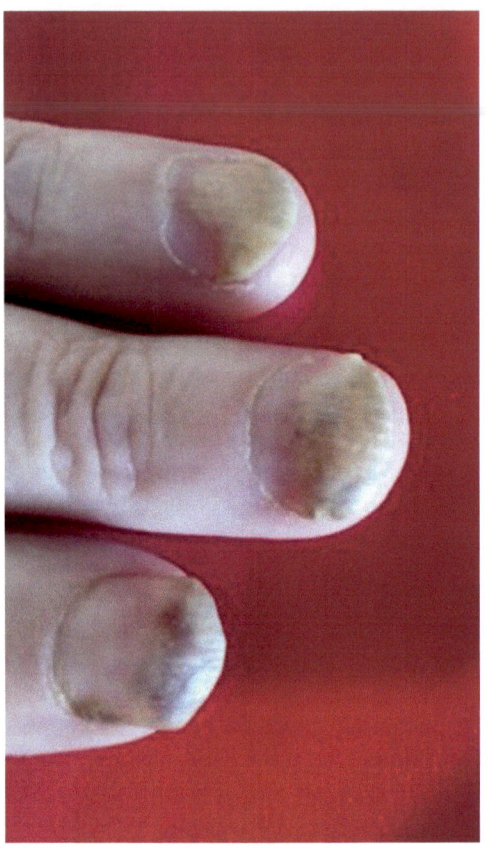

Extravasation Reactions

- **Incidence**: rare
- **Natural Course**: symptoms can be mild (e.g., localized pain, erythema, and edema) to severe (e.g., ulceration and necrosis) and typically develop immediately but can be delayed for days to weeks.
- **Clinical Features**: burning, pain, erythema, pruritus, edema at extravasation site; can progress to increased erythema, pain, induration, desquamation, and blistering; severe cases can cause ulceration and necrosis; typically followed by PIH.
- **Management**: see "General Principles of Management"; warm compresses; 1500 units of hyaluronidase diluted in 1 mL of sterile water of normal saline, injected in 5 × 0.2 mL around the periphery of the extravasation site, with a

needle changed for each injection; avoid topical steroids, which may worsen ulceration.

Eribulin

Indications
- Breast cancer
- Liposarcoma

Dermatologic Adverse Effects
- Maculopapular rash
- Alopecia
- Mucositis

Clinical Summary
- DAEs from eribulin therapy are not well-documented and include maculopapular rash, alopecia, and mucositis, which can be managed with conventional, symptomatic treatment (see "General Principles of Management").

Ixabepilone

Indications
- Breast cancer

Dermatologic Adverse Effects
- Maculopapular rash
- Alopecia
- Pruritus
- Edema
- Mucositis
- Nail changes (e.g., onycholysis) [92]

Clinical Summary
- DAEs from ixabepilone therapy are not well-documented and include maculopapular rash, alopecia, pruritus, edema, and mucositis, which can be managed with conventional, symptomatic treatment (see "General Principles of Management").
- Onycholysis from ixabepilone therapy has been reported and should be managed with conventional, symptomatic treatment (see "General Principles of Management").

Demethylating Agents, Arsenicals, and Retinoids

Agents

- Demethylating Agents
  - Azacitidine
  - Decitabine
- Arsenicals
  - Arsenic trioxide
- Retinoids
  - Bexarotene
  - Tretinoin (all-*trans* retinoid acid, ATRA)

Demethylating Agents

Azacitidine

Indications
- AML
- CML
- Myelodysplastic syndromes

Dermatologic Adverse Effects [93]
- Maculopapular rash
- Xerosis
- Injection site reactions
- AFND (Sweet syndrome) [94–98]

Clinical Summary
- DAEs from azacitidine therapy are uncommon and not well-documented. These DAEs include maculopapular rash and xerosis, which can be managed with conventional, symptomatic treatment (see "General Principles of Management").
- Injection site reactions and AFND from azacitidine therapy have been reported.

Decitabine

Indications
- AML
- Myelodysplastic syndromes

Dermatologic Adverse Effects
- Maculopapular rash
- Alopecia
- Xerosis
- Pruritus
- Rare: NEH [99]

Clinical Summary
- DAEs from decitabine therapy are uncommon and not well-documented. These DAEs include maculopapular rash and xerosis, which can be managed with conventional, symptomatic treatment (see "General Principles of Management").
- Neutrophilic eccrine hidradenitis from decitabine therapy has been reported.

Arsenic Trioxide

Indications
- APML

Dermatologic Adverse Effects
- Nonspecific rash
- Alopecia
- Xerosis
- Pruritus
- Nail changes (e.g., leukonychia, melanonychia)
- Reactivation of HSV/VZV [100]

Clinical Summary
- DAEs from arsenic trioxide therapy are uncommon and not well-documented. These DAEs include nonspecific rash, alopecia, xerosis, and pruritus, which can be managed with conventional, symptomatic treatment (see "General Principles of Management").

Bexarotene

Indications
- Cutaneous T-cell lymphoma

Dermatologic Adverse Effects
- Exfoliative dermatitis
- Xerosis
- Pruritus
- Rare: hypereosinophilic syndrome [101]

Clinical Summary
- DAEs from bexarotene therapy are uncommon and not well-documented (particularly because it is challenging to differentiate between complications of CTCL and true DAEs). Xerosis and pruritus can be managed with conventional, symptomatic treatment (see "General Principles of Management").

Tretinoin (All-*trans* Retinoid Acid, ATRA)

Indications
- APML

Dermatologic Adverse Effects
- Alopecia
- Xerosis
- Pruritus
- Ulcerations [102, 103]
- Rare: erythema nodosum [104], AFND [105], vasculitis [106]

Clinical Summary
- DAEs from systemic tretinoin therapy are uncommon and include alopecia, xerosis, and pruritus, which can be managed with conventional, symptomatic treatment (see "General Principles of Management").
- Scrotal ulcerations from systemic tretinoin therapy have been reported. This DAE can be painful and should be managed with local wound care and topical and/or systemic steroids.

ATRA-induced Scrotal Ulcerations
- **Incidence**: uncommon
- **Natural Course**: typically develops 1–4 week after initiating therapy; can occur with repeat administration of ATRA.
- **Clinical Features**: scrotal ulcerations that can manifest as single or multiple lesions; can be painful and associated with fever; preceding exfoliative dermatitis has been reported.
- **Management**: local wound care; topical and/or systemic steroids; consider dose reduction or drug discontinuation.

Review Questions and Answers

1. Which of the following is an appropriate and effective management strategy for platinum agent extravasation reactions?
   (a) Subcutaneous injection of DMSO, followed by topical application of DMSO and dry, cold compresses
   (b) Local injection of sodium thiosulfate
   (c) 1500 units of hyaluronidase diluted in 1 mL of sterile water of normal saline, injected in 5 × 0.2 mL around the periphery of the extravasation site, with a needle changed for each injection
   (d) Consider dose reduction or drug discontinuation
   (e) No treatment is necessary

   Answer: (b) Local injection of sodium thiosulfate

2. Which of the following medications pose of risk of drug induced alopecia for patients?
   (a) Pemetrexed
   (b) Chlorambucil
   (c) Cyclophosphamide
   (d) Azacitidine
   (e) Bexarotene

   Answer: (c) Cyclophosphamide

3. What are 2 common dermatologic adverse effects from methotrexate therapy?
   (a) Mucocutaneous ulceration and radiation recall dermatitis
   (b) Maculopapular rash and xerosis
   (c) Pruritus and edema
   (d) Exfoliative dermatitis and hypereosinophilic syndrome
   (e) DAEs from methotrexate therapy are uncommon and not well documented

   Answer: (a) Mucocutaneous ulceration and radiation recall dermatitis

4. A Patient comes to your office on daunorubicin therapy for ALL, what specific clinical manifestation of Anthracycline hyperpigmentation would you expect to find?
   (a) Predilection for flexural areas and palmar creases; can affect the tongue and buccal mucosa
   (b) Predilection for sun-exposed areas; can cause annular or polycyclic hyper-pigmentation of the scalp
   (c) Macular hyperpigmentation on the trunk, extremities, palms and soles
   (d) Reticular hyperpigmentation on the trunk and lower extremities

   Answer: (b) Predilection for sun-exposed areas; can cause annular or polycyclic hyperpigmentation of the scalp

5. Which of the following statement regarding taxane therapy is False?
   (a) Hypersensitivity reactions to taxane therapy are common, even with pre-medication; severe reactions have been reported.
   (b) Taxanes are considered irritants at lower concentrations and vesicants at higher concentrations; extravasation reactions from taxanes are typically mild but can be severe
   (c) Taxane induced nail changes occur in up to 90% of paients and include onychorrhexis, Beau lines, splinter hemorrhages, subungual hemorrhages, melanonychia and leukonychia, onychoschizia, onychomadesis, paronychia, and onycholysis
   (d) Periarticular thenar erythema and onycholysis (PATEO) is a clinically distinct form of hand-foot syndrome that, in contrast to the classical form, typically manifests on the dorsal aspects of the hands and feet, can involve the Achilles tendon region.
   (e) All of the above statements are true regarding taxane therapy.

Answer: (e) All of the above statements are true regarding taxane therapy

References

1. de Jonge ME, Mathôt RA, Dalesio O, Huitema AD, Rodenhuis S, Beijnen JH. Relationship between irreversible alopecia and exposure to cyclophosphamide, thiotepa and carboplatin (CTC) in high-dose chemotherapy. Bone Marrow Transplant. 2002;30(9):593–7.
2. Kim KJ, Chang SE, Choi JH, Sung KJ, Moon KC, Koh JK. Periungal hyperpigmentation induced by cisplatin. Clin Exp Dermatol. 2002;27(2):118–9.
3. Sroa N, Bartholomew DA, Magro CM. Lipodermatosclerosis as a form of vascular compromise-associated radiation recall dermatitis: case report and a review of literature. J Cutan Pathol. 2006;33(Suppl 2):55–9.
4. Rothe MJ, Bernstein ML, Grant-Kels JM. Life-threatening erythroderma: diagnosing and treating the "red man". Clin Dermatol. 2005;23(2):206–17.
5. Brodsky A, Aparici I, Argeri C, Goldenberg D. Stevens-Johnson syndrome, respiratory distress and acute renal failure due to synergic bleomycin-cisplatin toxicity. J Clin Pharmacol. 1989;29(9):821–3.
6. Broome CB, Schiff RI, Friedman HS. Successful desensitization to carboplatin in patients with systemic hypersensitivity reactions. Med Pediatr Oncol. 1996;26(2):105–10.
7. Altwerger G, Gressel GM, English DP, Nelson WK, Carusillo N, Silasi DA, Azodi M, Santin A, Schwartz PE, Ratner ES. Platinum desensitization in patients with carboplatin hypersensitivity: a single-institution retrospective study. Gynecol Oncol. 2017;144(1):77–82.
8. Kähler KC, Mustroph D, Hauschild A. Current recommendations for prevention and therapy of extravasation reactions in dermato-oncology. J Dtsch Dermatol Ges. 2009;7(1):21–8.
9. Alley E, Green R, Schuchter L. Cutaneous toxicities of cancer therapy. Curr Opin Oncol. 2002;14(2):212–6.
10. Brauer MJ, McEvoy BF, Mitus WJ. Hypersensitivity to nitrogen mustards in the form of erythema multiforme. A unique adverse reaction. Arch Intern Med. 1967;120(4):499–503.
11. Kovalyshyn I, Bijal AD, Lacouture ME, Brownell I. Cyclophosphamide-associated acneiform drug eruption in a patient with multiple myeloma. J Am Acad Dermatol. 2011;65(3):657–9.

12. Gottlieb D, Commens C. Recurrent "flare" of dermatitis herpetiformis after cytotoxic therapy for malignant lymphoma. Med J Aust. 1986;145(5):241.
13. Bargman HB, Fefferman I. Multiple dermatofibromas in a patient with myasthenia gravis treated with prednisone and cyclophosphamide. J Am Acad Dermatol. 1986;14:351–2.
14. Valks R, Fraga J, Porras-Luque J, Figuera A, Garcia-Diéz A, Fernández-Herrera J. Chemotherapy-induced eccrine squamous syringometaplasia. A distinctive eruption in patients receiving hematopoietic progenitor cells. Arch Dermatol. 1997;133(7):873–8.
15. Lienesch DW, Mutasim DF, Singh RR. Neutrophilic eccrine hidradenitis mimicking cutaneous vasculitis in a lupus patient: a complication of cyclophosphamide. Lupus. 2003;12(9):707–9.
16. Borroni G, Vassallo C, Brazzelli V, Martinoli S, Ardigò M, Alessandrino PE, Borroni RG, Franchini P. Radiation recall dermatitis, panniculitis, and myositis following cyclophosphamide therapy: histopathologic findings of a patient affected by multiple myeloma. Am J Dermatopathol. 2004;26(3):213–6.
17. Assier-Bonnet H, Aractingi S, Cadranel J, Wechsler J, Mayaud C, Saiag P. Stevens-Johnson syndrome induced by cyclophosphamide: report of two cases. Br J Dermatol. 1996;135(5):864–6.
18. Green RM, Schapel GJ, Sage RE. Cutaneous vasculitis due to cyclophosphamide therapy for chronic lymphocytic leukemia. Aust NZ J Med. 1989;19(1):55–7.
19. Landau M, Brenner S, Gat A, Klausner JM, Gutman M. Reticulate scleroderma after isolated limb perfusion with melphalan. J Am Acad Dermatol. 1998;39(6):1011–2.
20. Vaida I, Roszkiewicz F, Gruson B, Makdassi R, Damaj G. Drug rash with eosinophilia and systemic symptoms after chlorambucil treatment in chronic lymphocytic leukaemia. Pharmacology. 2009;83(3):148–9.
21. Hitchins RN, Hocker GA, Thomson DB. Chlorambucil allergy—a series of three cases. Aust NZ J Med. 1987;17(6):600–2.
22. Kilickap S, Kurt M, Aksoy S, Erman M, Turker A. Extensive exfoliative dermatitis induced by chlorambucil. Am J Hematol. 2006;81(11):891–2.
23. Clark E, Boffa M, Magri C, Muscat V. Chlorambucil-induced radiation recall dermatitis. Skinmed. 2015;13(4):317–9.
24. Barone C, Cassano A, Astone A. Toxic epidermal necrolysis during chlorambucil therapy in chronic lymphocytic leukaemia. Eur J Cancer (Oxford, England: 1990). 1990;26(11–12):1262.
25. Harber ID, Adams KV, Casamiquela K, Helms S, Benson BT, Herrin V. Bendamustine-induced acute generalized Exanthematous Pustulosis confirmed by patch testing. Dermatitis. 2017;28(4):292–3.
26. Carilli A, Favis G, Sundharkrishnan L, Hajdenberg J. Severe dermatologic reactions with bendamustine: a case series. Case Rep Oncol. 2014;7(2):465–70.
27. Gavini A, Telang GH, Olszewski AJ. Generalized purpuric drug exanthem with hemorrhagic plaques following bendamustine chemotherapy in a patient with B-prolymphocytic leukemia. Int J Hematol. 2012;95(3):311–4.
28. Rosman IS, Lloyd BM, Hayashi RJ, Bayliss SJ. Cutaneous effects of thiotepa in pediatric patients receiving high-dose chemotherapy with autologous stem cell transplantation. J Am Acad Dermatol. 2008;58(4):575–8.
29. Harben DJ, Cooper PH, Rodman OG. Thiotepa-induced leukoderma. Arch Dermatol. 1979;115(8):973–4.
30. Ritch PS, Louie AC. Skin rash following therapy with mitomycin C. Cancer. 1984;54(1):32–3.
31. Arikawa S, Uchida M, Ogoh E, Uozumi J, Yoshida S, Watanabe Y, Kaida H, Ishibashi N, Shirouzu K, Hayabuchi N. Drug eruption (erythema multiforme type) following chemoradiotherapy with mitomycin C and 5-fluorouracil administration for squamous cell carcinoma of the anal canal. Gan to kagaku ryoho. Cancer Chemother. 2010;37(4):727–30.
32. Spencer HJ. Local erythema multiforme-like drug reaction following intravenous mitomycin C and 5-fluorouracil. J Surg Oncol. 1984;26(1):47–50.
33. Andreu-Barasoain M, Gómez de la Fuente E, Pinedo F, Nuño A, López-Estebaranz JL. Intravesical mitomycin C-induced generalized pustular folliculitis. J Am Acad Dermatol. 2012;67(4):e142–3.

34. Tan SC, Tan JW. Symmetrical drug-related intertriginous and flexural exanthema. Curr Opin Allergy Clin Immunol. 2011;11(4):313–8.
35. Khong JJ, Muecke J. Complications of mitomycin C therapy in 100 eyes with ocular surface neoplasia. Br J Ophthalmol. 2006;90(7):819–22.
36. de Groot AC, Conemans JM. Systemic allergic contact dermatitis from intravesical instillation of the antitumor antibiotic mitomycin C. Contact Dermatitis. 1991;24(3):201–9.
37. Kunkeler L, Nieboer C, Bruynzeel DP. Type III and type IV hypersensitivity reactions due to mitomycin C. Contact Dermatitis. 2000;42(2):74–6.
38. Sanchis JM, Bagán JV, Gavaldá C, Murillo J, Diaz JM. Erythema multiforme: diagnosis, clinical manifestations and treatment in a retrospective study of 22 patients. J Oral Pathol Med. 2010;39(10):747–52.
39. Bronner AK, Hood AF. Cutaneous complications of chemotherapeutic agents. J Am Acad Dermatol. 1983;9(5):645–63.
40. Kyle RA, Dameshek W. Porthyria Cutanea Tarda associated with chronic granulocytic leukemia treated with Busulfan (Myleran). Blood. 1964;23:776–85.
41. Bandini G, Belardinelli A, Rosti G, Calori E, Motta MR, Rizzi S, Benini C, Tura S. Toxicity of high-dose busulphan and cyclophosphamide as conditioning therapy for allogeneic bone marrow transplantation in adults with haematological malignancies. Bone Marrow Transplant. 1994;13(5):577–81.
42. Granstein RD, Sober AJ. Drug- and heavy metal—induced hyperpigmentation. J Am Acad Dermatol. 1981;5(1):1–18.
43. Burns WA, McFarland W, Matthews MJ. Toxic manifestations of busulfan therapy. Med Ann Dist Columbia. 1971;40(9):567–72.
44. Shear NH, Knowles SR, Shapiro L, Poldre P. Dapsone in prevention of recurrent neutrophilic eccrine hidradenitis. J Am Acad Dermatol. 1996;35(5.2):819–22.
45. Rosen AC, Balagula Y, Raisch DW, Garg V, Nardone B, Larsen N, Sorrell J, West DP, Anadkat MJ, Lacouture ME. Life-threatening dermatologic adverse events in oncology. Anti-Cancer Drugs. 2014;25(2):225–34.
46. Wilkin JK. Flushing reactions in the cancer chemotherapy patient. The lists are longer but the strategies are the same. Arch Dermatol. 1992;128(10):1387–9.
47. Glovsky MM, Braunwald J, Opelz G, Alenty A. Hypersensitivity to procarbazine associated with angioedema, urticaria, and low serum complement activity. J Allergy Clin Immunol. 1976;57(2):134–40.
48. Giguere JK, Douglas DM, Lupton GP, Baker JR, Weiss RB. Procarbazine hypersensitivity manifested as a fixed drug eruption. Med Pediatr Oncol. 1988;16(6):378–80.
49. Treudler R, Georgieva J, Geilen CC, Orfanos CE. Dacarbazine but not temozolomide induces phototoxic dermatitis in patients with malignant melanoma. J Am Acad Dermatol. 2004;50(5):783–5.
50. Koehn GG, Balizet LB. Unusual local cutaneous reaction to dacarbazine. Arch Dermatol. 1982;118(12):1018–9.
51. Roider E, Schneider J, Flaig MJ, Ruzicka T, Kunte C, Berking C. Hypopigmentation in the sites of regressed melanoma metastases after successful dacarbazine therapy. Int J Dermatol. 2012;51(9):1142–4.
52. Johnson TM, Rapini RP, Duvic M. Inflammation of actinic keratoses from systemic chemotherapy. J Am Acad Dermatol. 1987;17(2.1):192–7.
53. Kennedy RD, McAleer JJ. Radiation recall dermatitis in a patient treated with dacarbazine. Clin Oncol (R Coll Radio). 2001;13(6):470–2.
54. Okada M, Miyake K, Shinomiya A, Kawai N, Tamiya T. Relapse of herpes encephalitis induced by temozolomide-based chemoradiation in a patient with malignant glioma. J Neurosurg. 2013;118(2):258–63.
55. Kanat O, Baskan BE, Kurt E, Evrensel T. Successful treatment of palmar-plantar erythrodysesthesia possibly due to temozolomide with dexamethasone. J Postgrad Med. 2007;53(2):146.
56. Sarma N. Stevens-Johnson syndrome and toxic epidermal necrolysis overlap due to oral temozolomide and cranial radiotherapy. Am J Clin Dermatol. 2009;10(4):264–7.

57. Dissemond J, Hoeft D, Knab J, Franckson T, Kroger K, Goos M. Leg ulcer in a patient associated with hydroxyurea therapy. Int J Dermatol. 2006;45(2):158–60.
58. Neill B, Ryser T, Neill J, Aires D, Rajpara A. A patient case highlighting the myriad of cutaneous adverse effects of prolonged use of hydroxyurea. Dermatol Online J. 2017;23(11):13030.
59. Jeevankumar B, Thappa DM. Blue lunula due to hydroxyurea. J Dermatol. 2003;30(8):628–30.
60. Zargari O, Kimyai-Asadi A, Jafroodi M. Cutaneous adverse reactions to hydroxyurea in patients with intermediate thalassemia. Pediatr Dermatol. 2004;21(6):633–5.
61. Kluger N, Naud M, Françès P. Toenails melanonychia induced by hydroxyurea. Presse Med (Paris, France: 1983). 2012;41(4):444–5.
62. Boyd AS, Neldner KH. Hydroxyurea therapy. J Am Acad Dermatol. 1991;25(3):518–24.
63. Sears ME. Erythema in areas of previous irradiation in patients treated with Hydroxyurea (NSC-32065). Cancer Chemother Rep. 1964;40:31–2.
64. Bouquet É, Urbanski G, Lavigne C, Lainé-Cessac P. Unexpected drug-induced Raynaud phenomenon: analysis from the French national pharmacovigilance database. Therapie. 2017;72(5):547–54.
65. Chowdhury MM, Patel GK, Inaloz HS, Holt PJ. Hydroxyurea-induced skin disease mimicking the baboon syndrome. Clin Exp Dermatol. 1999;24(4):336–7.
66. Kazlow DW, Federgrun D, Kurtin S, Lebwohl MG. Cutaneous ulceration caused by methotrexate. J Am Acad Dermatol. 2003;49(2 Suppl Case Reports):S197–8.
67. Del Pozo J, Martínez W, García-Silva J, Almagro M, Peña-Penabad C, Fonseca E. Cutaneous ulceration as a sign of methotrexate toxicity. Eur J Dermatol. 2001;11(5):450–2.
68. Khanfir K, Anchisi S. Pemetrexed-associated radiation recall dermatitis. Acta Oncol (Stockholm, Sweden). 2008;47(8):1607–8.
69. Ge J, Verma V, Hollander A, Langer C, Simone CB. Pemetrexed-induced radiation recall dermatitis in a patient with lung adenocarcinoma: case report and literature review. J Thorac Dis. 2016;8(12):E1589–93.
70. Kastalli S, Charfi O, Sahnoun R, Lakhoua G. Eyelid and feet edema induced by pemetrexed. Indian J Pharmacol. 2016;48(6):741–2.
71. Leitman SF, Tisdale JF, Bolan CD, Popovsky MA, Klippel JH, Balow JE, Boumpas DT, Illei GG. Transfusion-associated GVHD after fludarabine therapy in a patient with systemic lupus erythematosus. Transfusion. 2003;43(12):1667–71.
72. Baykal Y, Baykal C, Ozer S, Tulunay G. Topotecan induced nail pigmentation. J Dermatol. 2004;31(11):951–2.
73. Marini M, Wright D, Ropolo M, Abbruzzese M, Casas G. Neutrophilic eccrine hidradenitis secondary to topotecan. J Dermatolog Treat. 2002;13(1):35–7.
74. Ene-Stroescu D, Ellman MH, Peterson CE. Topotecan and the development of scleroderma or a scleroderma-like illness. Arthritis Rheum. 2002;46(3):844–5.
75. Lotem M, Hubert A, Lyass O, Goldenhersh MA, Ingber A, Peretz T, Gabizon A. Skin toxic effects of polyethylene glycol-coated liposomal doxorubicin. Arch Dermatol. 2000;136(12):1475–80.
76. Susser WS, Whitaker-Worth DL, Grant-Kels JM. Mucocutaneous reactions to chemotherapy. J Am Acad Dermatol. 1999;40(3):367–98. quiz 399–400.
77. Cohen PR. Paclitaxel-associated reticulate hyperpigmentation: report and review of chemotherapy-induced reticulate hyperpigmentation. World J Clin Cases. 2016;4(12):390–400.
78. Masson Regnault M, Gadaud N, Boulinguez S, Tournier E, Lamant L, Gladieff L, Roche H, Guenounou S, Recher C, Sibaud V. Chemotherapy-related reticulate hyperpigmentation: a case series and review of the literature. Dermatology (Basel, Switzerland). 2015;231(4):312–8.
79. Bertelli G. Prevention and management of extravasation of cytotoxic drugs. Drug Saf. 1995;12(4):245–55.
80. Luke E. Mitoxantrone-induced extravasation. Oncol Nurs Forum. 2005;32(1):27–9.
81. Zimmerman GC, Keeling JH, Burris HA, Cook G, Irvin R, Kuhn J, McCollough ML, Von Hoff DD. Acute cutaneous reactions to docetaxel, a new chemotherapeutic agent. Arch Dermatol. 1995;131(2):202–6.

82. Makdsi F, Deversa R. Inflammation of actinic keratosis with combination of alkylating and taxane agents: a case report. Cases J. 2009;2:6946.
83. Ji YZ, Geng L, Qu HM, Zhou HB, Xiao T, Chen HD, Wei HC. Acute generalized exanthematous pustulosis induced by docetaxel. Int J Dermatol. 2011;50(6):763–5.
84. Weinberg JM, Egan CL, Tangoren IA, Li LJ, Laughinghouse KA, Guzzo CA. Generalized pustular dermatosis following paclitaxel therapy. Int J Dermatol. 1997;36(7):559–60.
85. Kim SW, Lee UH, Jang SJ, Park HS, Kang YS. Acute localized exanthematous pustulosis induced by docetaxel. J Am Acad Dermatol. 2010;63(2):e44–6.
86. Feldweg AM, Lee CW, Matulonis UA, Castells M. Rapid desensitization for hypersensitivity reactions to paclitaxel and docetaxel: a new standard protocol used in 77 successful treatments. Gynecol Oncol. 2005;96(3):824–9.
87. Syrigou E, Dannos I, Kotteas E, Makrilia N, Tourkantonis I, Dilana K, Gkiozos I, Saif MW, Syrigos KN. Hypersensitivity reactions to docetaxel: retrospective evaluation and development of a desensitization protocol. Int Arch Allergy Immunol. 2011;156(3):320–4.
88. Scotté F, Tourani JM, Banu E, Peyromaure M, Levy E, Marsan S, Magherini E, Fabre-Guillevin E, Andrieu JM, Oudard S. Multicenter study of a frozen glove to prevent docetaxel-induced onycholysis and cutaneous toxicity of the hand. J Clin Oncol. 2005;23(19):4424–9.
89. Scotté F, Banu E, Medioni J, Levy E, Ebenezer C, Marsan S, Banu A, Tourani JM, Andrieu JM, Oudard S. Matched case-control phase 2 study to evaluate the use of a frozen sock to prevent docetaxel-induced onycholysis and cutaneous toxicity of the foot. Cancer. 2008;112(7):1625–31.
90. Gottschling S, Meyer S, Reinhard H, Krenn T, Graf N. First report of a vincristine dose-related Raynaud's phenomenon in an adolescent with malignant brain tumor. J Pediatr Hematol Oncol. 2004;26(11):768–9.
91. Arias D, Requena L, Hasson A, Gutierrez M, Domine M, Martin L, Barat A. Localized epidermal necrolysis (erythema multiforme-like reaction) following intravenous injection of vinblastine. J Cutan Pathol. 1991;18(5):344–6.
92. Alimonti A, Nardoni C, Papaldo P, Ferretti G, Caleno MP, Carlini P, Fabi A, Rasio D, Vecchione A, Cognetti F. Nail disorders in a woman treated with ixabepilone for metastatic breast cancer. Anticancer Res. 2005;25(5):3531–2.
93. Goldsmith SM, Sherertz EF, Powell BL, Hurd DD. Cutaneous reactions to azacitidine. Arch Dermatol. 1991;127(12):1847–8.
94. Tiwari SM, Caccetta T, Kumarasinghe SP, Harvey N. Azacitidine-induced sweet syndrome: two unusual clinical presentations. Australas J Dermatol. 2018;59(3):e224–5.
95. Kawano H, Suzuki T, Ishii S, Wakahashi K, Kawano Y, Sada A, Minagawa K, Ueno D, Yamasaki T, Itoh T, Yokozaki H, Katayama Y. Recurrence of abdominal large-vessel vasculitis and development of severe sweet syndrome after a single cycle of 5-azacytidine in a patient with myelodysplastic syndrome. Eur J Haematol. 2014;92(4):362–4.
96. Assenza B, Tripodi D, Scarano A, Perrotti V, Piattelli A, Iezzi G, D'Ercole S. Bacterial leakage in implants with different implant-abutment connections: an in vitro study. J Periodontol. 2012;83(4):491–7.
97. Trickett HB, Cumpston A, Craig M. Azacitidine-associated Sweet's syndrome. Am J Health Syst Pharm. 2012;69(10):869–71.
98. Tintle S, Patel V, Ruskin A, Halasz C. Azacitidine: a new medication associated with sweet syndrome. J Am Acad Dermatol. 2011;64(5):e77–9.
99. Ng ES, Aw DC, Tan KB, Poon ML, Yap ES, Liu TC, Tan LK, Chng WJ, Koh LP. Neutrophilic eccrine hidradenitis associated with decitabine. Leuk Res. 2010;34(5):e130–2.
100. Nouri K, Ricotti CA, Bouzari N, Chen H, Ahn E, Bach A. The incidence of recurrent herpes simplex and herpes zoster infection during treatment with arsenic trioxide. J Drugs Dermatol. 2006;5(2):182–5.
101. Ruiz-de-Casas A, Carrizosa-Esquivel A, Herrera-Saval A, Rios-Martín JJ, Camacho F. Sézary syndrome associated with granulomatous lesions during treatment with bexarotene. Br J Dermatol. 2006;154(2):372–4.
102. Shimizu D, Nomura K, Matsuyama R, Matsumoto Y, Ueda K, Masuda K, Taki T, Nishida K, Horiike S, Kishimoto S, Yanagisawa A, Taniwaki M. Scrotal ulcers arising during treatment

with all-trans retinoic acid for acute promyelocytic leukemia. Int Med (Tokyo, Japan). 2005;44(5):480–3.
103. Esser AC, Nossa R, Shoji T, Sapadin AN. All-trans-retinoic acid-induced scrotal ulcerations in a patient with acute promyelocytic leukemia. J Am Acad Dermatol. 2000;43(2 Pt 1):316–7.
104. Kuo MC, Dunn P, Wu JH, Shih LY. All- trans-retinoic acid-induced erythema nodosum in patients with acute promyelocytic leukemia. Ann Hematol. 2004;83(6):376–80.
105. Jagdeo J, Campbell R, Long T, Muglia J, Telang G, Robinson-Bostom L. Sweet's syndrome—like neutrophilic lobular panniculitis associated with all-trans-retinoic acid chemotherapy in a patient with acute promyelocytic leukemia. J Am Acad Dermatol. 2007;56(4):690–3.
106. Yanamandra U, Khadwal A, Saikia UN, Malhotra P. Genital vasculitis secondary to all-trans-retinoic-acid. BMJ Case Rep. 2016;2016:bcr2015212205.

Dermatologic Adverse Effects of Anticancer Therapy III: Targeted and Immunotherapies

10

Timothy Dang, Hannah Thompson, Vincent Liu, and Bernice Kwong

---

Learning Objectives

- To list the clinical indications (approved and off-label) for each anticancer therapy class.
- To recognize dermatologic adverse effects for each anticancer therapy class.
- To describe the incidence, natural course, clinical features, and management of DAEs from each therapy class summarized in the chapter.

Targeted Therapies

ALK Inhibitors (Table 10.1)

Agents
- Brigatinib
- Ceritinib
- Crizotinib

T. Dang (✉)
Santa Clara Valley Medical Center, San Jose, CA, USA
e-mail: timdang@stanford.edu

H. Thompson
University of Iowa Health Care, Iowa City, IA, USA
e-mail: hannah-j-thompson@uiowa.edu

V. Liu
Department of Dermatology, University of Iowa Health Care, Iowa, IA, USA
e-mail: Vincent-liu@uiowa.edu

B. Kwong
Stanford Health Care; Stanford Medicine, Palo Alto, CA, USA
e-mail: bernicek@stanford.edu

© Springer Nature Switzerland AG 2021
V. Liu (ed.), *Dermato-Oncology Study Guide*,
https://doi.org/10.1007/978-3-030-53437-0_10

Table 10.1 ALK inhibitors

Agent	Dermatologic adverse effects
Brigatinib	• Nonspecific rash
Ceritinib	• Nonspecific rash
	• Pruritus
	• Acneiform eruption
Crizotinib [1, 2]	• Nonspecific rash
	• Alopecia
	• Mucositis
	• Edema
	• Photosensitive, exfoliative rash
	• Rare: erythema multiforme

Main Indication
• Non-small cell lung cancer

Clinical Summary
• DAEs from ALK inhibitor therapy are not well-documented but include nonspecific rash, alopecia, mucositis, edema, and photosensitivity, which can be managed with conventional, symptomatic treatment (see "General Principles of Management").
• Rare DAEs including photosensitivity, exfoliative rash and erythema multiforme from ALK inhibitor therapy have been reported.

Epidermal Growth Factor Receptor Inhibitors (EGFR Inhibitors)
(Table 10.2)

Dermatologic Adverse Effects

• Acneiform rash
• Hair changes (e.g., hirsutism, hypertrichosis, trichomegaly, pili multigemini, alopecia)
• Nail changes (e.g., melanonychia, paronychia, onychorrhexis, onycholysis (Fig. 10.1)
• Xerosis
• Pruritus
• Telangiectasias

Clinical Summary
• Xerosis and pruritus are common DAEs from EGFR inhibitor therapy that can be managed with conventional, symptomatic treatment (see "General Principles of Management").
• EGFR inhibitor therapy often causes an acneiform rash that is typically self-limited but can cause significant morbidity. The development of acneiform rash correlates with overall response to therapy and should be managed with diligent

Table 10.2 EGFR inhibitors

Agent	Indications
Small molecule inhibitors	
Erlotinib	• Non-small cell lung cancer • Pancreatic cancer
Geftinib	• Non-small cell lung cancer
Afatinib (dual EGFR-Her 2)	• Non-small cell lung cancer
Lapatinib (dual EGFR-Her 2)	• Breast cancer
Neratinib (dual EGFR-Her 2)	• Breast cancer
Osimertinib	• Non-small cell lung cancer
Monoclonal antibodies	
Cetuximab	• Colorectal cancer • Head and neck cancer • Penile cancer • Squamous cell cancer of the skin
Panitumumab	• Colorectal cancer
Pertuzumab	• Breast cancer
Necitumumab	• Non-small cell lung cancer

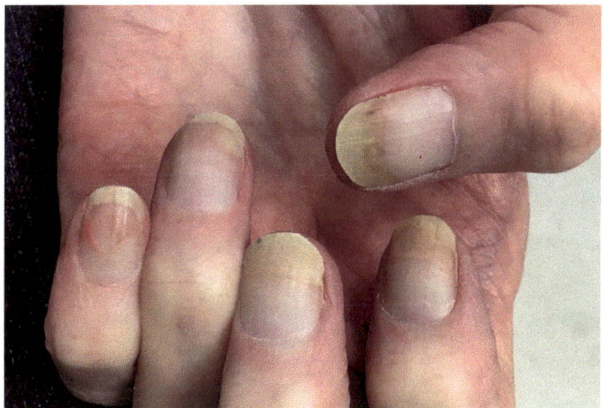

Fig. 10.1 EGFRi osimertinib-associated onycholysis

emolliation, sun protection, anti-inflammatory and antimicrobial strategies to avoid dose reduction or drug discontinuation if possible.
• Hair changes are common DAEs from EGFR inhibitor therapy and can cause significant distress and morbidity. Management strategies include treatment of underlying scalp inflammation and hair removal.
• Nail changes are common DAEs from EGFR inhibitor therapy and can cause significant distress and morbidity. Management strategies for paronychia include frequent emollition, avoidance of trauma, local wound care, antimicrobial therapy, and topical steroids.

EGFR Inhibitor-Induced Acneiform Rash

- **Incidence**: very common, >50% of patients
- **Natural Course**: onset of rash typically within 2–4 weeks of initiating therapy, with partial or complete resolution typically by 8 weeks after initiating therapy; associated xerosis typically persists; erythema and PIH can also persist
- **Clinical Features**: typically begins with tingling sensation, pain, burning, irritation, erythema, and/or edema with the eruption of an erythematous, folliculocentric, papulopustular rash without comedones on seborrheic-rich areas (e.g., face, scalp, upper chest, upper back); typically sterile but can become secondarily infected; associated with irritation, xerosis, and pruritus; sun exposure can trigger or exacerbate the rash; rash associated with lapatinib therapy [3] has a predilection for the trunk and infrequently affects the face
- **Additional Considerations**: there is an association between development of the acneiform rash and overall response rate to EGFR inhibitor therapy and survival; cutaneous toxicity may be a surrogate marker for efficacy of therapy
- **Management**: (Fig. 10.2)

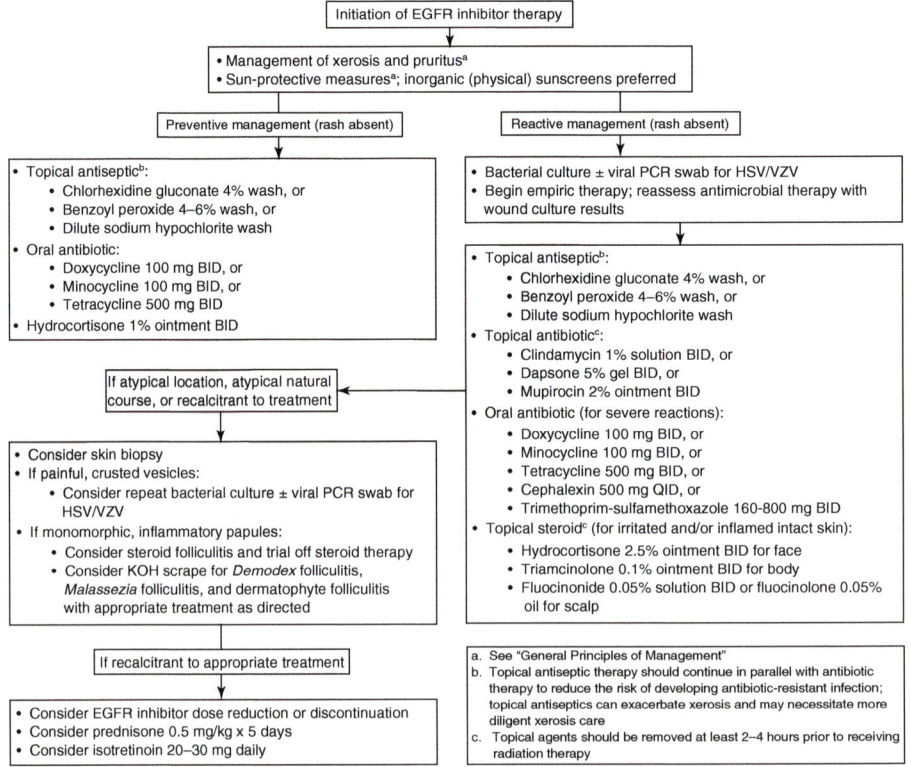

Fig. 10.2 Management of EGFR inhibitor-induced acneiform rash

EGFR Inhibitor-Induced Hair Changes

- **Incidence**: common; up to ~90% of patients
- **Natural Course**: typically develops after several months of therapy
- **Clinical Features**:

  - **Non-scarring, inflammatory alopecia**: patchy alopecia that often involves the crown of the scalp; can lead to persistent hair loss if underlying scalp inflammation is not adequately treated
  - **Hypertrichosis, hirsutism, trichomegaly, pili multigemini**
  - **Trichiasis**: can result in corneal trauma if eyelash growth contacts the cornea
  - **Color and texture changes**: hair can become darker or lighter; texture can change to straight or curly

- **Management**: treatment of underlying inflammation (see management for "Acneiform Rash"), hair removal (e.g., laser epilation, threading, plucking); ophthalmology referral for trichiasis

EGFR Inhibitor-Induced Paronychia

- **Incidence**: common; up to ~50% of patients
- **Natural Course**: often develops after trauma to the nails (e.g., from manicures, ill-fitting shoes); typically develops after several months of therapy
- **Clinical Features**: painful erythema and swelling of the proximal and lateral nail folds (Fig. 10.3); can have superficial abscess; initially sterile but often secondarily infected; may have associated ingrown toenail; may have associated pyogenic granuloma
- **Management**: Prevention strategies include avoidance of trauma (including wet work) and frequent emolliation
  - Treatment strategies include local wound care, antiseptic soaks (e.g., dilute sodium hypochlorite), topical and/or oral antimicrobials (e.g., mupirocin, ciclopirox), surgical incision and drainage for abscess, topical steroids for intact skin, and silver nitrate for pyogenic granuloma
  - Consider dose reduction or drug discontinuation
  - Consider partial or total nail avulsion for ingrown nail

Fig. 10.3 EGFRi erlotinib-associated paronychia

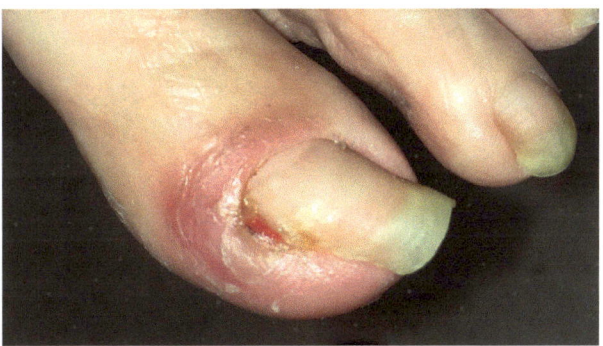

Table 10.3 CDK 4/6
inhibitor dermatologic
adverse effects

Agent	Dermatologic adverse effects
Palbociclib	• Nonspecific rash • Alopecia • Xerosis • Edema • Mucositis • Rare: SJS/TEN [5]
Abemaciclib	• Alopecia • Mucositis
Ribociclib	• Nonspecific rash • Alopecia • Xerosis • Edema • Mucositis • Pruritus • Rare: SJS/TEN [6]

CDK 4/6 Inhibitors [4] (Table 10.3)

Agents

- Palbociclib
- Abemaciclib
- Ribociclib

Main Indication
- Breast cancer

Clinical Summary
- DAEs from CDK 4/6 inhibitor therapy are not well-documented but include nonspecific rash, alopecia, xerosis, edema, mucositis, and pruritus, which can be managed with conventional, symptomatic treatment (see "General Principles of Management").
- SJS/TEN from CDK 4/6 inhibitor therapy has been reported.

Poly (ADP-Ribose) Polymerase (PARP) Inhibitors (Table 10.4)

Agents

- Olaparib
- Niraparib
- Rucaparib

Agent	Dermatologic adverse effects
Olaparib	• Nonspecific rash • Flushing reaction • Xerosis • Mucositis • Pruritus
Niraparib	• Nonspecific rash • Edema • Mucositis
Rucaparib	• Nonspecific rash • Alopecia • Photosensitivity • Mucositis • Pruritus

Table 10.4 PARP inhibitor-induced dermatologic adverse effects

Indications
- Gynecologic cancers (e.g., ovarian, fallopian tube, peritoneal)
- Breast cancer (olaparib only)

Clinical Summary
- DAEs from PARP inhibitor therapy are not well-documented but include non-specific rash, alopecia, xerosis, edema, mucositis, pruritus, and photosensitivity, which can be managed with conventional, symptomatic treatment (see "General Principles of Management").

BRAF Inhibitors (Table 10.5)

Agents

- Dabrafenib
- Vemurafenib
- Encorafenib

Dermatologic Adverse Effects
- Hyperproliferative epidermal neoplasms: verrucal keratoses, plantar hyperkeratosis, transient acantholytic dermatosis (Grover disease), actinic keratoses, cutaneous SCCs, keratoacanthomas
- Maculopapular rash
- Keratosis pilaris-like rash
- Seborrheic dermatitis-like rash
- Panniculitis (e.g., erythema nodosum)
- AFND
- Photosensitivity
- Pyogenic granulomas
- Alopecia
- Xerosis

Table 10.5 BRAF-inhibitor indications

Agent	Indications
Dabrafenib	• Melanoma • Non-small cell lung cancer • Thyroid cancer
Vemurafenib	• Melanoma • Non-small cell lung cancer
Encorafenib	• Melanoma

- Hair changes (e.g., texture change, color change) [7]
- Nail changes (e.g., onychorrhexis, onycholysis) [7]
- Radiation recall dermatitis
- Primary melanocytic lesions
- Hand-foot syndrome
- Rare: AGEP [8], DRESS [8], Kaposi varicelliform eruption [9], cutis certicis gyrate [10], SJS/TEN

Clinical Summary

- DAEs from BRAF inhibitor therapy are very common. The incidence of BRAF inhibitor-induced DAEs reduces with concurrent MEK inhibitor therapy.
- The most common DAEs are hyperproliferative epidermal neoplasms (e.g., verrucal keratoses, actinic keratoses, cutaneous SCCs), which should be managed with regular dermatologic evaluations, biopsies, and destructive therapies as needed. Dose reduction or drug discontinuation is not typically necessary.
- BRAF inhibitor-induced melanocytic lesions, including dysplastic nevi and primary melanoma, have been reported.

BRAF Inhibitor-Induced Verrucal Keratoses

- **Incidence**: common; up to ~70% with BRAF inhibitor monotherapy
- **Natural Course**: can develop as early as 1 week after initiating therapy but typically develop 6–12 weeks after initiating therapy; not considered malignant but may rarely progress to cutaneous SCCs [11]
- **Clinical Features**: white, hyperkeratotic, verrucous lesions can develop at any body site; on histopathologic examination, lesions typically lack viral inclusions (no association with HPV) [12]
- **Management**: cryotherapy; consider biopsy and/or excision of larger lesions

BRAF Inhibitor-Induced Cutaneous SCCs and KAs

- **Incidence**: common; up to ~10% with dabrafenib monotherapy; up to ~30 with vemurafenib monotherapy [13]

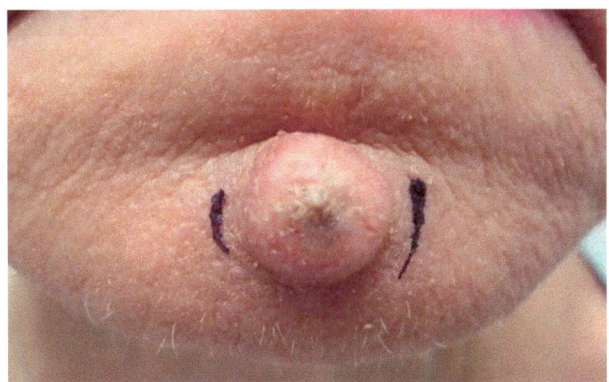

Fig. 10.4 BRAF inhibitor vemurafinib-induced squamous cell carcinoma

- **Natural Course**: median time to development after initiating therapy: 16 weeks for dabrafenib, 8 weeks for vemurafenib [13]
- **Clinical Features**: typical manifestations of cutaneous SCCs and KAs (Fig. 10.4); not typically aggressive
- **Management**: full fusiform excisions not always necessary; can treat with deep shave biopsy, electrodessication and curettage, aggressive cryotherapy, topical or intralesional fluorouracil, PDT, and/or low-dose acitretin; dose reduction or drug discontinuation is not typically necessary

BRAF Inhibitor-Induced Melanocytic Lesions [14]

- **Incidence**: ~2% of patients on BRAF inhibitor monotherapy
- **Natural Course**: median time to development after initiating therapy: 8 weeks for primary melanoma (range: 4–27 weeks); 17.5 weeks for dysplastic nevi (range: 2–42 weeks)
- **Clinical Features**: BRAF-wild type melanocytic lesions, including common melanocytic nevi, dysplastic nevi, and primary melanoma
- **Management**: sun-protective measures and regular dermatologic examinations; biopsy and excision per standard management

MEK Inhibitors (Table 10.6)

Agents

- Trametinib
- Cobimetinib
- Binimetinib

Table 10.6 MEK inhibitors

Agent	Indications
Trametinib	• Melanoma • Non-small cell lung cancer • Thyroid cancer
Cobimetinib	• Melanoma
Binimetinib	• Melanoma

Dermatologic Adverse Effects

- Acneiform rash
- Maculopapular rash
- Pruritus
- Xerosis
- Alopecia
- Edema
- Paronychia
- Hyperpigmentation
- Hair changes (e.g., depigmentation, trichomegaly, hypertrichosis)
- Telangiectasias
- Mucositis
- Photosensitivity

Clinical Summary

- DAEs from MEK inhibitor therapy are common and include maculopapular rash, pruritus, xerosis, alopecia, edema, mucositis, and photosensitivity, which can be managed with conventional, symptomatic treatment (see "General Principles of Management").
- The incidence of MEK inhibitor-induced DAEs reduces with concurrent BRAF inhibitor therapy.
- MEK inhibitor therapy often causes an acneiform rash similar to that induced by EGFR inhibitor therapy; management strategies are similar.

MEK Inhibitor-Induced Acneiform Rash [15]

- **Incidence**: very common, up to 90% of patients
- **Natural Course**: onset of rash typically within 2–4 weeks of initiating therapy (similar to EGFR inhibitor-induced acneiform rash); incidence and time to onset of rash is reduced in patients who receive concurrent BRAF inhibitor therapy
- **Clinical Features**: similar to EGFR inhibitor-induced acneiform rash
- **Management**: similar to EGFR inhibitor-induced acneiform rash

Histone Deacetylase Inhibitors

Agents

- Histone Deacetylase Inhibitors
  - Vorinostat
  - Romidepsin

Vorinostat

Indications
- Cutaneous T-cell lymphoma

Dermatologic Adverse Effects
- Alopecia
- Nonspecific rash
- Edema
- Pruritus
- Leukonychia
- Rare: AFND [16]

Clinical Summary
- DAEs from vorinostat therapy are uncommon and not well-documented. These DAEs include alopecia, edema, and pruritus, which can be managed with conventional, symptomatic treatment (see "General Principles of Management").
- Leukonychia [17] from vorinostat therapy has been reported; this DAE does not require any specific therapy and resolves with time after drug discontinuation.
- Acute febrile neutrophilic dermatosis from vorinostat therapy (in conjunction with azacitidine) has been reported.

Romidepsin

Indications
- Cutaneous T-cell lymphoma
- Peripheral T-cell lymphoma

Dermatologic Adverse Effects
- Nonspecific rash
- Pruritus
- Rare: hypersensitivity reactions [18]

Clinical Summary
- DAEs from romidepsin therapy are uncommon and not well-documented. These DAEs include nonspecific rash and pruritus, which can be managed with conventional, symptomatic treatment (see "General Principles of Management").
- Hypersensitivity reaction to romidepsin has been reported. This reaction should be managed with standard resuscitation (e.g., epinephrine, antihistamines, bronchodilators) if required. Systemic steroid premedication has been unsuccessfully tried.

Proteasome Inhibitors

Agents

- Bortezomib
- Carfilzomib
- Ixazomib

Bortezomib

Indications
- Mantle cell lymphoma
- Multiple myeloma
- Cutaneous T-cell lymphoma
- Peripheral T-cell lymphoma
- Follicular lymphoma
- Waldenström macroglobulinemia

Dermatologic Adverse Effects
- Maculopapular rash
- Pruritus
- Injection site reactions [19]
- Cutaneous nodules
- VZV reactivation [20]
- AFND [21–25]
- Rare: vasculitis [26], lupus erythematosus tumidus [27]

Clinical Summary
- DAEs from bortezomib therapy are uncommon and not well-documented. These DAEs include maculopapular rash and pruritus, which can be managed with conventional, symptomatic treatment (see "General Principles of Management").
- AFND from bortezomib therapy has been reported.

Carfilzomib

Indications
- Multiple myeloma
- Waldenström macroglobulinemia

Dermatologic Adverse Effects
- Nonspecific rash [28]
- Edema
- HSV reactivation [29]
- VZV reactivation [29]
- Infusion reactions (premed not recommended)

Clinical Summary
- DAEs from carfilzomib therapy are uncommon and not well-documented. These DAEs include nonspecific rash and edema, which can be managed with conventional, symptomatic treatment (see "General Principles of Management").
- Viral reactivation (e.g., of HSV and VZV) after carfilzomib therapy has been reported.
- AFND from bortezomib therapy has been reported.
- Infusion reactions from carfilzomib therapy have been reported. Standard premedication (e.g., acetaminophen, systemic steroids, H_1 blockade, and H_2 blockade) is prudent. Severe cases should be managed with standard resuscitation. Refer to section on infusion reactions from rituximab therapy for additional considerations.

Ixazomib

Indications
- Multiple myeloma

Dermatologic Adverse Effects
- Nonspecific rash
- Edema
- Xerosis
- Pruritus
- Rare: HSV reactivation, alopecia, acneiform rash [30], vasculitis [31], AFND [32], erythema multiforme [32], exfoliative dermatitis [32], SJS/TEN [32]

Clinical Summary
- DAEs from ixazomib therapy are uncommon and not well-documented. These DAEs include nonspecific rash, edema, xerosis, and pruritus, which can be managed with conventional, symptomatic treatment (see "General Principles of Management").

- Viral reactivation of HSV after ixazomib therapy has been reported.
- Rare DAEs from ixazomib therapy have been reported and include vasculitis, AFND, and SJS/TEN.

Hedgehog Pathway Inhibitors (SMO Inhibitors) (Table 10.7)

Clinical Summary

- DAEs from hedgehog pathway inhibitor therapy are not well-documented but include alopecia, pruritus, and mucositis, which can be managed with conventional, symptomatic treatment (see "General Principles of Management").
- DRESS and development of keratoacanthomas have been reported from vismodegib therapy.

Phosphatidylinositol 3-Kinase (PI3K) Inhibitors (Table 10.8)

Agents

- Idelalisib
- Copanlisib
- Duvelisib

Table 10.7 SMO inhibitors

Agent	Indication	Dermatologic adverse effects
Sonidegib	• Basal cell carcinom	• Alopecia • Pruritus
Vismodegib [33, 34]		• Alopecia [35] • Mucositis [36] • Rare: keratoacanthomas [37], DRESS [38]

Table 10.8 PI3K inhibitors

Agent	Indications
Idelalisib	• CLL • SLL • Follicular B-cell non-Hodgkin lymphoma
Copanlisib	• Follicular lymphoma
Duvelisib	• CLL • SLL • Follicular lymphoma

Idelalisib

Dermatologic Adverse Effects
- Nonspecific rash [39–41]
- Edema [39]
- Mucositis
- Rare: exfoliative dermatitis [42], SJS/TEN [40]

Clinical Summary
- DAEs from idelalisib therapy are not well-documented and include nonspecific rash, edema, and mucositis, which can be managed with conventional, symptomatic treatment (see "General Principles of Management").
- Rare DAEs of exfoliative dermatitis and SJS/TEN have been reported.

Copanlisib

Dermatologic Adverse Effects [43]
- Maculopapular rash
- Mucositis

Clinical Summary
- DAEs from copanlisib therapy are not well-documented and include maculopapular rash and mucositis, which can be managed with conventional, symptomatic treatment (see "General Principles of Management").

Duvelisib

Dermatologic Adverse Effects [44, 45]
- Maculopapular rash
- Mucositis
- Edema

Clinical Summary
- DAEs from duvelisib therapy are not well-documented and include maculopapular rash, mucositis, and edema, which can be managed with conventional, symptomatic treatment (see "General Principles of Management").

Other Small Molecule Inhibitors (Table 10.9)

Imatinib

Dermatologic Adverse Effects
- Edema
- Maculopapular rash
- Xerosis

Table 10.9 Small molecule inhibitors

Agent	Target(s)	Indications
Imatinib	• Bcr-Abl • PDGFR • SCF • c-KIT	• ALL • Systemic mastocytosis • CML • Dermatofibrosarcoma protuberans • Gastrointestinal stromal tumors • Hypereosinophilic syndrome • Chronic eosinophilic leukemia • Myeloproliferative diseases • Desmoid tumors • Melanoma
Nilotinib	• Bcr-Abl • PDGFR • c-KIT	• CML • ALL • Gastrointestinal stromal tumors
Dasatinib	• Bcr-Abl • SRC family • c-KIT • EPHA2 • PDGFR	• ALL • CML • Gastrointestinal stromal tumors
Ponatinib	• Bcr-Abl • VEGFR • FGFR • PDGFR • EPH • SRC • KIT • RET • TIE2 • FLT3	• ALL • CML
Bosutinib	• Bcr-Abl • SRC family	• CML
Sorafenib	• Raf kinases (e.g., CRAF, BRAF) • VEGFR • PDGFR • c-KIT • FLT-3 • RET • RET/PTC	• Hepatocellular carcinoma • Renal cell carcinoma • Thyroid carcinoma • Angiosarcoma • Gastrointestinal stromal tumors

Table 10.9 (continued)

Agent	Target(s)	Indications
Sunitinib	• PDGFR • VEGFR • FLT3 • CSF-1R • RET	• Gastrointestinal stromal tumors • Pancreatic neuroendocrine tumors • Renal cell carcinoma • Thyroid carcinoma • Soft tissue sarcomas
Axitinib	• VEGFR	• Renal cell carcinoma • Thyroid carcinoma
Regorafenib	• VEGFR • KIT • PDGFR • RET • FGFR • TIE2 • DDR2 • TrkA • Eph2A • RAF-1 • BRAF • BRAF^{V600E} • SAPK2 • PTK5 • Abl	• Colorectal carcinoma • Gastrointestinal stromal tumors • Hepatocellular carcinoma
Pazopanib	• VEGFR • PDGFR • FGFR • c-KIT • ITK • Lck • c-Fms	• Renal cell carcinoma • Soft tissue sarcomas • Thyroid carcinoma
Vandetanib	• EGFR • VEGFR • RET • BRK • EPH • SRC	• Thyroid carcinoma
Cabozantinib	• AXL • FLT-3 • KIT • MER • MET • RET • ROS1 • TIE-2 • TRKB • TYRO3 • VEGFR	• Renal cell carcinoma • Medullary thyroid carcinoma

(continued)

Table 10.9 (continued)

Agent	Target(s)	Indications
Everolimus	• mTOR • VEGFR • HIF-1	• Breast carcinoma • Neuroendocrine tumors • Renal cell carcinoma • Tuberous sclerosis complex-associated renal angiomyolipoma • Tuberous sclerosis complex-associated subependymal giant cell astrocytoma • Carcinoid tumors • Hodgkin lymphoma • Thymoma and thymic carcinomas • Waldenström macroglobulinemia
Temsirolimus	• mTOR • HIF-1 • HIF-2 • VEGFR	• Renal cell carcinoma • Endometrial carcinoma
Ibrutinib	• BTK	• CLL/SLL • Mantle cell lymphoma • Marginal zone lymphoma • Waldenström macroglobulinemia • Chronic GVHD
Acalabrutinib	• BTK	• Mantle cell lymphoma
Lenvatinib	• VEGFR • FGFR • PDGFR • KIT • RET	• Hepatocellular carcinoma • Renal cell carcinoma • Thyroid carcinoma
Midostaurin	• FLT3 • ITD • TKD • KIT • PDGFR • VEGFR • PKC family	• AML • Mast cell leukemia • Systemic mastocytosis
Alectinib	• ALK • RET	• Non-small cell lung cancer
Enasidenib	• IDH2	• AML

- Pruritus
- Mucositis
- Alopecia
- Nail changes (e.g., melanonychia, onychodystrophy)
- Hyperhidrosis
- Psoriasiform eruption
- Lichenoid dermatitis
- Hypersensitivity reactions
- Photosensitivity
- Pigmentation changes

- Rare: AFND [46], AGEP, NEH, follicular mucinosis, hand-foot syndrome, vasculitis, erythema nodosum, lichen planus-like eruption, GVHD-like eruption, exfoliative dermatitis, DRESS, SJS/TEN

Clinical Summary

- DAEs from imatinib therapy are common but typically mild in severity. These DAEs include edema, maculopapular rash, xerosis, pruritus, mucositis, alopecia, and photosensitivity, which can be managed with conventional, symptomatic treatment (see "General Principles of Management").
- Edema associated with imatinib therapy often precedes the development of a maculopapular rash. Edema has a predilection for the face and typically resolves with drug discontinuation. Maculopapular rashes typically affect the trunk/limbs and resolve over time with symptomatic management of pruritus and inflammation.
- Pigmentary changes (including both depigmentation and hyperpigmentation) from imatinib therapy are common; depigmentation is more common in darker skin types. These changes typically resolve with drug discontinuation.

Imatinib-Induced Edema

- **Incidence**: common; dose-dependent; up to ~80% of patients
- **Natural Course**: typically develops ~4–6 weeks after initiating therapy (and often before a maculopapular rash) and can persist with drug continuation
- **Clinical Features**: edema of the face (particularly periorbital area) and dependent parts of the body; can be associated with epiphora and visual obstruction from severe eyelid edema
- **Management**: see "General Principles of Management"; does not typically respond to diuretic therapy

Imatinib-Induced Maculopapular Rash

- **Incidence**: common; dose-dependent; up to ~50% of patients
- **Natural Course**: self-limited; typically develops ~6–8 weeks after initiating therapy and resolves with time and symptomatic treatment (even with drug continuation)
- **Clinical Features**: typically affects the trunk and limbs; can be associated with pruritus; histopathologic examination is not typically necessary but shows non-specific perivascular mononuclear cell infiltrates
- **Management**: see "General Principles of Management"; severe cases may require systemic steroids and/or drug discontinuation with reintroduction at a lower dose

Imatinib-Induced Pigmentation Changes

- **Incidence**: common; up to ~40% of patients; more common among patients with darker skin types
- **Natural Course**: self-limited; typically develops ~4 weeks after initiating imatinib therapy and resolves upon drug discontinuation

- **Clinical Features**: localized or diffuse depigmentation or hyperpigmentation of skin, hair, and/or nails
- **Management**: see "General Principles of Management"

Imatinib-Induced Lichenoid Dermatitis [40, 47]

- **Incidence**: uncommon
- **Natural Course**: self-limited; typically develops 2–3 months after initiating imatinib therapy and responds to symptomatic management
- **Clinical Features**: lichenoid dermatitis on any body surface; can affect mucosal surfaces
- **Management**: topical and/or systemic steroids; successful resolution of the rash with acitretin therapy has been reported; consider dose reduction or drug discontinuation

Nilotinib

Dermatologic Adverse Effects

- Maculopapular rash
- Pruritus
- Hyperhidrosis
- Alopecia
- Xerosis
- Acneiform rash
- Edema
- Pigmentation changes
- Rare: AFND, exfoliative dermatitis, hypersensitivity reactions
- Nilotinib: rash, pruritus, sweats, alopecia, xerosis, acne, exfoliative dermatitis, allergic reactions, folliculitis, urticaria

Clinical Summary

- DAEs from nilotinib therapy are common but typically mild. These DAEs include edema, maculopapular rashes, xerosis, alopecia, and pruritus, which can be managed with conventional, symptomatic treatment (see "General Principles of Management").

Dasatinib

Dermatologic Adverse Effects

- Maculopapular rashes
- Pruritus
- Mucositis
- Edema
- Xerosis

- Rare: AFND, alopecia, erythema nodosum, exfoliative dermatitis, flushing reactions, acneiform rash, hyperhidrosis, alopecia, photosensitivity, pigmentation changes, nail changes, panniculitis, hand-foot syndrome

Clinical Summary
- DAEs from dasatinib therapy are common but typically mild. These DAEs include maculopapular rash, pruritus, mucositis, and xerosis, which can be managed with conventional, symptomatic treatment (see "General Principles of Management").
- Panniculitis is a rare DAE from dasatinib therapy that may require drug discontinuation; systemic steroids can facilitate successful drug reintroduction.

Dasatinib-Induced Panniculitis [48]
- **Incidence**: rare
- **Natural Course**: has been reported to develop 1–3 months after initiating therapy; typically resolves with drug discontinuation and recurs upon reintroduction
- **Clinical Features**: painful, subcutaneous nodules with overlying erythema on the extremities; can involve the genitals
- **Management**: drug discontinuation; in some cases, systemic steroids can prevent rash recurrence from drug reintroduction

Ponatinib

Dermatologic Adverse Effects
- Maculopapular rash
- Acneiform rash
- Edema
- Xerosis
- Mucositis
- Rare: acneiform rash, flushing reaction, panniculitis [49, 50], erythema nodosum [51], pityriasis rubra pilaris-like eruption

Clinical Summary
- DAEs from ponatinib therapy are common but typically mild. These DAEs include maculopapular rash, edema, mucositis, and xerosis, which can be managed with conventional, symptomatic treatment (see "General Principles of Management").
- Pityriasis rubra pilaris-like eruptions are rare DAEs from ponatinib therapy; retinoid therapy can result in resolution of the rash even with drug continuation.

Ponatinib-Induced Pityriasis Rubra Pilaris-Like Eruption [50, 52]
- **Incidence**: rare
- **Natural Course**: has been reported to develop 1–3 months after initiating therapy

- **Clinical Features**: heterogeneous; includes xerotic, atrophic, ichthyosiform, and erythematous plaques on the face, scalp, extremities and trunk with islands of sparing; histopathologic examination can show alternating orthokeratosis, parakeratosis and a sparse superficial and/or perivascular lymphocytic infiltrate
- **Management**: topical retinoids (e.g., tretinoin 0.025% cream, acitretin 25 mg daily) and nbUVB can result in resolution of the rash within weeks; topical steroids may not be effective

Bosutinib

Dermatologic Adverse Effects
- Maculopapular rash
- Pruritus
- Acneiform rash
- Exfoliative dermatitis
- Urticaria
- Edema

Clinical Summary
- DAEs from bosutinib therapy are common but typically mild. These DAEs include maculopapular rash, edema, pruritus, and xerosis, which can be managed with conventional, symptomatic treatment (see "General Principles of Management").

Sorafenib

Dermatologic Adverse Effects [53]
- Acneiform rash
- Maculopapular rash
- Inflammation of actinic keratoses
- Keratoacanthomas [54] and squamous cell carcinomas
- Alopecia
- Hair changes (e.g., texture change, depigmentation)
- Splinter hemorrhages
- Xerosis
- Mucositis
- Pruritus
- Edema [55]
- Hyperpigmentation
- Hand foot skin reaction
- Facial erythema
- Angioedema [56]
- Genital rash (erythematous, desquamating lesions)

- Eruptive nevi
- Keratosis pilaris [57]
- Follicular cysts [58]
- Perforating folliculitis [56]
- Radiation recall dermatitis [59]
- Rare: erythema multiforme [60], AGEP [61, 62], DRESS [63], SJS/TEN [64], PRP-like reaction [65]

Clinical Summary

- DAEs from sorafenib therapy are common and variable. These DAEs include edema, maculopapular rash, xerosis, pruritus, mucositis, and alopecia, which can be managed with conventional, symptomatic treatment (see "General Principles of Management").
- **Hand-foot skin reaction** (HFSR) is a common DAE from sorafenib therapy that is clinically distinct from hand-foot syndrome; HFSR typically causes hyperkeratotic lesions in pressure-bearing and trauma-prone areas and should be managed with gentle exfoliation, topical keratolytics, frequent emolliation, topical steroids for inflammation, and pain control. Severe HFSR may require dose reduction or drug discontinuation.
- Facial erythema is a common DAE from sorafenib therapy that is typically self-limited and can be managed symptomatically with frequent emolliation and topical steroids.
- Splinter hemorrhages are a common DAE from sorafenib due to VEGF inhibition; this DAE is asymptomatic and typically resolves upon drug discontinuation.
- Patients who receive sorafenib therapy have a higher risk of developing keratoacanthomas and nonmelanoma skin cancers; patients should practice diligent sun-protective measures and receive regular dermatologic examinations.

Hand-Foot Skin Reaction (HFSR)

- **Incidence**: common, up to ~60% of patients
- **Natural Course**: typically develops 2–4 weeks after initiating therapy and worsen with continuing sorafenib therapy
- **Clinical Features**: predilection for pressure-bearing and trauma-prone areas (e.g., heels, sides of the feet); typically has prodromal tingling, numbness, and/or pain; progresses to hyperkeratosis; can have associated inflammation and desquamation; can also develop as bullous lesions
- **Management**: gentle exfoliation and/or paring of hyperkeratotic lesions; avoidance of friction and trauma; topical keratolytics (e.g., urea cream); frequent emolliation; topical steroids for inflammation; pain control; consider dose reduction or drug discontinuation
- **Additional Considerations**: development of HFSR is associated with overall survival; HFSR is clinically distinct from the hand-foot syndrome associated with other types of anticancer therapy (e.g., capecitabine) in that the characteristic lesions are more hyperkeratotic than inflammatory

Sorafenib-Induced Facial Erythema
- **Incidence**: common, up to ~60% of patients
- **Natural Course**: typically develops days to weeks after initiating therapy; typically self-limited; up to 60% of patients in some series
- **Clinical Features**: erythema of the centrofacial area and scalp; typically associated with sensitivity and pain
- **Management**: topical steroids; frequent use of emollients

Sorafenib-Induced Splinter Hemorrhages
- **Incidence**: not well-documented; up to 70% of patients in some series; likely underreported given that this DAE is asymptomatic
- **Natural Course**: typically develops 2–4 weeks after initiating therapy and resolve after drug discontinuation
- **Clinical Features**: painless, longitudinal, dark lines on the distal nail plate; thought to be related to microtrauma and impaired wound healing in the setting of VEGF inhibition
- **Management**: none necessary
- **Additional Considerations**: other drugs that inhibit VEGF (e.g., bevacizumab) can also cause splinter hemorrhages

Sorafenib-Induced Keratoacanthomas and Nonmelanoma Skin Cancers
- **Incidence**: up to 10% of patients
- **Natural Course**: typically develop weeks to months after initiating therapy
- **Clinical Features**: KAs develop as fast-growing, dome-shaped nodules with central keratotic crust; SCCs typically have clinical and histopathologic similarities to KAs (e.g., nests of atypical cells that invade the dermis; crateriform pattern with bulging borders; not always in sun-exposed areas); not typically aggressive
- **Management**: sun-protective measures and regular dermatologic examinations
- **Additional Considerations**: though KAs can regress spontaneously, these lesions should be resected completely

Sunitinib

Dermatologic Adverse Effects
- Mucositis
- Splinter hemorrhages
- Alopecia
- Xerosis
- Pruritus
- Hair depigmentation [66, 67]
- Facial edema
- Nonspecific, erythematous rash

- Genital rash (erythematous, desquamating lesions)
- Hand-foot skin reaction
- Yellow skin discoloration
- Rare: pyoderma gangrenosum [68–70]

Clinical Summary

- DAEs from sunitinib therapy are common and variable. These DAEs include edema, xerosis, pruritus, mucositis, and alopecia, which can be managed with conventional, symptomatic treatment (see "General Principles of Management").
- Hand-foot skin reaction (HFSR) occurs less frequently (up to 20% of patients) from sunitinib therapy than from sorafenib therapy; this DAE should be managed similarly to HFSR from sorafenib therapy.
- Hair depigmentation is a benign, uncommon DAE that typically occurs 5–6 weeks after initiating sunitinib therapy and resolves with 2–3 weeks off of therapy. As a result, patients on periodic schedule of sunitinib therapy (e.g., 4 weeks on, 2 weeks off) can develop bands of hair that alternate from normal to depigmented in color. This DAE does not require specific management.
- Yellow skin discoloration is a benign, common DAE that typically occurs within weeks after initiating sunitinib therapy and does not require specific management.

Sunitinib-Induced Hair Depigmentation

- **Incidence**: up to 15% of patients
- **Natural Course**: typically occurs 5–6 weeks after initiating sunitinib therapy and resolves with 2–3 weeks off therapy
- **Clinical Features**: typically manifests with alternating bands of normal and depigmented hair (due to periodic administration of sunitinib with 4 weeks on therapy and 2 weeks off of therapy); can affect hair at any site on the body
- **Management**: no specific management required

Sunitinib-Induced Yellow Skin Discoloration

- **Incidence**: common, up to 30% of patients
- **Natural Course**: self-limited; typically occurs within weeks after initiating sunitinib therapy and resolves with time
- **Clinical Features**: mild, generalized, yellowish skin discoloration
- **Management**: no specific management required

Axitinib

Dermatologic Adverse Effects

- Maculopapular rash
- Pruritus
- Alopecia
- Mucositis
- Hand-foot skin reaction

Clinical Summary
- DAEs from axitinib therapy are uncommon and include maculopapular rash, pruritus, mucositis, and alopecia, which can be managed with conventional, symptomatic treatment (see "General Principles of Management").
- Hand-foot skin reaction (HFSR) occurs less frequently (up to 25% of patients) from axitinib therapy than from sorafenib therapy; this DAE should be managed similarly to HFSR from sorafenib therapy.

Regorafenib

Dermatologic Adverse Effects
- Desquamating, erythematous rash
- Maculopapular rash
- Pruritus
- Alopecia
- Mucositis
- Hand-foot skin reaction
- Rare: hair depigmentation, eruptive nevi [71], erythema multiforme, SJS/TEN [71]

Clinical Summary
- DAEs from regorafenib therapy are uncommon and include maculopapular rash, erythematous and desquamating rash, pruritus, mucositis, and alopecia, which can be managed with conventional, symptomatic treatment (see "General Principles of Management").
- Hand-foot skin reaction (HFSR) is a common DAE from regorafenib therapy (up to 60% of patients) and should be managed similarly to HFSR from sorafenib therapy.
- Hair depigmentation (similar to from sunitinib therapy) and eruptive nevi (similar to sorafenib therapy) have been reported in association with regorafenib therapy.

Pazopanib

Dermatologic Adverse Effects
- Changes in hair color [72]
- Skin hypopigmentation
- Splinter hemorrhages
- Nonspecific, erythematous rash
- Alopecia
- Xerosis
- Genital rash (erythematous, desquamating lesions)
- Hand-foot skin reaction

- Edema [73]
- Rare: leukocytoclastic vasculitis [74], pyoderma gangrenosum [75]

Clinical Summary
- DAEs from pazopanib therapy are common and include an erythematous rash, alopecia, xerosis, and edema, which can be managed with conventional, symptomatic treatment (see "General Principles of Management").
- Hand-foot skin reaction (HFSR) is an uncommon DAE from pazopanib therapy (up to 5% of patients) and should be managed similarly to HFSR from sorafenib therapy.
- Changes in hair color and skin hypopigmentation have been reported in association with pazopanib therapy.

Vandetanib

Dermatologic Adverse Effects
- Acneiform rash
- Alopecia
- Xerosis
- Pruritus
- Photosensitivity
- Splinter hemorrhages
- Hand-foot skin reaction
- Genital rash (erythematous, desquamating lesions)
- Nail changes (e.g., paronychia, onycholysis) [76]
- Rare: erythema multiforme [77], lichenoid drug rash [78], SCLE, SJS/TEN [79]

Clinical Summary
- DAEs from vandetanib therapy are common and include acneiform rash, alopecia, xerosis, pruritus, and photosensitivity, which can be managed with conventional, symptomatic treatment (see "General Principles of Management").
- Hand-foot skin reaction (HFSR) has been reported in association with vandetanib therapy, but its incidence is not well-documented. This DAE should be managed similarly to HFSR from sorafenib therapy.

Cabozantinib

Dermatologic Adverse Effects
- Alopecia
- Xerosis
- Mucositis
- Hair changes (e.g., color change, depigmentation)
- Skin hypopigmentation

- Photosensitivity
- Splinter hemorrhages
- Genital rash (erythematous, desquamating lesions)
- Hand-foot skin reaction

Clinical Summary
- DAEs from cabozantinib therapy are common and include alopecia, xerosis, mucositis, and photosensitivity, which can be managed with conventional, symptomatic treatment (see "General Principles of Management").
- Hand-foot skin reaction (HFSR) is a common DAE from cabozantinib therapy (up to 50% of patients) and should be managed similarly to HFSR from sorafenib therapy.
- Changes in hair color and skin hypopigmentation have been reported in association with cabozantinib therapy.

Everolimus

Dermatologic Adverse Effects
- Acneiform rash
- Maculopapular rash
- Edema
- Mucositis
- Pruritus
- Xerosis
- Nail changes (e.g., paronychia, onychorrhexis)
- Radiation recall dermatitis [80]
- Rare: SDRIFE (Fig. 10.5) [81], bullous pemphigoid [82], leukocytoclastic vasculitis [83]

Clinical Summary
- DAEs from everolimus therapy are common and include maculopapular rash, edema, mucositis, pruritus, and xerosis, which can be managed with conventional, symptomatic treatment (see "General Principles of Management").
- Acneiform rash is a common DAE from everolimus therapy and should be managed similarly to acneiform rash from EGFR inhibitor therapy
- Rare DAEs from everolimus therapy include SDRIFE, bullous pemphigoid, and leukocytoclastic vasculitis.

Temsirolimus

Dermatologic Adverse Effects [84]
- Acneiform rash [85]
- Maculopapular rash
- Mucositis

Fig. 10.5 Representative
SDRIFE reaction

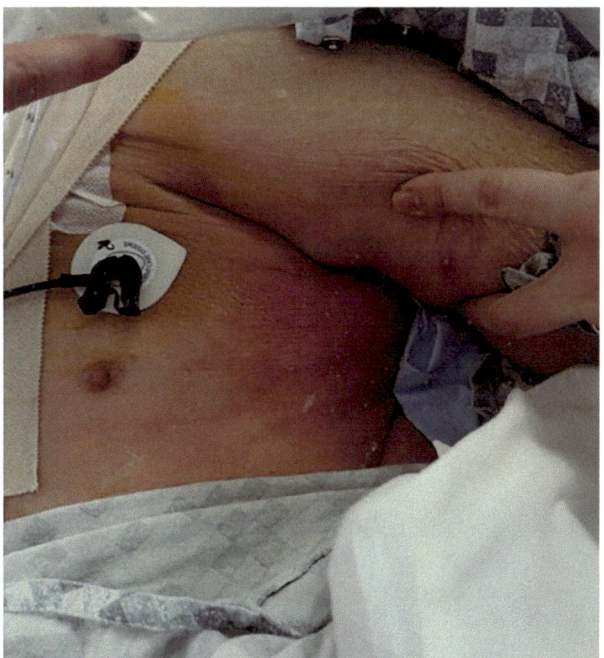

- Pruritus
- Edema
- Xerosis
- Nail changes [86, 87] (e.g., paronychia, onychorrhexis)
- Hypersensitivity reactions

Clinical Summary
- DAEs from temsirolimus therapy are common and include maculopapular rash, edema, mucositis, pruritus, edema, and xerosis, which can be managed with conventional, symptomatic treatment (see "General Principles of Management").
- Acneiform rash is a common DAE from temsirolimus therapy and should be managed similarly to acneiform rash due to EGFR inhibitor therapy.

Ibrutinib

Dermatologic Adverse Effects [88]
- Petechiae and purpura [89, 90]
- Edema [91]
- Mucositis
- Hair changes (e.g., texture change)
- Nail changes (e.g., onychoschizia, onychorrhexis)
- Panniculitis [91, 92]
- Rare: pyoderma gangrenosum [93]

Clinical Summary
- DAEs from ibrutinib therapy are common and include maculopapular rash, edema, mucositis, pruritus, edema, and xerosis, which can be managed with conventional, symptomatic treatment (see "General Principles of Management").
- Acneiform rash is a common DAE from ibrutinib therapy and should be managed similarly to acneiform rash due to EGFR inhibitor therapy.

Ibrutinib-Induced Petechiae and Purpura
- **Incidence**: common, up to 50% of patients
- **Natural Course**: petechial rash typically develops ~3 months after initiating ibrutinib therapy; purpuric rash typically develops weeks after initiating ibrutinib therapy and resolves with symptomatic treatment or within weeks after drug discontinuation
- **Clinical Features**: petechial rash is mild and associated with thrombocytopenia; purpuric rash can be severe and associated with allergic symptoms (e.g., lip tingling, tongue swelling)
- **Management**: no specific management required for petechial rash; purpuric rash can be managed with topical steroids and oral antihistamines; consider dose reduction or drug discontinuation in severe cases; successful rechallenge of ibrutinib with systemic steroids has been reported

Ibrutinib-Induced Panniculitis
- **Incidence**: rare
- **Natural Course**: typically develops weeks after initiating ibrutinib therapy
- **Clinical Features**: painful, erythematous nodules on the extremities (Fig. 10.6); not typically associated with systemic symptoms; histopathologic examination typically shows lymphohistiocytic, lobular panniculitis with prominent leukocytoclasis and occasional eosinophils

Fig. 10.6 Ibrutinib-induced panniculitis

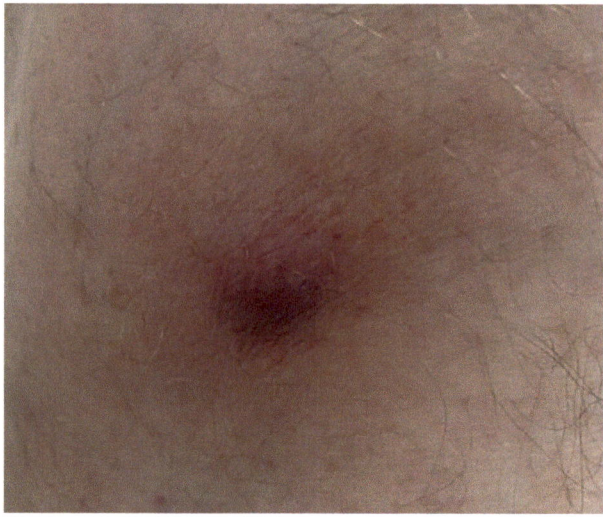

- **Management**: topical and systemic steroids; pain control; consider dose reduction or drug discontinuation in severe cases; successful rechallenge of ibrutinib with systemic steroids has been reported

Acalabrutinib

Dermatologic Adverse Effects
- Edema
- Petechiae
- Rare: nonspecific rash, purpura

Clinical Summary
- DAEs from ibrutinib therapy are uncommon and include nonspecific rash and edema, which can be managed with conventional, symptomatic treatment (see "General Principles of Management").
- Petechiae and purpura from acalabrutinib therapy occur with much less frequency than from ibrutinib therapy.

Lenvatinib

Dermatologic Adverse Effects [94]
- Maculopapular rash
- Edema
- Xerosis
- Alopecia
- Mucositis
- Hand-foot skin reaction

Clinical Summary
- DAEs from lenvatinib therapy are common and include maculopapular rash, edema, xerosis, mucositis, and alopecia, which can be managed with conventional, symptomatic treatment (see "General Principles of Management").
- Hand-foot skin reaction (HFSR) is a common DAE from lenvatinib therapy (up to 30% of patients) and should be managed similarly to HFSR from sorafenib therapy.

Midostaurin

Dermatologic Adverse Effects
- Nonspecific rash
- Petechiae
- Pruritus

- Edema
- Mucositis
- Xerosis
- Photosensitivity [95]

Clinical Summary
- DAEs from midostaurin therapy are not well-documented and include nonspecific rash, pruritus, edema, mucositis, xerosis, and photosensitivity, which can be managed with conventional, symptomatic treatment (see "General Principles of Management").

Alectinib

Dermatologic Adverse Effects [94]
- Maculopapular rash [96]
- Edema
- Photosensitivity
- Rare: alopecia, erythema multiforme [97]

Clinical Summary
- DAEs from alectinib therapy (similar to DAEs from other ALK inhibitors such as crizotinib) are not well-documented but include maculopapular rash, edema, photosensitivity, and alopecia, which can be managed with conventional, symptomatic treatment (see "General Principles of Management").
- Erythema multiforme from alectinib therapy has been reported.

Enasidenib

Dermatologic Adverse Effects
- Maculopapular rash [98]

Clinical Summary
DAEs from enasidenib therapy are not well-documented and include maculopapular rash, which can be managed with conventional, symptomatic treatment (see "General Principles of Management").

Immunotherapy

Antibodies Against CTLA-4, PD-1, and PD-L1

Agents
- Anti-CTLA-4 antibody

- Ipilimumab
- Anti-PD-1 antibodies

  - Pembrolizumab
  - Nivolumab

- Anti-PD-L1 antibodies

  - Atezolizumab
  - Avelumab
  - Durvalumab

Anti-CTLA-4 Antibodies

Agents
- **Ipilimumab**

Indications
- Colorectal cancer
- Melanoma
- Renal cell carcinoma
- Small cell lung cancer

Dermatologic Adverse Effects
- Maculopapular rash
- Pruritus
- Xerosis
- Vitiligo-like melanoma-associated depigmentation
- Rare: prurigo nodularis, transient acantholytic dermatosis (Grover disease) [99], lichenoid dermatitis, vasculitis [100], pyoderma gangrenosum, photosensitivity, radiation recall dermatitis, alopecia, sarcoidosis [101], Sjögren syndrome [102], AFND [103, 104], DRESS, SJS/TEN

Clinical Summary
- Common DAEs from ipilimumab therapy include maculopapular rash, pruritus, and xerosis, which can be managed with conventional, symptomatic treatment (see "General Principles of Management").
- Vitiligo-like melanoma-associated depigmentation is an uncommon DAE from ipilimumab therapy that is associated with response to therapy; no specific management is required for this DAE.
- Several rare DAEs from ipilimumab have been reported, including AFND, DRESS, and SJS/TEN.

Ipilimumab-Induced Vitiligo-Like Melanoma-Associated Depigmentation [105, 106]

- **Incidence**: uncommon, up to ~10% of patients
- **Natural Course**: typically develops weeks after initiating therapy; does not resolve upon drug discontinuation
- **Clinical Features**: depigmented macules and patches, often near the sites of a treated melanoma or metastases
- **Management**: no specific management required; can be distinguished from hypopigmentation with Wood lamp examination
- **Additional Considerations**: associated with response to therapy

Anti-PD-1 and Anti-PD-L1 Antibodies (Tables 10.10 and 10.11)

Table 10.10 Anti-PD-1 antibodies

Agent	Indications
Pembrolizumab	• Microsatellite instability-high cancers • Cervical cancer • Gastric cancer • Head and neck cancer • Hodgkin lymphoma • Melanoma • Non-small cell lung cancer • Primary mediastinal large B-cell lymphoma • Urothelial carcinoma • Merkel cell carcinoma
Nivolumab	• Colorectal cancer • Head and neck cancer • Hepatocellular carcinoma • Hodgkin lymphoma • Melanoma • Lung cancers (e.g., non-small cell lung cancer, small cell lung cancer) • Renal cell cancer • Urothelial carcinoma

Table 10.11 Antibodies against PD-L1

Agent	Indications	Dermatologic adverse effects
Atezolizumab	• Lung cancers (e.g., non-small cell lung cancer, small cell lung cancer) • Urothelial carcinoma	• Maculopapular rash • Halo nevi [107] • Pruritus • Rare: Sjögren syndrome [102]
Avelumab [107]	• Merkel cell carcinoma • Urothelial carcinoma	• Lichenoid dermatitis
Durvalumab [108, 109]	• Non-small cell lung cancer • Urothelial carcinoma	• Maculopapular rash • Autoimmune bullous dermatitis

Fig. 10.7 Morbilliform eruption secondary to pembrolizumab

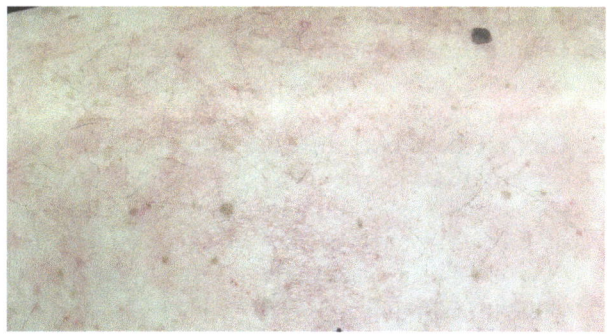

Dermatologic Adverse Effects
- Maculopapular rash (Fig. 10.7)
- Lichenoid dermatitis
- Vitiligo
- Pruritus
- Xerosis
- Mucositis
- Autoimmune blistering dermatitis
- Rare: alopecia, paronychia [110], photosensitivity, hyperhidrosis, hair pigmentation changes, sarcoidosis, erythema multiforme [111], erythema nodosum [112], sarcoidosis [113], psoriasis, keratoacanthomas, granuloma annulare [114], lupus-like eruption [115], panniculitis [116], scleroderma [117, 118], AGEP [119], AFND, Sjögren syndrome [102, 120], SJS/TEN [121]

Clinical Summary
- Common DAEs from anti-PD-1 antibody therapy include maculopapular rash, pruritus, xerosis, and mucositis, which can be managed with conventional, symptomatic treatment (see "General Principles of Management").
- Vitiligo is an uncommon DAE from anti-PD-1 antibody therapy that, similar to ipilimumab-induced vitiligo, is associated with response to therapy; no specific management is required for this DAE.
- Lichenoid dermatitis is a common DAE of anti-PD-1 antibody therapy that is clinically heterogeneous and should be managed with topical and/or systemic steroids. Dose reduction and/or drug discontinuation should be considered in severe cases.
- Autoimmune blistering dermatitis is an uncommon DAE of anti-PD-1 antibody therapy that should be managed with topical and/or systemic steroids. Dose reduction and/or drug discontinuation should be considered in severe cases, though the DAE can persist for months even after discontinuation of therapy.

Anti-PD-1-Induced Lichenoid Dermatitis [122, 123]

- **Incidence**: common; incidence not well-characterized
- **Natural Course**: variable time of onset (range: 3 weeks to 2 years after initiation of therapy) [47]
- **Clinical Features**: clinically heterogeneous; includes erythematous to violaceous, pruritic papules and plaques on the trunk oral lesions include flat-topped papules and erosions; histopathologic examination typically shows an abundance of CD3+ lymphocytes, few PD-1 positive T cells, and CD163+ histiocytes (which are not typically seen in cases of lichen planus)
- **Management**: topical and/or systemic steroids; consider dose reduction or drug discontinuation

Anti-PD-1-Induced Autoimmune Blistering Dermatitis [47, 108]

- **Incidence**: uncommon; incidence not well-characterized
- **Natural Course**: typically begins 3–4 months after initiating therapy
- **Clinical Features**: initial phase includes a pruritic, maculopapular rash that progresses to localized or generalized tense blisters and erosions; associated with oral mucosal involvement in 10–30% of cases; can persist for months after drug discontinuation
- **Management**: skin biopsy of lesion and perilesional skin for H&E staining and DIF; IIF and/or serum studies for autoantibodies (e.g., anti-BP180 antibodies, anti-BP230 antibodies); topical and/or systemic steroids; other therapies (e.g., methotrexate, omalizumab) have been successfully tried

Clinical Summary

- DAEs from anti-PD-L1antibody therapy are not well-documented but include maculopapular rash and pruritus, which can be managed with conventional, symptomatic treatment (see "General Principles of Management").
- Lichenoid dermatitis (Fig. 10.8) and autoimmune bullous dermatitis from anti-PD-L1antibody therapy have been reported and should be managed similarly to these DAEs from anti-PD-1antibody therapy.
- The similarities between the mechanisms of action of and anti-PD-1antibody and anti-PD-L1antibody therapy likely result in shared DAEs between these agents.

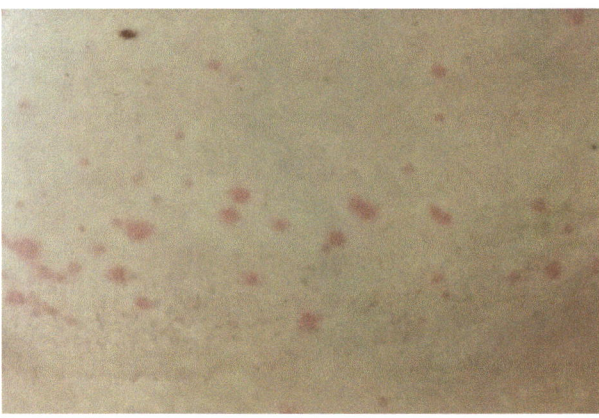

Fig. 10.8 Lichenoid dermatitis secondary to avelumab

Review Questions and Answers

1. Which of the following is the first step in reactive management of EGFR inhibitor therapy induced rash?

   (a) Consider skin biopsy
   (b) Consider prednisone 0.5 mg/kg × 5 days
   (c) Consider isotretinoin 20–30 mg daily
   (d) Bacterial culture, consider viral PCR swab for HSV/VZV and begin empiric therapy
   (e) Consider EGFR dose reduction or discontinuation

   Answer: (d) Bacterial culture, consider viral PCR swab for HSV/VZV and begin empiric therapy

2. Which of the following medications pose a risk of drug induced hyperproliferative epidermal neoplasms?

   (a) PD-1 inhibitors
   (b) Sunitinib
   (c) Copanlisib
   (d) Binimetinib
   (e) BRAF inhibitors

   Answer: (e) BRAF inhibitors

3. What are two common dermatologic adverse effects from bortezomib (protease inhibitor) therapy?

   (a) Mucositis and xerosis
   (b) Injection site reaction and VZV reactivation
   (c) Photosensitivity and edema
   (d) Hair and nail changes
   (e) Splinter hemorrhages and genital rash

   Answer: (b) Injection site reaction and VZV reactivation

4. Which of the following is false regarding hand-foot skin reactions resulting from sorafenib therapy?

   (a) Incidence is common, with up to 60% of patients reporting this DAE
   (b) Typically develops 2–4 weeks after sorafenib initiation
   (c) Hand-foot skin reactions have a predilection for non-pressure bearing areas
   (d) Manage with gentle exfoliation or pairing of hyperkeratotic lesions and avoid friction or trauma to the area
   (e) The reaction is distinct from chemotherapy-related hand-foot syndrome

   Answer: (c) Hand-foot skin reactions have a predilection for non-pressure bearing areas

5. A patient undergoing anti-PD-1 therapy might be expected to experience which of the following DAE?

   (a) Vitiligo-like melanoma-associated depigmentation
   (b) Panniculitis
   (c) Lichenoid dermatitis
   (d) Petechiae and purpura
   (e) DAE's are uncommon with PD-1 therapy

   Answer: (c) Lichenoid dermatitis

References

1. Rothenstein JM, Letarte N. Managing treatment-related adverse events associated with Alk inhibitors. Curr Oncol. 2014;21(1):19–26.
2. Oser MG, Jänne PA. A severe photosensitivity dermatitis caused by crizotinib. J Thorac Oncol. 2014;9(7):e51–3.
3. Lacouture ME, Laabs SM, Koehler M, Sweetman RW, Preston AJ, Di Leo A, Gomez HL, Salazar VM, Byrne JA, Koch KM, Blackwell KL. Analysis of dermatologic events in patients with cancer treated with lapatinib. Breast Cancer Res Treat. 2009;114(3):485–93.
4. Spring LM, Zangardi ML, Moy B, Bardia A. Clinical management of potential toxicities and drug interactions related to cyclin-dependent kinase 4/6 inhibitors in breast cancer: practical considerations and recommendations. Oncologist. 2017;22(9):1039–48.
5. Karagounis T, Vallurupalli M, Nathan N, Nazarian R, Vedak P, Spring L, Chen ST. Stevens-Johnson syndrome-like eruption from palbociclib in a patient with metastatic breast cancer. JAAD Case Rep. 2018;4(5):452–4.
6. Widmer S, Grossman M. Chemotherapy patient with Stevens-Johnson syndrome presents to the emergency department: a case report. Am J Emerg Med. 2018;36(7):1325.e3–4.
7. Dika E, Patrizi A, Ribero S, Fanti PA, Starace M, Melotti B, Sperandi F, Piraccini BM. Hair and nail adverse events during treatment with targeted therapies for metastatic melanoma. Eur J Dermatol. 2016;26(3):232–9.
8. Gey A, Milpied B, Dutriaux C, Mateus C, Robert C, Perro G, Taieb A, Ezzedine K, Jouary T. Severe cutaneous adverse reaction associated with vemurafenib: DRESS, AGEP or overlap reaction. J Eur Acad Dermatol Venereol. 2016;30(1):178–9.
9. Gupta M, Huang V, Linette G, Cornelius L. Unusual complication of vemurafenib treatment of metastatic melanoma: exacerbation of acantholytic dyskeratosis complicated by Kaposi varicelliform eruption. Arch Dermatol. 2012;148(8):966–8.
10. Harding JJ, Barker CA, Carvajal RD, Wolchok JD, Chapman PB, Lacouture ME. Cutis verticis gyrata in association with vemurafenib and whole-brain radiotherapy. J Clin Oncol. 2014;32(14):e54–6.
11. Anforth R, Tembe V, Blumetti T, Fernandez-Peñas P. Mutational analysis of cutaneous squamous cell carcinomas and verrucal keratosis in patients taking BRAF inhibitors. Pigment Cell Melanoma Res. 2012;25(5):569–72.
12. Anforth R, Fernandez-Penas P. BRAF inhibitor induced verrucal keratosis. Am J Dermatopathol. 2014;36(2):192.
13. Anforth R, Fernandez-Peñas P, Long GV. Cutaneous toxicities of RAF inhibitors. Lancet Oncol. 2013;14(1):e11–8.
14. Zimmer L, Hillen U, Livingstone E, Lacouture ME, Busam K, Carvajal RD, Egberts F, Hauschild A, Kashani-Sabet M, Goldinger SM, Dummer R, Long GV, McArthur G, Scherag A, Sucker A, Schadendorf D. Atypical melanocytic proliferations and new primary melanomas in patients with advanced melanoma undergoing selective BRAF inhibition. J Clin Oncol. 2012;30(19):2375–83.

15. Anforth R, Liu M, Nguyen B, Uribe P, Kefford R, Clements A, Long GV, Fernandez-Peñas P. Acneiform eruptions: a common cutaneous toxicity of the MEK inhibitor trametinib. Australas J Dermatol. 2014;55(4):250–4.

16. Pang A, Tan KB, Aw D, Hsieh WS, Goh BC, Lee SC. A case of Sweet's syndrome due to 5-Azacytidine and vorinostat in a patient with NK/T cell lymphoma. Cutan Ocul Toxicol. 2012;31(1):64–6.

17. Anderson KA, Bartell HL, Olsen EA. Leukonychia related to vorinostat. Arch Dermatol. 2009;145(11):1338–9.

18. Kakar R, Rommel J, McKinley-Grant L, Shenoy AG, DeSimone JA. Hypersensitivity to romidepsin. J Am Acad Dermatol. 2014;70(1):e21–2.

19. Obeid KM, Ferrara R, Sharma M. Cutaneous lesion induced by a subcutaneous administration of bortezomib. Clin Lymphoma Myeloma Leuk. 2012;12(4):284–6.

20. Pour L, Adam Z, Buresova L, Krejci M, Krivanova A, Sandecka V, Zahradova L, Buchler T, Vorlicek J, Hajek R. Varicella-zoster virus prophylaxis with low-dose acyclovir in patients with multiple myeloma treated with bortezomib. Clin Lymphoma Myeloma. 2009;9(2):151–3.

21. Kim JS, Roh HS, Lee JW, Lee MW, Yu HJ. Distinct variant of Sweet's syndrome: bortezomib-induced histiocytoid Sweet's syndrome in a patient with multiple myeloma. Int J Dermatol. 2012;51(12):1491–3.

22. Murase JE, Wu JJ, Theate I, Cole GW, Barr RJ, Dyson SW. Bortezomib-induced histiocytoid Sweet syndrome. J Am Acad Dermatol. 2009;60(3):496–7.

23. Tanguy-Schmidt A, Avenel-Audran M, Croué A, Lissandre S, Dib M, Zidane-Marinnes M, Moles MP, Hunault-Berger M. [Bortezomib-induced acute neutrophilic dermatosis]. Annales de dermatologie et de venereologie 2009;136(5):443–6.

24. Thuillier D, Lenglet A, Chaby G, Royer R, Vaida I, Viseux V, Dadban A, Billet A, Christophe O, Chatelain D, Marolleau JP, Lok C, Damaj G. [Bortezomib-induced eruption: sweet syndrome? Two case reports]. Annales de dermatologie et de venereologie 2009;136(5):427–30.

25. Knoops L, Jacquemain A, Tennstedt D, Theate I, Ferrant A, Van den Neste E. Bortezomib-induced Sweet syndrome. Br J Haematol. 2005;131(2):142.

26. Hayaishi O. Prostaglandin D2 and sleep. Adv Prostaglandin Thromboxane Leukot Res. 1989;19:26–33.

27. Böckle BC, Baltaci M, Weyrer W, Sepp NT. Bortezomib-induced lupus erythematosus tumidus. Oncologist. 2009;14(6):637–9.

28. Liu L, Zhao N, Xu W, Sheng Z, Wang L. Pooled analysis of the reports of carfilzomib, panobinostat, and elotuzumab combinations in patients with refractory/relapsed multiple myeloma. J Hematol Oncol. 2016;9(1):54.

29. Nooka AK. Management of hematologic adverse events in patients with relapsed and/or refractory multiple myeloma treated with single-agent carfilzomib. Oncology. 2013;27(Suppl 3):11–8.

30. Smith DC, Kalebic T, Infante JR, Siu LL, Sullivan D, Vlahovic G, Kauh JS, Gao F, Berger AJ, Tirrell S, Gupta N, Di Bacco A, Berg D, Liu G, Lin J, Hui AM, Thompson JA. Phase 1 study of ixazomib, an investigational proteasome inhibitor, in advanced non-hematologic malignancies. Investig New Drugs. 2015;33(3):652–63.

31. Alloo A, Khosravi H, Granter SR, Jadeja SM, Richardson PG, Castillo JJ, LeBoeuf NR. Ixazomib-induced cutaneous necrotizing vasculitis. Support Care Cancer. 2018;26(7):2247–50.

32. Kumar SK, Bensinger WI, Zimmerman TM, Reeder CB, Berenson JR, Berg D, Hui AM, Gupta N, Di Bacco A, Yu J, Shou Y, Niesvizky R. Phase 1 study of weekly dosing with the investigational oral proteasome inhibitor ixazomib in relapsed/refractory multiple myeloma. Blood. 2014;124(7):1047–55.

33. Dummer R, Guminski A, Gutzmer R, Dirix L, Lewis KD, Combemale P, Herd RM, Kaatz M, Loquai C, Stratigos AJ, Schulze HJ, Plummer R, Gogov S, Pallaud C, Yi T, Mone M, Chang AL, Cornélis F, Kudchadkar R, Trefzer U, Lear JT, Sellami D, Migden MR. The 12-month analysis from basal cell carcinoma outcomes with LDE225 treatment (BOLT): a phase II,

randomized, double-blind study of sonidegib in patients with advanced basal cell carcinoma. J Am Acad Dermatol. 2016;75(1):113–125.e5.

34. Sanmartín O. Skin manifestations of targeted antineoplastic therapy. Curr Probl Dermatol. 2018;53:93–104.

35. Chang AL, Solomon JA, Hainsworth JD, Goldberg L, McKenna E, Day BM, Chen DM, Weiss GJ. Expanded access study of patients with advanced basal cell carcinoma treated with the Hedgehog pathway inhibitor, vismodegib. J Am Acad Dermatol. 2014;70(1):60–9.

36. Berlin J, Bendell JC, Hart LL, Firdaus I, Gore I, Hermann RC, Mulcahy MF, Zalupski MM, Mackey HM, Yauch RL, Graham RA, Bray GL, Low JA. A randomized phase II trial of vismodegib versus placebo with FOLFOX or FOLFIRI and bevacizumab in patients with previously untreated metastatic colorectal cancer. Clin Cancer Res. 2013;19(1):258–67.

37. Aasi S, Silkiss R, Tang JY, Wysong A, Liu A, Epstein E, Oro AE, Chang AL. New onset of keratoacanthomas after vismodegib treatment for locally advanced basal cell carcinomas: a report of 2 cases. JAMA Dermatol. 2013;149(2):242–3.

38. Lam T, Wolverton SE, Davis CL. Drug hypersensitivity syndrome in a patient receiving vismodegib. J Am Acad Dermatol. 2014;70(3):e65–6.

39. Falchi L, Baron JM, Orlikowski CA, Ferrajoli A. BCR signaling inhibitors: an overview of toxicities associated with Ibrutinib and Idelalisib in patients with chronic lymphocytic leukemia. Mediterr J Hematol Infect Dis. 2016;8(1):e2016011.

40. Greenwell IB, Ip A, Cohen JB. PI3K inhibitors: understanding toxicity mechanisms and management. Oncology. 2017;31(11):821–8.

41. Miller BW, Przepiorka D, de Claro RA, Lee K, Nie L, Simpson N, Gudi R, Saber H, Shord S, Bullock J, Marathe D, Mehrotra N, Hsieh LS, Ghosh D, Brown J, Kane RC, Justice R, Kaminskas E, Farrell AT, Pazdur R. FDA approval: idelalisib monotherapy for the treatment of patients with follicular lymphoma and small lymphocytic lymphoma. Clin Cancer Res. 2015;21(7):1525–9.

42. Gabriel JG, Kapila A, Gonzalez-Estrada A. A severe case of cutaneous adverse drug reaction secondary to a novice drug: idelalisib. J Investig Med High Impact Case Rep. 2017;5(2):2324709617711463.

43. Dreyling M, Morschhauser F, Bouabdallah K, Bron D, Cunningham D, Assouline SE, Verhoef G, Linton K, Thieblemont C, Vitolo U, Hiemeyer F, Giurescu M, Garcia-Vargas J, Gorbatchevsky I, Liu L, Koechert K, Peña C, Neves M, Childs BH, Zinzani PL. Phase II study of copanlisib, a PI3K inhibitor, in relapsed or refractory, indolent or aggressive lymphoma. Ann Oncol. 2017;28(9):2169–78.

44. Flinn IW, Patel M, Oki Y, Horwitz S, Foss FF, Allen K, Douglas M, Stern H, Sweeney J, Kharidia J, Kelly P, Kelly VM, Kahl B. Duvelisib, an oral dual PI3K-δ, γ inhibitor, shows clinical activity in indolent non-Hodgkin lymphoma in a phase 1 study. Am J Hematol. 2018;93(11):1311–7.

45. Horwitz SM, Koch R, Porcu P, Oki Y, Moskowitz A, Perez M, Myskowski P, Officer A, Jaffe JD, Morrow SN, Allen K, Douglas M, Stern H, Sweeney J, Kelly P, Kelly V, Aster JC, Weaver D, Foss FM, Weinstock DM. Activity of the PI3K-δ,γ inhibitor duvelisib in a phase 1 trial and preclinical models of T-cell lymphoma. Blood. 2018;131(8):888–98.

46. Ayirookuzhi SJ, Ma L, Ramshesh P, Mills G. Imatinib-induced sweet syndrome in a patient with chronic myeloid leukemia. Arch Dermatol. 2005;141(3):368–70.

47. Ransohoff JD, Kwong BY. Cutaneous adverse events of targeted therapies for hematolymphoid malignancies. Clin Lymphoma Myeloma Leuk. 2017;17(12):834–51.

48. Assouline S, Laneuville P, Gambacorti-Passerini C. Panniculitis during dasatinib therapy for imatinib-resistant chronic myelogenous leukemia. N Engl J Med. 2006;354(24):2623–4.

49. Zhang M, Hassan KM, Musiek A, Rosman IS. Ponatinib-induced neutrophilic panniculitis. J Cutan Pathol. 2014;41(7):597–601.

50. Eber AE, Rosen A, Oberlin KE, Giubellino A, Romanelli P. Ichthyosiform Pityriasis Rubra pilaris-like eruption secondary to Ponatinib therapy: case report and literature review. Drug Saf. 2017;4(1):19.

51. Butler TW, Waddell JA, Solimando DA. Drug monographs: pomalidomide and ponatinib. Hosp Pharm. 2013;48(8):636–41.
52. Jack A, Mauro MJ, Ehst BD. Pityriasis rubra pilaris-like eruption associated with the multi-kinase inhibitor ponatinib. J Am Acad Dermatol. 2013;69(5):e249–50.
53. Lee WJ, Lee JL, Chang SE, Lee MW, Kang YK, Choi JH, Moon KC, Koh JK. Cutaneous adverse effects in patients treated with the multitargeted kinase inhibitors sorafenib and suni-tinib. Br J Dermatol. 2009;161(5):1045–51.
54. Arnault JP, Wechsler J, Escudier B, Spatz A, Tomasic G, Sibaud V, Aractingi S, Grange JD, Poirier-Colame V, Malka D, Soria JC, Mateus C, Robert C. Keratoacanthomas and squamous cell carcinomas in patients receiving sorafenib. J Clin Oncol. 2009;27(23):e59–61.
55. Uraizee I, Cheng S, Moslehi J. Reversible cardiomyopathy associated with sunitinib and sorafenib. N Engl J Med. 2011;365(17):1649–50.
56. Wolber C, Udvardi A, Tatzreiter G, Schneeberger A, Volc-Platzer B. Perforating folliculitis, angioedema, hand-foot syndrome—multiple cutaneous side effects in a patient treated with sorafenib. J Dtsch Dermatol Ges. 2009;7(5):449–52.
57. Autier J, Escudier B, Wechsler J, Spatz A, Robert C. Prospective study of the cuta-neous adverse effects of sorafenib, a novel multikinase inhibitor. Arch Dermatol. 2008;144(7):886–92.
58. Kong HH, Turner ML. Array of cutaneous adverse effects associated with sorafenib. J Am Acad Dermatol. 2009;61(2):360–1.
59. Chung C, Dawson LA, Joshua AM, Brade AM. Radiation recall dermatitis triggered by multi-targeted tyrosine kinase inhibitors: sunitinib and sorafenib. Anti Cancer Drugs. 2010;21(2):206–9.
60. Namba M, Tsunemi Y, Kawashima M. Sorafenib-induced erythema multiforme: three cases. Eur J Dermatol. 2011;21(6):1015–6.
61. Mancano MA. Pancreatitis-associated with riluzole; linezolid-induced hypoglycemia; sorafenib-induced acute generalized exanthematous pustulosis; creatine supplementation-induced thrombotic events; acute pancreatitis associated with quetiapine; hypomagnesemia and seizure associated with rabeprazole. Hosp Pharm. 2014;49(11):1004–8.
62. Liang CP, Yang CS, Shen JL, Chen YJ. Sorafenib-induced acute localized exanthematous pustulosis in a patient with hepatocellular carcinoma. Br J Dermatol. 2011;165(2):443–5.
63. Kim DK, Lee SW, Nam HS, Jeon DS, Park NR, Nam YH, Lee SK, Baek YH, Han SY, Lee SW. A case of sorafenib-induced DRESS syndrome in hepatocelluar carcinoma. Korean J Gastroenterol. 2016;67(6):337–40.
64. Ikeda M, Fujita T, Amoh Y, Mii S, Matsumoto K, Iwamura M. Stevens-Johnson syndrome induced by sorafenib for metastatic renal cell carcinoma. Urol Int. 2013;91(4):482–3.
65. Paz C, Querfeld C, Shea CR. Sorafenib-induced eruption resembling pityriasis rubra pilaris. J Am Acad Dermatol. 2011;65(2):452–3.
66. Hartmann JT, Kanz L. Sunitinib and periodic hair depigmentation due to temporary c-KIT inhibition. Arch Dermatol. 2008;144(11):1525–6.
67. McLellan B, Kerr H. Cutaneous toxicities of the multikinase inhibitors sorafenib and suni-tinib. Dermatol Ther. 2011;24(4):396–400.
68. Dean SM, Zirwas M. A second case of Sunitinib-associated pyoderma Gangrenosum. J Clin Aesthet Dermatol. 2010;3(8):34–5.
69. ten Freyhaus K, Homey B, Bieber T, Wilsmann-Theis D. Pyoderma gangrenosum: another cutaneous side-effect of sunitinib. Br J Dermatol. 2008;159(1):242–3.
70. Nadauld LD, Miller MB, Srinivas S. Pyoderma gangrenosum with the use of sunitinib. J Clin Oncol. 2011;29(10):e266–7.
71. Mihara Y, Yamaguchi K, Nakama T, Nakayama G, Kamei H, Ishibashi N, Uchida S, Akagi Y, Ogata Y. [Stevens-Johnson syndrome induced by regorafenib in a patient with progressive recurrent rectal carcinoma]. Cancer Chemother 2015;42(2):233–236.
72. Sternberg CN, Davis ID, Mardiak J, Szczylik C, Lee E, Wagstaff J, Barrios CH, Salman P, Gladkov OA, Kavina A, Zarbá JJ, Chen M, McCann L, Pandite L, Roychowdhury DF,

Hawkins RE. Pazopanib in locally advanced or metastatic renal cell carcinoma: results of a randomized phase III trial. J Clin Oncol. 2010;28(6):1061–8.

73. Kawai A, Araki N, Hiraga H, Sugiura H, Matsumine A, Ozaki T, Ueda T, Ishii T, Esaki T, Machida M, Fukasawa N. A randomized, double-blind, placebo-controlled, phase III study of pazopanib in patients with soft tissue sarcoma: results from the Japanese subgroup. Jpn J Clin Oncol. 2016;46(3):248–53.

74. Alpuim Costa D, Baptista de Almeida S, Coelho Barata P, Quintela A, Cabral P, Afonso A, Maia Silva J. Pazopanib-induced cutaneous leukocytoclastic vasculitis: an exclusion diagnosis of a multidisciplinary approach. Case Rep Oncol. 2017;10(3):1041–9.

75. Usui S, Otsuka A, Kaku Y, Dainichi T, Kabashima K. Pyoderma gangrenosum of the penis possibly associated with pazopanib treatment. J Eur Acad Dermatol Venereol. 2016;30(7):1222–3.

76. Negulescu M, Zerdoud S, Boulinguez S, Tournier E, Delord JP, Baran R, Sibaud V. Development of photoonycholysis with vandetanib therapy. Skin Append Disord. 2017;2(3–4):146–51.

77. Caro-Gutiérrez D, Floristán Muruzábal MU, de la Fuente EG, Franco AP, López Estebaranz JL. Photo-induced erythema multiforme associated with vandetanib administration. J Am Acad Dermatol. 2014;71(4):e142–4.

78. Giacchero D, Ramacciotti C, Arnault JP, Brassard M, Baudin E, Maksimovic L, Mateus C, Tomasic G, Wechsler J, Schlumberger M, Robert C. A new spectrum of skin toxic effects associated with the multikinase inhibitor vandetanib. Arch Dermatol. 2012;148(12):1418–20.

79. Yoon J, Oh CW, Kim CY. Stevens-Johnson syndrome induced by vandetanib. Ann Dermatol. 2011;23(Suppl 3):S343–5.

80. Ioannidis G, Gkogkou P, Charalampous P, Diamandi M, Ioannou R. Radiation-recall dermatitis with the everolimus/exemestane combination ten years after adjuvant whole-breast radiotherapy. Radiother Oncol. 2014;112(3):449–50.

81. Kurtzman DJ, Oulton J, Erickson C, Curiel-Lewandrowski C. Everolimus-induced symmetrical drug-related intertriginous and flexural exanthema (SDRIFE). Dermatitis. 2016;27(2):76–7.

82. Atzori L, Conti B, Zucca M, Pau M. Bullous pemphigoid induced by m-TOR inhibitors in renal transplant recipients. J Eur Acad Dermatol Venereol. 2015;29(8):1626–30.

83. Yee KW, Hymes SR, Heller L, Prieto VG, Welch MA, Giles FJ. Cutaneous leukocytoclastic vasculitis in a patient with myelodysplastic syndrome after therapy with the rapamycin analogue everolimus: case report and review of the literature. Leuk Lymphoma. 2006;47(5):926–9.

84. Bhojani N, Jeldres C, Patard JJ, Perrotte P, Suardi N, Hutterer G, Patenaude F, Oudard S, Karakiewicz PI. Toxicities associated with the administration of sorafenib, sunitinib, and temsirolimus and their management in patients with metastatic renal cell carcinoma. Eur Urol. 2008;53(5):917–30.

85. Okuno S, Bailey H, Mahoney MR, Adkins D, Maples W, Fitch T, Ettinger D, Erlichman C, Sarkaria JN. A phase 2 study of temsirolimus (CCI-779) in patients with soft tissue sarcomas: a study of the Mayo phase 2 consortium (P2C). Cancer. 2011;117(15):3468–75.

86. Peuvrel L, Quéreux G, Brocard A, Saint-Jean M, Dréno B. Onychopathy induced by temsirolimus, a mammalian target of rapamycin inhibitor. Dermatology. 2012;224(3):204–8.

87. Sibaud V, Dalenc F, Mourey L, Chevreau C. Paronychia and pyogenic granuloma induced by new anticancer mTOR inhibitors. Acta Derm Venereol. 2011;91(5):584–5.

88. Bitar C, Farooqui MZ, Valdez J, Saba NS, Soto S, Bray A, Marti G, Wiestner A, Cowen EW. Hair and nail changes during long-term therapy with ibrutinib for chronic lymphocytic leukemia. JAMA Dermatol. 2016;152(6):698–701.

89. Lipsky AH, Farooqui MZ, Tian X, Martyr S, Cullinane AM, Nghiem K, Sun C, Valdez J, Niemann CU, Herman SE, Saba N, Soto S, Marti G, Uzel G, Holland SM, Lozier JN, Wiestner A. Incidence and risk factors of bleeding-related adverse events in patients with chronic lymphocytic leukemia treated with ibrutinib. Haematologica. 2015;100(12):1571–8.

90. Iberri DJ, Kwong BY, Stevens LA, Coutre SE, Kim J, Sabile JM, Advani RH. Ibrutinib-associated rash: a single-centre experience of clinicopathological features and management. Br J Haematol. 2018;180(1):164–6.
91. Fabbro SK, Smith SM, Dubovsky JA, Gru AA, Jones JA. Panniculitis in patients undergoing treatment with the Bruton tyrosine kinase inhibitor ibrutinib for lymphoid leukemias. JAMA Oncol. 2015;1(5):684–6.
92. Stewart J, Bayers S, Vandergriff T. Self-limiting ibrutinib-induced neutrophilic panniculitis. Am J Dermatopathol. 2018;40(2):e28–9.
93. Sławińska M, Barańska-Rybak W, Sobjanek M, Wilkowska A, Mital A, Nowicki R. Ibrutinib-induced pyoderma gangrenosum. Polskie Archiwum Medycyny Wewnetrznej. 2016;126(9):710–1.
94. Haddad RI, Schlumberger M, Wirth LJ, Sherman EJ, Shah MH, Robinson B, Dutcus CE, Teng A, Gianoukakis AG, Sherman SI. Incidence and timing of common adverse events in lenvatinib-treated patients from the SELECT trial and their association with survival outcomes. Endocrine. 2017;56(1):121–8.
95. Chandesris MO, Damaj G, Canioni D, Brouzes C, Lhermitte L, Hanssens K, Frenzel L, Cherquaoui Z, Durieu I, Durupt S, Gyan E, Beyne-Rauzy O, Launay D, Faure C, Hamidou M, Besnard S, Diouf M, Schiffmann A, Niault M, Jeandel PY, Ranta D, Gressin R, Chantepie S, Barete S, Dubreuil P, Bourget P, Lortholary O, Hermine O. Midostaurin in advanced systemic mastocytosis. N Engl J Med. 2016;374(26):2605–7.
96. Shirasawa M, Kubotaa M, Harada S, Niwa H, Kusuhara S, Kasajima M, Hiyoshi Y, Ishihara M, Igawa S, Masuda N. Successful oral desensitization against skin rash induced by alectinib in a patient with anaplastic lymphoma kinase-positive lung adenocarcinoma: a case report. Lung Cancer. 2016;99:66–8.
97. Kimura T, Sowa-Osako J, Nakai T, Ohyama A, Kawaguchi T, Tsuruta D, Ohsawa M, Hirata K. Alectinib-induced erythema multiforme and successful rechallenge with alectinib in a patient with anaplastic lymphoma kinase-rearranged lung cancer. Case Rep Oncol. 2016;9(3):826–32.
98. Eytan M. Stein et al. "Ivosidenib or Enasidenib combined with standard induction chemotherapy is well tolerated and active in patients with newly diagnosed AML with an IDH1 or IDH2 mutation: initial results from a phase 1 trial". Blood. 2017;130(Suppl 1):726.
99. Koelzer VH, Buser T, Willi N, Rothschild SI, Wicki A, Schiller P, Cathomas G, Zippelius A, Mertz KD. Grover's-like drug eruption in a patient with metastatic melanoma under ipilimumab therapy. J Immunother Cancer. 2016;4:47.
100. Arellano K, Mosley JC, Moore DC. Case report of ipilimumab-induced diffuse, non-necrotizing granulomatous lymphadenitis and granulomatous vasculitis. J Pharm Pract. 2018;31(2):227–9.
101. Firwana B, Ravilla R, Raval M, Hutchins L, Mahmoud F. Sarcoidosis-like syndrome and lymphadenopathy due to checkpoint inhibitors. J Oncol Pharm Prac. 2017;23(8):620–4.
102. Calabrese C, Kirchner E, Kontzias A, Velcheti V, Calabrese LH. Rheumatic immune-related adverse events of checkpoint therapy for cancer: case series of a new nosological entity. RMD Open. 2017;3(1):e000412.
103. Gormley R, Wanat K, Elenitsas R, Giles J, McGettigan S, Schuchter L, Takeshita J. Ipilimumab-associated Sweet syndrome in a melanoma patient. J Am Acad Dermatol. 2014;71(5):e211–3.
104. Pintova S, Sidhu H, Friedlander PA, Holcombe RF. Sweet's syndrome in a patient with metastatic melanoma after ipilimumab therapy. Melanoma Res. 2013;23(6):498–501.
105. Lacouture ME, Wolchok JD, Yosipovitch G, Kähler KC, Busam KJ, Hauschild A. Ipilimumab in patients with cancer and the management of dermatologic adverse events. J Am Acad Dermatol. 2014;71(1):161–9.
106. Choi JN. Dermatologic adverse events to chemotherapeutic agents, part 2: BRAF inhibitors, MEK inhibitors, and ipilimumab. Semin Cutan Med Surg. 2014;33(1):40–8.
107. Shen J, Chang J, Mendenhall M, Cherry G, Goldman JW, Kulkarni RP. Diverse cutaneous adverse eruptions caused by anti-programmed cell death-1 (PD-1) and anti-programmed cell

death ligand-1 (PD-L1) immunotherapies: clinical features and management. Therap Adv Med Oncol. 2018;10:1758834017751634.

108. Naidoo J, Schindler K, Querfeld C, Busam K, Cunningham J, Page DB, Postow MA, Weinstein A, Lucas AS, Ciccolini KT, Quigley EA, Lesokhin AM, Paik PK, Chaft JE, Segal NH, D'Angelo SP, Dickson MA, Wolchok JD, Lacouture ME. Autoimmune bullous skin disorders with immune checkpoint inhibitors targeting PD-1 and PD-L1. Cancer Immunol Res. 2016;4(5):383–9.

109. Yang H, Shen K, Zhu C, Li Q, Zhao Y, Ma X. Safety and efficacy of durvalumab (MEDI4736) in various solid tumors. Drug Des Devel Ther. 2018;12:2085–96.

110. Khokhar MO, Kettle J, Palla AR. Debilitating skin toxicity associated with Pembrolizumab therapy in an 81-year-old female with malignant melanoma. Case Rep Oncol. 2016;9(3):833–9.

111. Nomura H, Takahashi H, Suzuki S, Kurihara Y, Chubachi S, Kawada I, Yasuda H, Betsuyaku T, Amagai M, Funakoshi T. Unexpected recalcitrant course of drug-induced erythema multiforme-like eruption and interstitial pneumonia sequentially occurring after nivolumab therapy. J Dermatol. 2017;44(7):818–21.

112. Laroche A, Alarcon Chinchilla E, Bourgeault E, Doré MA. Erythema nodosum as the initial presentation of nivolumab-induced sarcoidosis-like reaction. J Cutan Med Surg. 2018;22(6):627–9.

113. Lomax AJ, McGuire HM, McNeil C, Choi CJ, Hersey P, Karikios D, Shannon K, van Hal S, Carr U, Crotty A, Gupta SK, Hollingsworth J, Kim H, Fazekas de St Groth B, McGill N. Immunotherapy-induced sarcoidosis in patients with melanoma treated with PD-1 checkpoint inhibitors: case series and immunophenotypic analysis. Int J Rheum Dis. 2017;20(9):1277–85.

114. Charollais R, Aubin F, Roche-Kubler B, Puzenat E. [Two cases of granuloma annulare under anti-PD1 therapy]. Annales de dermatologie et de venereologie 2018;145(2):116–119.

115. Shao K, McGettigan S, Elenitsas R, Chu EY. Lupus-like cutaneous reaction following pembrolizumab: an immune-related adverse event associated with anti-PD-1 therapy. J Cutan Pathol. 2018;45(1):74–7.

116. Burillo-Martinez S, Morales-Raya C, Prieto-Barrios M, Rodriguez-Peralto JL, Ortiz-Romero PL. Pembrolizumab-induced extensive panniculitis and nevus regression: two novel cutaneous manifestations of the post-immunotherapy granulomatous reactions spectrum. JAMA Dermatol. 2017;153(7):721–2.

117. Barbosa NS, Wetter DA, Wieland CN, Shenoy NK, Markovic SN, Thanarajasingam U. Scleroderma induced by Pembrolizumab: a case series. Mayo Clin Proc. 2017;92(7):1158–63.

118. Tjarks BJ, Kerkvliet AM, Jassim AD, Bleeker JS. Scleroderma-like skin changes induced by checkpoint inhibitor therapy. J Cutan Pathol. 2018;45(8):615–8.

119. Zhao CY, Consuegra G, Chou S, Fernández-Peñas P. Intracorneal pustular drug eruption, a novel cutaneous adverse event in anti-programmed cell death-1 patients that highlights the effect of anti-programmed cell death-1 in neutrophils. Melanoma Res. 2017;27(6):641–4.

120. Ghosn J, Vicino A, Michielin O, Coukos G, Kuntzer T, Obeid M. A severe case of neuro-Sjögren's syndrome induced by pembrolizumab. J Immunother Cancer. 2018;6(1):110.

121. Saw S, Lee HY, Ng QS. Pembrolizumab-induced Stevens-Johnson syndrome in non-melanoma patients. Eur J Cancer. 2017;81:237–9.

122. Joseph RW, Cappel M, Goedjen B, Gordon M, Kirsch B, Gilstrap C, Bagaria S, Jambusaria-Pahlajani A. Lichenoid dermatitis in three patients with metastatic melanoma treated with anti-PD-1 therapy. Cancer Immunol Res. 2015;3(1):18–22.

123. Collins LK, Chapman MS, Carter JB, Samie FH. Cutaneous adverse effects of the immune checkpoint inhibitors. Curr Probl Cancer. 2017;41(2):125–8.

Dermatologic Adverse Effects of Anticancer Therapy IV: Endocrine, Biologic, Radiation and Other Therapies

11

Timothy Dang, Hannah Thompson, Vincent Liu, and Bernice Kwong

---

Learning Objectives

- To recognize clinical indications (approved and off-label) for each therapy class.
- To identify the dermatologic adverse effects of each therapy class.
- To describe the incidence, natural course, clinical features, and management of DAEs from each therapy class summarized in the chapter.

Endocrine Therapy (Table 11.1)

Agents

- Selective estrogen receptor modulators (SERMs)
  - Tamoxifen
  - Toremifene
  - Raloxifene

T. Dang (✉)
Santa Clara Valley Medical Center, San Jose, CA, USA
e-mail: timdang@stanford.edu

H. Thompson
University of Iowa Health Care, Iowa City, IA, USA
e-mail: hannah-j-thompson@uiowa.edu

V. Liu
Department of Dermatology, University of Iowa Health Care, Iowa, IA, USA
e-mail: Vincent-liu@uiowa.edu

B. Kwong
Stanford Health Care; Stanford Medicine, Palo Alto, CA, USA
e-mail: bernicek@stanford.edu

© Springer Nature Switzerland AG 2021
V. Liu (ed.), *Dermato-Oncology Study Guide*,
https://doi.org/10.1007/978-3-030-53437-0_11

Table 11.1 Endocrinologic anti-cancer agents

Agent	Indications
Tamoxifen	• Breast cancer • Desmoid tumors • Ovarian cancer • Endometrial cancer
Toremifene	• Breast cancer • Desmoid tumors
Raloxifene	• Breast cancer
Anastrozole	• Breast cancer • Ovarian cancer • Endometrial cancer • Uterine cancer
Letrozole	• Breast cancer • Ovarian cancer
Exemestane	• Breast cancer
Fulvestrant	• Breast cancer
Leuprolide	• Prostate cancer • Uterine leiomyomata • Breast cancer
Flutamide Bicalutamide Nilutamide Enzalutamide Apalutamide Abiraterone (17α-hydroxylase)	• Prostate cancer
Fluoxymesterone	• Breast cancer
Estradiol	• Breast cancer • Prostate cancer
Octreotide	• Carcinoid tumors • Vasoactive intestinal peptide-secreting tumors • Diarrhea associated with GVHD, chemotherapy • Gastroenteropancreatic neuroendocrine tumors • Zollinger-Ellison syndrome • Thymoma/thymic malignancies
Lanreotide	• Carcinoid tumors • Gastroenteropancreatic neuroendocrine tumors
Megestrol acetate	• Breast cancer • Endometrial cancer
Medroxyprogesterone acetate	• Endometrial cancer
Pamidronate disodium	• Hypercalcemia of malignancy • Osteolytic bone metastases of breast cancer • Osteolytic lesions of multiple myeloma • Bone loss associated with androgen deprivation therapy • Symptomatic bone metastases of thyroid cancer
Zoledronic acid	• Bone metastases from solid tumors • Hypercalcemia of malignancy • Multiple myeloma • Bone loss associated with androgen deprivation therapy • Bone loss associated with aromatase inhibitor therapy
Goserelin	• Breast cancer • Prostate cancer
Degarelix	• Prostate cancer

- Aromatase Inhibitors
  - Nonsteroidal
    - Anastrozole
    - Letrozole
  - Steroidal
    - Exemestane
- Estrogen receptor downregulator
  - Fulvestrant
- Luteinizing-hormone releasing hormone agonist
  - Leuprolide
- Antiandrogens
  - Flutamide
  - Bicalutamide
  - Nilutamide
  - Enzalutamide
  - Apalutamide
  - Abiraterone (17α-hydroxylase)
- Androgen
  - Fluoxymesterone
- Estrogen
  - Estradiol
- Somatostatin Analogs
  - Octreotide
  - Lanreotide
- Progestational Agents
  - Megestrol acetate
  - Medroxyprogesterone acetate
- Bisphosphonate
  - Pamidronate disodium
  - Zoledronic acid
- GnRH agonist
  - Goserelin
- GnRH antagonist
  - Degarelix

Selective Estrogen Receptor Modulators (SERMs)

Agents

- Tamoxifen
- Toremifene
- Raloxifene

Dermatologic Adverse Effects [1]

- Maculopapular rash
- Alopecia
- Xerosis
- Pruritus
- Edema
- Hot flashes
- Vaginal dryness
- Radiation recall dermatitis [2, 3]
- Rare: hair color changes [4], hirsutism [5], vasculitis [6], dermatomyositis [7], eccrine squamous syringometaplasia [8], SCLE [9], SJS/TEN (Fig. 11.1), porphyria cutanea tarda, melasma

Clinical Summary

- DAEs from SERM therapy are common and include maculopapular rash, alopecia, xerosis, pruritus, and edema, which can be managed with conventional, symptomatic treatment (see "General Principles of Management"). DAEs from tamoxifen therapy are better documented than those from toremifene and raloxifene.
- Hot flashes are most common DAE from SERM therapy and can be managed with a combination of non-pharmacologic (e.g., loose-fitting clothing) and pharmacologic (e.g., citalopram therapy) measures.

SERM-Induced Hot Flashes

- **Incidence**: common; up to 80% of patients who receive tamoxifen therapy; less prevalent among patients who receive toremifene or raloxifene therapy
- **Natural Course**: Resolves upon treatment discontinuation, however for patients unable to discontinue use, there are effective management options

Fig. 11.1 Representative TEN

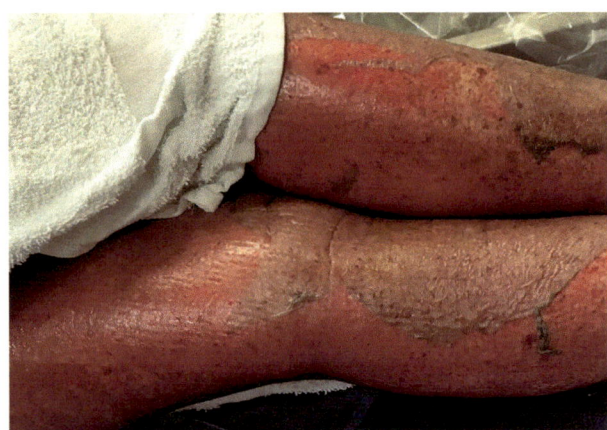

- **Clinical Features**: episodic sensation of intense heat in the chest, neck, and/or face; often accompanied by flushing, diaphoresis, palpitations, and anxiety; episodes typically last 2–4 min; can occur several times daily
- **Management**: non-pharmacologic measures (e.g., loose-fitting clothing, avoidance of alcohol); SSRIs/SNRIs (e.g., venlafaxine, citalopram, desvenlafaxine, escitalopram); gabapentin, pregabalin
- **Additional Considerations**: paroxetine and fluoxetine therapy should be avoided because these agents inhibit CYP2D6, which converts tamoxifen to its most active metabolite

Aromatase Inhibitors

Agents

- Anastrozole
- Letrozole
- Exemestane

Dermatologic Adverse Effects
- Maculopapular rash [10]
- Hair thinning [10]
- Hyperhidrosis
- Xerosis
- Pruritus
- Edema
- Mucositis
- Xerostomia
- Hot flashes
- Vaginal dryness
- Radiation recall dermatitis [11, 12]
- Rare: erythema multiforme [13], SCLE [14, 15], erythema nodosum [16], vasculitis [17–21] transient acantholytic dermatosis (Grover disease) [22], SJS/TEN [23]

Clinical Summary
- DAEs from aromatase inhibitor therapy are common and include maculopapular rash, hair thinning, xerosis, pruritus, edema, and mucositis, which can be managed with conventional, symptomatic treatment (see "General Principles of Management").

- Hot flashes are a common DAE from aromatase inhibitor therapy and can be managed with a combination of nonpharmacologic and pharmacologic measures. Refer to section on hot flashes from SERM therapy for additional considerations.
- Several rare DAEs from aromatase inhibitor therapy have been reported, including erythema multiforme, SCLE, and vasculitis.

Fulvestrant

Dermatologic Adverse Effects

- Pruritus
- Mucositis
- Hot flashes
- Hyperhidrosis
- Bromhidrosis
- Injection site reactions
- Rare: alopecia

Clinical Summary
- DAEs from fulvestrant therapy are uncommon and include pruritus, mucositis, and alopecia, which can be managed with conventional, symptomatic treatment (see "General Principles of Management").
- Hot flashes are a common DAE from fulvestrant therapy and can be managed with a combination of nonpharmacologic and pharmacologic measures. Refer to section on hot flashes from SERM therapy for additional considerations.

Leuprolide

Dermatologic Adverse Effects

- Maculopapular rash
- Hyperhidrosis
- Edema
- Photosensitivity
- Xerosis
- Alopecia
- Hot flashes
- Hyperhidrosis
- Injection site reactions

- Rare: SCLE [24] and lupus-like syndrome [25], psoriasiform eruption [26], granulomas [27–29], hypersensitivity reactions [30], serum sickness [31], dermatitis herpetiformis [32, 33], fixed drug eruption [34], vasculitis [35]

Clinical Summary
- DAEs from leuprolide therapy are common and include maculopapular rash, edema, photosensitivity, xerosis, and alopecia, which can be managed with conventional, symptomatic treatment (see "General Principles of Management").
- Hot flashes are a common DAE from leuprolide therapy and can be managed with a combination of nonpharmacologic and pharmacologic measures. Refer to section on hot flashes from SERM therapy for additional considerations.
- Several rare DAEs from leuprolide therapy have been reported, including SCLE, serum sickness, dermatitis herpetiformis, and vasculitis.

Antiandrogens

Agents

- Flutamide
- Bicalutamide
- Nilutamide
- Enzalutamide
- Apalutamide
- Abiraterone (17α-hydroxylase)

Dermatologic Adverse Effects
- Maculopapular rash
- Xerosis
- Pruritus
- Edema
- Hot flashes
- Photosensitivity [36, 37] and photosensitive drug reactions [38]
- Rare: photoleukomelanoderma [36, 39], lupus-like reaction [40], AGEP [41]

Clinical Summary
- DAEs from antiandrogen therapy are common and include maculopapular rash, xerosis, pruritus, edema, and photosensitivity, which can be managed with conventional, symptomatic treatment (see "General Principles of Management").
- Hot flashes are a common DAE from antiandrogen therapy and can be managed with a combination of nonpharmacologic and pharmacologic

measures. Refer to section on hot flashes from SERM therapy for additional considerations.

- Several rare DAEs from antiandrogen therapy have been reported, including photoleukomelanoderma, lupus-like reaction, and AGEP.

Fluoxymesterone

Dermatologic Adverse Effects

- Acne vulgaris [42]
- Androgenetic alopecia
- Hirsutism

Clinical Summary
- Fluoxymesterone therapy causes androgenic effects that lead to DAEs such as acne vulgaris, androgenetic alopecia, and hirsutism. These DAEs may require dose reduction or drug discontinuation.

Estradiol

Dermatologic Adverse Effects

- Rare: erythema nodosum [43], chronic dermatitis with transformation to cutaneous T-cell lymphoma [44]

Clinical Summary
- DAEs from estradiol therapy are rare. Erythema nodosum and chronic dermatitis that transformed into cutaneous T-cell lymphoma at the site of a transdermal estradiol patch have been reported.

Somatostatin Analogs

Agents

- Octreotide
- Lanreotide

Dermatologic Adverse Effects
- Rare: nail changes (e.g., Beau lines) [45], granulomas [46], hypersensitivity reactions [47], eruptive nevi [48], alopecia [49–51], radiation recall dermatitis [52], injection site reactions

Clinical Summary

- DAEs from somatostatin analog therapy are rare. Granulomas, alopecia, radiation recall dermatitis, and injection site reactions have been reported.

Megestrol Acetate

Dermatologic Adverse Effects

- Hot flashes
- Rare: alopecia, hirsutism

Clinical Summary

- DAEs from megestrol acetate therapy are uncommon. Hot flashes can be managed with a combination of nonpharmacologic and pharmacologic measures. Refer to section on hot flashes from SERM therapy for additional considerations.

Medroxyprogesterone Acetate

Dermatologic Adverse Effects

- Hot flashes
- Acne vulgaris [53]
- Injection site reactions
- Rare: AGEP [54], erythema multiforme [55]

Clinical Summary

- DAEs from megestrol acetate therapy are uncommon. Hot flashes can be managed with a combination of nonpharmacologic and pharmacologic measures. Refer to section on hot flashes from SERM therapy for additional considerations.
- Rare DAEs from megestrol acetate therapy have been reported such as AGEP and erythema multiforme.

Pamidronate Disodium

Dermatologic Adverse Effects

- Rare: nonspecific rash, nail changes (e.g., subungual hemorrhage, pincer nails)

Clinical Summary
- DAEs from pamidronate disodium therapy are rare. Nonspecific rash and nail changes have been reported.

Zoledronic Acid

Dermatologic Adverse Effects

- Injection site reactions
- Rare: maculopapular rash, dermatomyositis [56, 57], lip ulceration [58], cutaneous B-cell pseudolymphoma [59]

Clinical Summary
- DAEs from zoledronic acid therapy are rare. Dermatomyositis and cutaneous B-cell pseudolymphoma have been reported.

Goserelin

Dermatologic Adverse Effects

- Maculopapular rash [60]
- Xerosis
- Pruritus
- Acne vulgaris
- Hot flashes
- Rare: hypersensitivity reactions [61]

Clinical Summary
- DAEs from goserelin therapy are uncommon and include xerosis and pruritus, which can be managed with conventional, symptomatic treatment (see "General Principles of Management").
- Hot flashes can be managed with a combination of nonpharmacologic and pharmacologic measures. Refer to section on hot flashes from SERM therapy for additional considerations.
- Rare hypersensitivity reactions from goserelin therapy have been reported.

Degarelix

Dermatologic Adverse Effects

- Hot flashes

Clinical Summary

- DAEs from degarelix therapy are rare and not well-documented.
- Hot flashes can be managed with a combination of nonpharmacologic and pharmacologic measures. Refer to section on hot flashes from SERM therapy for additional considerations.

Biotherapy

Agents

- Interferon α (Table 11.2)
  - Interferon alfa-2b
  - Pegylated interferon (peginterferon) alfa-2b
- Interleukins
  - Interleukin-2 (IL-2, aldesleukin)

Dermatologic Adverse Effects [62, 63]

- Maculopapular rash
- Alopecia
- Pruritus
- Xerosis
- Urticaria
- Hair changes (e.g., alopecia, hair discoloration, texture change) [64]
- Psoriasis (exacerbations and new-onset) [65, 66]
- Injection site reactions
- Radiation recall dermatitis [67]

Table 11.2 Interferon α

Agents	Indications
Interferon alfa-2b	• AIDS-related Kaposi sarcoma • Follicular lymphoma • Hairy cell leukemia • Melanoma • CML • Cutaneous T-cell lymphoma • Desmoid tumors • Multiple myeloma • Neuroendocrine tumors (e.g., carcinoid syndrome, islet cell tumor) • Non-Hodgkin lymphoma • Renal cell carcinoma
Pegylated interferon (peginterferon) alfa-2b	• Melanoma

- Rare: halo dermatitis [68], lipoatrophy [69], livedo reticularis [70], eosinophilic fasciitis, autoimmune blistering disorders [71, 72], SLE, Sjögren syndrome [73], AFND [74], livedoid dermatitis [75], granuloma annulare [76], panniculitis [77], Raynaud phenomenon [62], sarcoidosis [78], pyoderma gangrenosum [79], vasculitis, dermatomyositis [80], Kaposi sarcoma [81], SJS/TEN

Clinical Summary
- DAEs from interferon alfa-2b therapy are common and heterogeneous; however, attribution of DAEs specifically to interferon alfa-2b therapy is challenging because of frequent administration with other agents (e.g., ribavirin and telaprevir in the treatment of hepatitis C infection).
- Common DAEs include alopecia, maculopapular rash, and injection site reactions, which can be managed with conventional, symptomatic treatment (see "General Principles of Management").
- Exacerbations of psoriasis from interferon alfa-2b therapy have been reported and should be managed with conventional treatment for psoriasis (e.g., topical and/or systemic steroids, topical calcineurin inhibitors). Consider dose reduction or drug discontinuation in severe in cases.
- Rare DAEs from interferon alfa-2b therapy such as AFND and autoimmune blistering disorders have been described.

Interferon Alfa-2b-Related Psoriasis

- **Incidence**: uncommon; up to 5% of patients
- **Natural Course**: typically begins 3–4 months after initiating therapy
- **Clinical Features**: includes new-onset and exacerbations of psoriasis
- **Management**: topical and/or systemic steroids; topical calcineurin inhibitors; consider dose reduction or drug discontinuation

Interleukin-2 (IL-2, Aldesleukin)

Indications

- Melanoma
- Renal cell cancer

Dermatologic Adverse Effects
- Maculopapular rash
- Urticaria
- Edema (from capillary leak syndrome)

- Alopecia
- Xerosis
- Pruritus
- Mucositis
- Erythema nodosum
- Bullous eruptions [82, 83] (e.g., autoimmune blistering disorders) [84, 85]
- Psoriasis (exacerbations and new-onset) [86, 87]
- Injection site reactions
- Rare: AFND [88], AGEP [89], hypersensitivity reactions [90], erythema multiforme [91], erythema nodosum [92], Kaposi sarcoma [93], sarcoidosis [94], scleroderma [95]

Clinical Summary
- DAEs from interleukin-2 therapy are common and heterogeneous, These DAEs tend to be mild and do not require dose reduction or drug discontinuation.
- Common DAEs from interleukin-2 therapy include maculopapular rash, alopecia, xerosis, pruritus, mucositis, and injection site reactions, which can be managed with conventional, symptomatic treatment (see "General Principles of Management").
- Rare DAEs from interleukin-2 therapy such as AFND and autoimmune blistering disorders have been described.

Radiation Therapy

Radiopharmaceuticals

Agents
- Lutetium Lu-177 Dotatate
- Radium 223 Dichloride

Lutetium Lu-177 Dotatate

Main Indication
- Gastroenteropancreatic neuroendocrine tumors

Dermatologic Adverse Effects [96]
- Alopecia
- Edema
- Flushing reaction

Clinical Summary
- DAEs from lutetium Lu-177 dotatate therapy are not well-documented and include alopecia and edema, which can be managed with conventional, symptomatic treatment (see "General Principles of Management").

Radium Ra-223 Dichloride

Main Indication
- Pancreatic cancer

Dermatologic Adverse Effects
- Nonspecific rash [97]
- Edema [98, 99]

Clinical Summary
- DAEs from radium Ra-223 dichloride therapy are not well-documented and include nonspecific rash and edema, which can be managed with conventional, symptomatic treatment (see "General Principles of Management").

Radiation Dermatitis

Acute Radiation Dermatitis [100, 101]

- **Incidence**: Common, occurs in 85–95% of individuals undergoing radiotherapy treatment
- **Natural Course**: Acute radiation dermatitis occurs within 90 days of exposure to radiation therapy, most commonly after 2–4 weeks; time to resolution of DAE is usually 2–4 weeks after the end of treatment
- **Clinical Features**: Skin changes can vary, ranging from reddening and faint erythema to desquamation (Fig. 11.2), ulceration, and skin necrosis depending on the patients severity of reaction (see Table 11.3)
- **Management**: Wash area with non-irritant soap and water, use topical corticosteroids and silver nylon dressing; careful assessment of radiation dose and distribution is essential, consider discontinuing any concomitant medications that may have contributed, consider alternative explanation for skin reactions and investigate, avoid sun exposure and topical irritants as well as trauma to affected area to limit damage; no consensus exists for treatment but many treatments (herbal, hygienic, topical vitamins, endogenous agents, pharmaceuticals, metallic ointments and dressings) have been implemented

Fig. 11.2 Acute radiation dermatitis

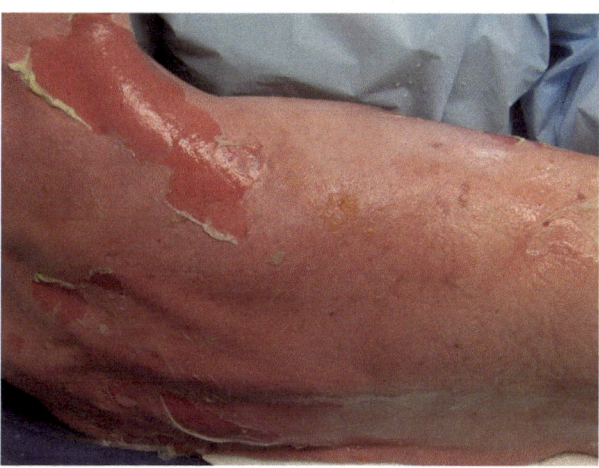

Table 11.3 Grading criteria for acute radiation dermatitis

Grade 1	Faint erythema or dry desquamation
Grade 2	Moderate erythema or desquamation confined to skin folds and creases, moderate swelling
Grade 3	Confluent moist desquamation > 1.5 cm diameter, not confined to skin folds, pitting edema
Grade 4	Skin ulceration or necrosis of full thickness dermis

Chronic Radiation Dermatitis [100, 102]

- **Incidence**: No precise data for chronic radiation dermatitis; up to 95% of patients receiving radiotherapy have some form of skin toxicity
- **Natural Course**: Skin changes that occur more than 90 days after starting therapy; considered an irreversible and progressive condition
- **Clinical Features**: Disappearance of follicular structures, damage to elastic fibers in the dermis, increase in collagen, fragile epidermis, telangiectasias (Fig. 11.3); secondary skin cancers (BCC) have also been reported from chronic radiation exposure; most severe complication includes radiation-induced fibrosis (RIF) which manifests as skin induration or retraction, lymphedema and possibly restriction of joint movement
- **Management**: If unclear manifestation, use of biopsy with histopathological examination is necessary to rule out secondary malignancy; most effective strategy is prophylaxis using proper radiation therapy technique; use of pulse dye laser therapy may be used to treat telangiectasias; chronic wounds can be managed symptomatically (see "General Principles of Management"); RIF can be addressed with rehabilitation, pharmacotherapy, hyperbaric oxygen and laser therapy

Fig. 11.3 Chronic
radiation dermatitis

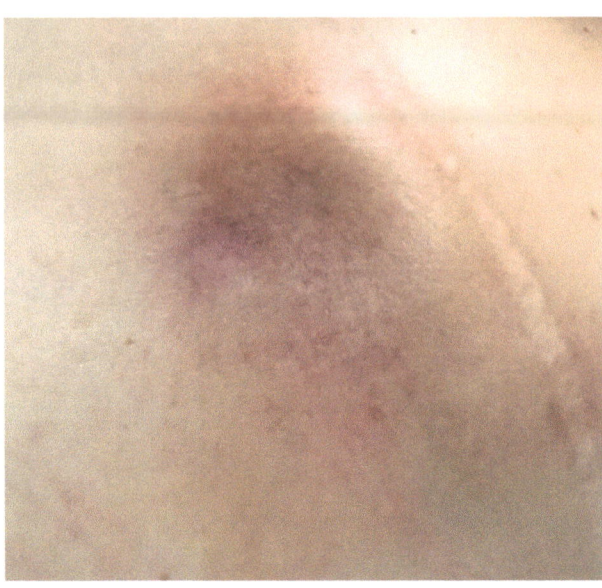

Radiation Recall Dermatitis [103]

- **Incidence**: Rare
- **Natural Course**: Reaction may occur weeks to years after completion of radiation therapy
- **Clinical Features**: Radiation recall dermatitis occurs at sites of previous skin exposure to radiation therapy after a patient initiates chemotherapy agents; clinical features vary and range from erythema, bleeding, ulceration, and necrosis (Fig. 11.4)
- **Management**: Radiation recall dermatitis usually resolves without treatment; steroids and antihistamines have been used to control pruritus/pain, but time to resolution is not impacted

Clinical Summary
- Radiation dermatitis is most commonly caused by radiotherapy for underlying malignancies, leading to various patterns of direct tissue damage to the epidermis and endothelial/support cells of the dermis.
- Clinical features of radiation dermatitis manifest within days to weeks after the start of radiation therapy. Exact timing depends on radiation dose intensity and sensitivity of tissue.
- Clinical features include moderate to severe skin reactions such as erythema, ulceration, and necrosis that is sharply demarcated and confined to skin areas that have received radiation.

Fig. 11.4 Radiation recall dermatitis

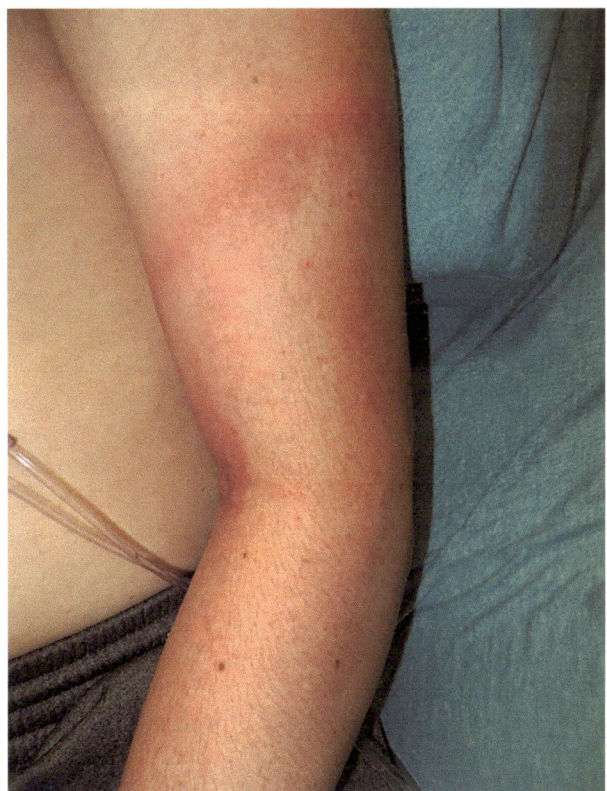

- Risk factors include poor nutrition, pre-existing skin disease (SLE, SS, MCTD), overlapping skin folds/obesity, diabetes mellitus, prolonged exposure, concurrent cetuximab therapy, application of skin creams immediately before treatment; limiting risk factors can help manage disease severity.

Stem Cell Transplantation

Acute Graft Versus Host Disease [104–106]

- **Incidence**: Of patients receiving allogenic hematopoietic-cell transplant, up to 50% have acute manifestations with ~14% having severe disease
- **Natural Course**: Manifests within 100 days following stem cell transplantation, skin involvement is usually the earliest clinical manifestation
- **Clinical Features**: Erythematous, maculopapular morbilliform rash (Fig. 11.5) starting on the face/palms/soles, erythroderma, and jaundice from resulting hyperbilirubinemia; other features include nausea, vomiting, anorexia and watery/bloody diarrhea

Fig. 11.5 Acute graft-versus-host disease

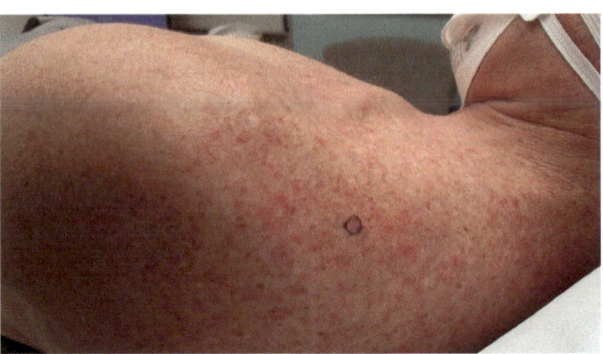

- **Management**: In addition to optimization of systemic immunosuppression (e.g. dose adjustment of calcineurin inhibitors (CNI)), topical therapy and oral anti-histamines to relieve symptoms may be needed; topical steroids are considered first line but systemic steroid therapy may be needed in more severe disease; second line therapy often includes extracorporeal photopheresis (ECP), myco-phenolate mofetil and TNF antagonists

Chronic Graft Versus Host Disease [106, 107]

- **Incidence**: Common, 60–70% of patients surviving beyond day 100 develop chronic GVHD
- **Natural Course**: Manifests more than 100 days following stem cell transplanta-tion, skin is commonly the most affected organ
- **Clinical Features**: Non-sclerotic eruptions consist of lichen planus-like and poi-kiloderma features occurring on the face/palms/soles; sclerotic eruptions cause fibrosis and may resemble lichen sclerosus and morphea (Fig. 11.6), or possibly systemic scleroderma; Chronic GVHD can affect dermis and subcutaneous tis-sue causing adnexal loss, alopecia and ulcers
- **Management**: Mild cases are treated with topical therapies (barrier protection and steroid); with systemic therapy (steroid combined with CNI) considered if organ involvement occurs; second line therapy also includes ECP, mycopheno-late mofetil, imatinib and rituximab

Clinical Summary
- Cutaneous manifestations are the most common sign of GVHD and often times the presenting symptom.
- Preventative measures with cyclosporine and methotrexate have been used; con-trol of additional risk factors for disease includes minimizing the degree of HLA

Fig. 11.6 Sclerodermoid chronic graft-versus-host disease

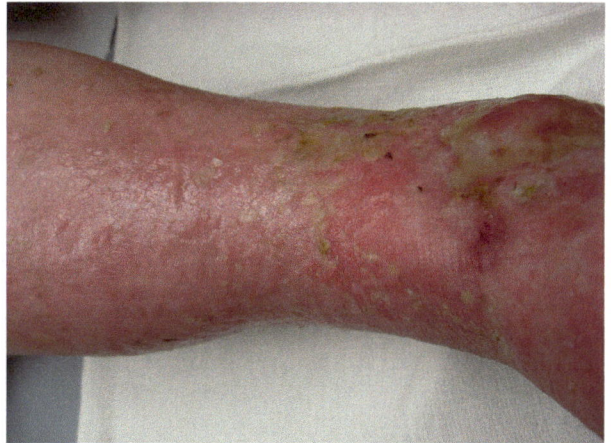

mismatch, avoiding transplantation from an unrelated donor, avoiding female donor for male recipient, and effective prophylaxis.
- Management strategies include prophylaxis as a mainstay of treatment. Treatment depends on severity of disease and multiple modalities are available.

Other Anticancer Therapies (Table 11.4)

Venetoclax [108, 109]

Dermatologic Adverse Effects
- Nonspecific rash
- Edema
- Mucositis

Clinical Summary
- DAEs from Venetoclax therapy do occur and include nonspecific rash, mucositis, and edema, which can be managed with conventional, symptomatic treatment (see "General Principles of Management").

Omacetaxine Mepesuccinate [110]

Dermatologic Adverse Effects
- Nonspecific skin rash
- Alopecia

Table 11.4 Other anticancer therapies

Agent	Indication
Venetoclax	• AML • CLL/SLL • Mantle cell lymphoma
Omacetaxine mepesuccinate	• CML
Panobinostat	• Multiple myeloma (MM)
Plerixafor	• Peripheral Stem cell mobilization for patients with NHL and MM
Pomalidomide	• Multiple myeloma
Axicabtagene ciloleucel	• Large B-cell lymphoma
Belinostat	• Peripheral T-Cell lymphoma
Pralatrexate	• Peripheral T-cell lymphoma • Cutaneous T-cell lymphoma
Eltrombopag	• Anaplastic anemia • Hepatitis C infection associated thrombocytopenia • Chronic immune thrombocytopenia

- Edema
- Injection site reaction
- Hyperhidrosis
- Hyperpigmentation
- Pruritus
- Xeroderma

Clinical Summary
- DAEs from Omacetaxine mepesuccinate therapy have been documented and include alopecia, edema, injection site reaction, pruritus, and xeroderma, which can be managed with conventional, symptomatic treatment (see "General Principles of Management").

Panobinostat

Dermatologic Adverse Effects
- Nonspecific rash
- Cheilitis
- Erythema
- Edema

Clinical Summary
- DAEs from Panobinostat therapy are not well-documented and include cheilitis, erythema, edema, and nonspecific rash, which can be managed with conventional, symptomatic treatment (see "General Principles of Management").

Plerixafor [111]

Dermatologic Adverse Effects
- Injection site reaction
- Erythema
- Hyperhidrosis
- Rare- periorbital edema

Clinical Summary
- Clinical profile is overall favorable with DAEs from Plerixafor therapy, and not often reported. Uncommon DAEs include injection site reaction, erythema, and hyperhidrosis, which can be managed with conventional, symptomatic treatment (see "General Principles of Management").
- Rare reports of periorbital edema have been described.

Pomalidomide

Dermatologic Adverse Effects
- Nonspecific skin rash
- Edema
- Pruritus
- Xeroderma
- Hyperhidrosis
- Cellulitis
- Rare- angioedema, BCC, DRESS, HSV, SJS/TEN

Clinical Summary
- DAEs from Pomalidomide therapy are not well-documented and include non-specific rash, edema, pruritus, and xeroderma, which can be managed with conventional, symptomatic treatment (see "General Principles of Management").
- Rare DAEs of DRESS and SJS/TEN have been reported.

Axicabtagene Ciloleucel [112]

Dermatologic Adverse Effects
- Non-specific skin rash
- Edema

Clinical Summary
- DAEs from Axicabtagene ciloleucel therapy have been documented and are mild. These include non-specific rash and edema, which can be managed

with conventional, symptomatic treatment (see "General Principles of Management").

Belinostat

Dermatologic Adverse Effects
- Non-specific skin rash
- Edema
- Pruritus
- Injection site reaction

Clinical Summary
- DAEs from Belinostat therapy are not well-documented and include nonspecific rash, edema, and pruritus, which can be managed with conventional, symptomatic treatment (see "General Principles of Management").

Pralatrexate [113]

Dermatologic Adverse Effects
- Non-specific skin rash
- Pruritus
- Edema
- Mucositis
- Night sweats
- Rare- Dermal ulcers, desquamation, SJS/TEN, HSV

Clinical Summary
- DAEs from Palatrexate therapy have been reported and include non-specific rash, edema, and pruritus, which can be managed with conventional, symptomatic treatment (see "General Principles of Management").
- Rare DAEs of exfoliative dermatitis and SJS/TEN have been reported.

Eltrombopag [109, 114]

Dermatologic Adverse Effects
- Non-specific rash
- Alopecia
- Rare- hyperpigmentation

Clinical Summary

- DAEs from Eltrombopag therapy are typically mild and include non-specific rash, and alopecia, which can be managed with conventional, symptomatic treatment (see "General Principles of Management").
- Rare reports of skin hyperpigmentation have been documented.

Alternating Electric Field Therapy (Tumor Treating Fields) [115]

Indications

- Glioblastoma multiforme
- Solid tumors

Dermatologic Adverse Effects
- Irritant/contact dermatitis
- Allergic contact dermatitis
- Ulcers

Clinical Summary
- DAEs from TT Fields treatments can be avoided with adequate skin preparation to ensure effective contact between the array transducer and the skin.
- Long term skin integrity can be maintained with good scalp hygiene and array placement.
- Most DAEs can be treated with topical therapy and slight adjustments of the arrays to minimize skin irritation.

CAR-T- Tisagenlecleucel [116]

Indications

- ALL
- Diffuse large B-cell lymphoma

Dermatologic adverse effects
- Skin rash
- Edema
- Dermatitis

Clinical summary
- DAEs from Tisagenlecleucel therapy have been reported and include non-specific rash, and edema, which can be managed with conventional, symptomatic treatment (see "General Principles of Management").

Immunostimulant—Sipuleucel-T [117]

Indications

- Metastatic prostate cancer

Dermatologic Adverse Effects
- Nonspecific rash
- Hyperhidrosis
- Infusion reaction

Clinical Summary
- DAEs from Sipuleucel-T therapy are not well-documented and include non-specific rash, hyperhidrosis and infusion reaction, which can be managed with conventional, symptomatic treatment (see "General Principles of Management").

Transcription factor interference—Trabectedin [118]

Indications

- Soft tissue sarcoma (liposarcoma, leiomyosarcoma)
- Ovarian cancer

Dermatologic adverse effects
- Mucositis
- Hand-foot syndrome
- Edema
- Injection site reactions
- Non-specific rash

Clinical summary
- DAEs from Trabectedin therapy have been documented and include injection site reaction, mucositis and hand-foot syndrome, which can be managed with conventional, symptomatic treatment (see "General Principles of Management").

Talimogene Laherparepvec (Oncolytic Herpes Virus) [119]

Indications

- Melanoma (non-resectable)

Dermatologic Adverse Effects
- Panniculitis [120]
- Injection site reaction
- Cellulitis
- Psoriasis exacerbation
- Vitiligo

Clinical Summary
- DAEs from Talimogene laherparepvec therapy have been documented and include injection site reactioncellulitis and exacerbation of psoriasis, which can be managed with conventional, symptomatic treatment (see "General Principles of Management").

Other Chemotherapy Medications

- L-Asparaginase
- Pegaspargase
- Bleomycin
- Thalidomide
- Lenalidomide

L-Asparaginase

Indications
- ALL

Dermatologic Adverse Effects
- Mucositis
- Injection site reaction

Clinical Summary
- DAEs from L-Asparaginase therapy have been documented and include injection site reaction (common >30%), and less commonly mucositis, which can be managed with conventional, symptomatic treatment (see "General Principles of Management").

- Appropriate monitoring of treatment is essential with possible modification if needed in patients experiencing intolerable L-Asparaginase related toxicities.

Pegaspargase [121]

Indications
- ALL
- Non-Hodgkin lymphoma

Dermatologic Adverse Effects
- Injection site reactions
- Edema
- Perioral edema
- Petechiae/purpura
- Mucositis

Clinical Summary
- DAEs from pegaspargase therapy are uncommon and include injection site reaction, mucositis, petechial rash and perioral edema, which can be managed with conventional, symptomatic treatment (see "General Principles of Management").
- Appropriate monitoring of treatment is essential with possible modification if needed in patients experiencing intolerable pegaspargase related toxicities.

Bleomycin

Indications
- Hodgkin lymphoma
- Head and neck cancers
- Malignant pleural effusion
- Testicular cancer
- Malignant germ cell tumors

Dermatologic Adverse Effects
- Hyperpigmentation
- Atrophic striae
- Erythema
- Flagellate dermatitis [122]
- Desquamation
- Hyperkeratosis
- Localized vesicle formation
- Sclerosis
- Alopecia
- Nailbed changes

Clinical Summary

- DAEs from Bleomycin therapy are relatively common (up to 50%) and include hyperpigmentation, atrophic striae, desquamation (most common palms and soles), and hyperkeratosis, which can be managed with conventional, symptomatic treatment (see "General Principles of Management").
- Bleomycin-induced flagellated dermatitis has been reported and can be managed clinically.
- Alopecia and changes to the nailbed may be dose related and reversible with discontinuation of medication.
- Appropriate monitoring of treatment is essential with possible modification if needed in patients experiencing intolerable Bleomycin related toxicities.

Bleomycin-induced Flagellate Dermatitis

- **Incidence**: Reports ~8–20% of patients on Bleomycin therapy
- **Natural Course**: Dermatitis appears days to weeks after induction of therapy; heals over time (weeks) after discontinuation of therapy
- **Clinical Features**: Pruritus with concurrent erythematous linear/whip-like streaks found on back and flanks; after resolution of itch and erythema, brown pigmentation is left until fully resolved
- **Management**: Rash usually resolves spontaneously; address pruritus with general treatment management; systemic steroids may speed resolution; rash will worsen with heat exposure, therefore cooling before chemotherapy administration may prevent worsening

Thalidomide

Indications

- Erythema nodosum leprosum
- Multiple myeloma
- AIDS-related aphthous stomatitis
- Chronic GVHD
- Systemic light chain amyloidosis
- Waldenström macroglobulinemia

Dermatologic Adverse Effects
- Pruritus
- Edema
- Fungal dermatitis
- Maculopapular rash (Fig. 11.7)
- Nail disease
- Rare- erythema multiforme, myxedema, petechiae/purpura, SJS/TEN, VZV

Fig. 11.7 Thalidomide cutaneous hypersensitivity reaction

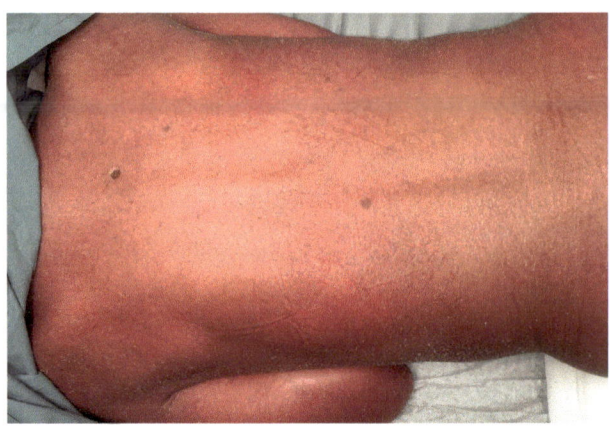

Clinical summary

- DAEs from thalidomide therapy are reported and include pruritus, edema, fungal dermatitis, and maculopapular rash (commonly on trunk, back, arms and legs), which can be managed with conventional, symptomatic treatment (see "General Principles of Management").
- Rare DAEs of SJS/TEN have been reported.
- Appropriate monitoring of treatment is essential with possible modification if needed in patients experiencing intolerable thalidomide related toxicities.

Lenalidomide [109, 123]

Indications

- Follicular lymphoma
- Mantle cell lymphoma
- Marginal zone lymphoma
- Multiple myeloma
- Myelodysplastic syndromes
- CLL
- DLBCL
- Systemic light chain amyloidosis

Dermatologic Adverse Effects
- Pruritus
- Xeroderma
- Nonspecific skin rash
- Rare- DRESS [124]

Clinical Summary

- DAEs from Lenalidomide therapy are common (>10%) and include nonspecific rash, pruritus (up to 42% in patients with MDS), and xeroderma, which can be managed with conventional, symptomatic treatment (see "General Principles of Management").
- Onset of variable rash (morbilliform, urticarial, dermatitis, acneiform) most commonly occurred in the first month, however delayed onset also observed.
- Appropriate monitoring of treatment is essential with possible modification if needed in patients experiencing intolerable lenalidomide-related toxicities.

Other Hematologic Agents (Tables 11.5 and 11.6)

Clinical Summary

- DAEs from erythropoietic growth factor therapy are common (>10%) and include non-specific rash, and pruritis, which can be managed with conventional, symptomatic treatment (see "General Principles of Management").
- Although this medication is not used to treat malignancy, it provides support for patients suffering from anemia due to a disease state (or treatment of disease).
- DAEs are usually mild and reversible with completion of treatment.

Clinical Summary

- DAEs from granulocyte colony-stimulating factor therapy are uncommon and include maculopapular rash, mucositis, erythema, and injection site reactions, which can be managed with conventional, symptomatic treatment (see "General Principles of Management").
- Although rare, Sweet syndrome (Fig. 11.8) has been reported with use of GCSF medications. Most occur within days to weeks of treatment. Management typically includes systemic steroids and discontinuation of treatment if severe.

Table 11.5 Erythropoietic growth factors

Agent	Indication	DAEs
Epoetin alfa	• Supportive medication for disease induced anemia	• Non-specific Rash • Pruritus • Injection site reactions
Darbepoetin alfa	• Supportive medication foe disease-induced anemia	• Edema • Erythema • Injection site reactions

Table 11.6 Granulocyte colony-stimulating factors

Agent	Indication	DAEs
Filgrastim	• Chemotherapy-induced myelosuppression • AML • Bone marrow transplant • Hematopoietic radiation injury syndrome • Peripheral blood progenitor cell collection/therapy • Severe chronic neutropenia • Alcoholic hepatitis • Anemia in myelodysplastic syndrome • Neutropenia in advanced HIV infection	• Maculopapular rash • Edema • Mucositis • Alopecia • Erythema • Sweet syndrome [125]
Pegfilgrastim (Pegylated filgrastim)	• Hematopoietic radiation injury syndrome • Prevention of chemotherapy-induced neutropenia	• Erythema • Injection site reaction • Non-specific rash • Sweet syndrome [125] • Urticaria

Fig. 11.8 Sweet syndrome

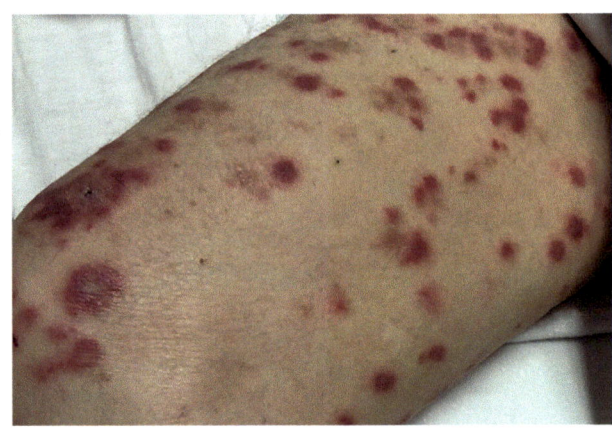

Granulocyte–Macrophage Colony-Stimulating Factors—Sargramostim

Indications

- AML
- Allogenic bone marrow transplantation

- Autologous peripheral blood progenitor cell mobilization/collection
- Hematopoietic radiation injury syndrome
- Prophylaxis of neutropenia in patients receiving chemotherapy

Dermatologic Adverse Effects
- Non-specific skin rash
- Pruritus
- Edema
- Injection site reaction
- Urticaria

Clinical Summary
- DAEs from sargramostim therapy are common and include non-specific skin rash, pruritus, edema, injection site reactions, and rarely urticaria, which can be managed with conventional, symptomatic treatment (see "General Principles of Management").
- DAEs are almost always reversible with discontinuation of treatment.

Thrombopoietic Growth Factor—Oprelvekin (Interleukin 11)

Indications

- Chemotherapy-associated thrombocytopenia

Dermatologic Adverse Effects
- Urticaria
- Erythema
- Pruritus
- Edema
- Mucositis
- Nonspecific rash

Clinical Summary
- DAEs from Oprelvekin therapy are documented and include non-specific rash, edema, and mucositis, which can be managed with conventional, symptomatic treatment (see "General Principles of Management").

Topical Anticancer Therapies

Agents

- Chemotherapy (Table 11.7)

Table 11.7 Topical chemotherapy agents

Agent	Indication	DAEs
Carmustine [126]	• Mycosis Fungoides	• Erythema • Alopecia • Hyperpigmentation • Telangiectasias • Irritant dermatitis
Mechlorethamine [127]	• Cutaneous T-cell lymphoma	• Dermatitis • Pruritus • Bacterial skin infection • Secondary malignancies • Dermal ulcer • Skin hyperpigmentation
Fluorouracil [128]	• Actinic or solar keratosis • BCC • SCC • Malignant melanoma • Keratoacanthoma • Vitiligo • Verruca vulgaris • Darier's disease	• Application site irritation • Skin erosion • Erythema • Pruritus • Edema • Alopecia

   – Carmustine
   – Mechlorethamine
   – Fluorouracil

Clinical Summary

- DAEs from topical Carmustine therapy are common and include erythema with burning sensation that is accentuated in body fold locations. Management includes topical steroids, emollients, and cool compresses, with the reaction subsiding within a few weeks.
- DAEs from topical Mechlorethamine (nitrogen mustard) are well documented and include dermatitis (up to 50% of patients) (Fig. 11.9)
- Other DAEs from topical Mechlorethamine include bacterial skin infection, and less commonly (<10%) dermal ulcers, and skin hyperpigmentation. Careful monitoring and discontinuation of treatment if serious cutaneous toxicities arise. Manage with typical therapy and symptomatic treatment.
- DAEs from topical fluorouracil therapy are extremely common on the application site. Up to 95% of patients report application site scaling, dryness, erythema,

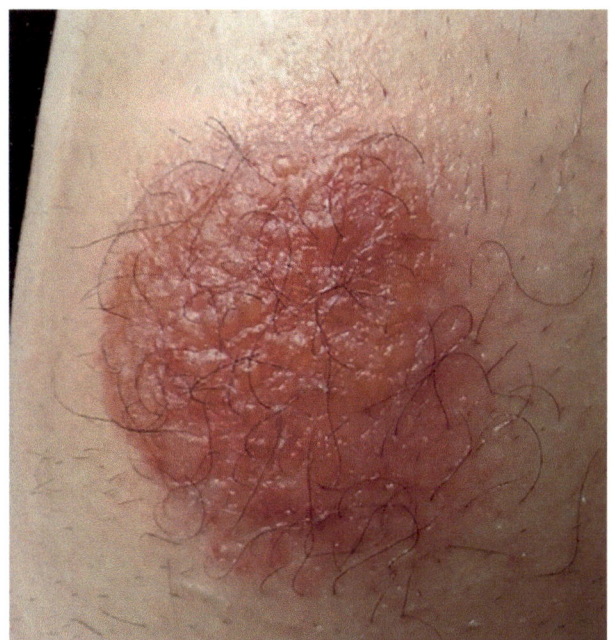

Fig. 11.9 Contact dermatitis secondary to mechlorethamine

pruritus, pain, and edema. Management strategies include avoidance of sunlight, emollients, and topical steroids with resolution within 2–4 weeks.
- **Retinoids** (Table 11.8)
  - Alitretinoin
  - Bexarotene

Clinical Summary
- DAEs from retinoid therapy are variable, but mild and include pigmentation changes, erythema, desquamation, photosensitivity, and pruritus, which can be managed with conventional, symptomatic treatment (see "General Principles of Management").
- Most cutaneous toxicities (most common of which included nonspecific rash in ~70% of patients) were limited to application site, are mild to moderate, and reversible with treatment discontinuation or symptomatic treatment.
- **Biologic Response Modifiers**
  - Imiquimod (Table 11.9)

Table 11.8 Topical retinoid agents

Agent	Indication	DAEs
Alitretinoin [129]	• Kaposi sarcoma	• Hyper/hypopigmentation • Erythema • Desquamation • Edema • Pruritus • Irritant dermatitis • Photosensitivity
Bexarotene [130]	• Cutaneous T-cell lymphoma	• Non-specific rash • Pruritus • Contact dermatitis • Edema • Desquamation

Table 11.9 Topical biologic response modifiers

Agent	Indication	DAEs
Imiquimod [131]	• Actinic keratosis • Genital and perianal warts • Superficial BCC • Cutaneous flat warts • Acyclovir resistant HSV	• Erythema • Xeroderma • Sclerosis • Dermal ulcer • Vesicles • Excoriation • Edema • Hypertrophy • Cellulitis • Eczema • Seborrheic keratosis • Tinea • Desquamation • Erythema multiforme • HSV • SCC

Clinical Summary

- DAEs from imiquimod therapy are varied and well-documented and include erythema, sclerosis, excoriation, cellulitis, and seborrheic keratosis, which can be managed with conventional, symptomatic treatment (see "General Principles of Management").
- Rare DAEs of desquamation, erythema multiforme, HSV infection, and SCC have been reported with imiquimod use.

Review Questions and Answers

1. Which medication should be avoided in patients taking tamoxifen as it can convert tamoxifen to a more active metabolite, worsening side effects?
   (a) Venlafaxine
   (b) Fluoxetine
   (c) Desvenlafaxine
   (d) Gabapentin
   (e) Pregabalin

   Answer: (b) Fluoxetine

2. Which of the following are common dermatologic adverse effects of antiandrogen therapy?
   (a) Edema and photosensitivity
   (b) Alopecia and flushing
   (c) Urticaria
   (d) Mucositis and erythema nodosum
   (e) Bullous eruption

   Answer: (a) Edema and photosensitivity

3. Which of the following statements regarding Interferon-α therapy is false?
   (a) DAEs from interferon-alfa-2b therapy are common and heterogeneous; however, attribution of DAEs specifically to interferon alfa-2b therapy is challenging because of frequent administration with other agents
   (b) Exacerbations of psoriasis from interferon-alfa-2b therapy have been reported and should be managed with conventional treatment for psoriasis
   (c) Rare DAEs from interferon-alfa-2b therapy such as AFND and autoimmune blistering disorders have been described.
   (d) DAEs occur within days upon initiation of interferon-α therapy
   (e) Alopecia is an established DAE from interferon-α therapy

   Answer: (d) DAEs occur within days upon initiation of interferon-α therapy

4. Which of the following medications pose a risk of drug-induced psoriasis for patients?
   (a) Lutetium Lu-177 Dotatate
   (b) Degarelix
   (c) Zoledronic Acid
   (d) Aldesleukin
   (e) Letrozole

   Answer: (d) Aldesleukin

5. A patient comes into the office after three weeks of radiation therapy with moderate erythema and desquamation near skin folds of the treatment area. Which of the following is correct?
   (a) This would be classified as grade 1 acute radiation dermatitis
   (b) Acute radiation dermatitis is uncommon in patients, and an alternative cause should be investigated in this patient
   (c) Although there is no consensus on exact treatment, use of non-irritant soap with topical steroids or silver nitrate dressing are commonly used
   (d) All of the above statements are incorrect

Answer: (c) Although there is no consensus on exact treatment, use of non-irritant soap with topical steroids or silver nitrate dressing are commonly used

References

1. Andrew P, Valiani S, MacIsaac J, Mithoowani H, Verma S. Tamoxifen-associated skin reactions in breast cancer patients: from case report to literature review. Breast Cancer Res Treat. 2014;148(1):1–5.
2. Ledet JJ, Grafton LH. Tamoxifen-induced ultraviolet, recall dermatitis. J Drugs Dermatol. 2009;8(8):761–2.
3. Salah E. Tamoxifen-induced radiation recall dermatitis: three calls from Egypt. J Eur Acad Dermatol Venereol. 2017;31(9):e386–8.
4. Hampson JP, Donnelly A, Lewis-Jones MS, Pye JK. Tamoxifen-induced hair colour change. Br J Dermatol. 1995;132(3):483–4.
5. Al-Niaimi F, Lyon C. Tamoxifen-induced hirsutism. J Drugs Dermatol. 2011;10(7):799–801.
6. Betto P, Gennari E, Germi L, Bonoldi E, Scalco G, Tosti A. Tamoxifen and purpuric vasculitis: a case report. J Eur Acad Dermatol Venereol. 2008;22(6):762–3.
7. Harris AL, Smith IE, Snaith M. Tamoxifen-induced tumour regression associated with dermatomyositis. Br J Dermatol. 1982;284(6330):1674–5.
8. Teoh DC, Aw DC, Jaffar H, Ling W, Yong WP, Lee YS, Choo SN, Tan KB. Tamoxifen-induced eccrine squamous syringometaplasia. J Cutan Pathol. 2012;39(5):554–7.
9. Fumal I, Danchin A, Cosserat F, Barbaud A, Schmutz JL. Subacute cutaneous lupus erythematosus associated with tamoxifen therapy: two cases. Dermatology. 2005;210(3):251–2.
10. Bremec T, Demsar J, Luzar B, Pavlović MD. Drug-induced pruritic micropapular eruption: anastrozole, a commonly used aromatase inhibitor. Dermatol Online J. 2009;15(7):14.
11. Foster LM, Mahoney ME, Harmon MW, Allen JW, Luh JY. Radiation recall reaction with letrozole therapy in breast cancer. Clin Breast Cancer. 2014;14(3):e95–7.
12. Marchand A, Georgin-Mège M, Cellier P, Martin L, Avenel-Audran M, Le Corre Y. Exemestane-induced radiation recall dermatitis and morbilliform rash. J Dermatol. 2016;43(5):575–6.
13. Wollina U, Schönlebe J, Heinig B, Tchernev G, França K, Lotti T. Segmental erythema multiforme-like drug eruption by aromatase inhibitor anastrozole—first case report and another example of an immunocompromised district. Open Access Macedonian J Med Sci. 2018;6(1):79–81.
14. Trancart M, Cavailhes A, Balme B, Skowron F. Anastrozole-induced subacute cutaneous lupus erythematosus. Br J Dermatol. 2008;158(3):628–9.

15. Zarkavelis G, Kollas A, Kampletsas E, Vasiliou V, Kaltsonoudis E, Drosos A, Khaled H, Pavlidis N. Aromatase inhibitors induced autoimmune disorders in patients with breast cancer: a review. J Adv Res. 2016;7(5):719–26.

16. Jhaveri K, Halperin P, Shin SJ, Vahdat L. Erythema nodosum secondary to aromatase inhibitor use in breast cancer patients: case reports and review of the literature. Breast Cancer Res Treat. 2007;106(3):315–8.

17. Shoda H, Inokuma S, Yajima N, Tanaka Y, Setoguchi K. Cutaneous vasculitis developed in a patient with breast cancer undergoing aromatase inhibitor treatment. Ann Rheum Dis. 2005;64(4):651–2.

18. Conti-Beltraminelli M, Pagani O, Ballerini G, Richetti A, Graffeo R, Ruggeri M, Forni V, Pianca S, Schönholzer C, Mainetti C, Cavalli F, Goldhirsch A. Henoch-Schönlein purpura (HSP) during treatment with anastrozole. Ann Oncol. 2007;18(1):205–7.

19. Pathmarajah P, Shah K, Taghipour K, Ramachandra S, Thorat MA, Chaudhry Z, Patkar V, Peters F, Connor T, Spurrell E, Tobias JS, Vaidya JS. Letrozole-induced necrotising leukocytoclastic small vessel vasculitis: first report of a case in the UK. Int J Surg Case Rep. 2015;16:77–80.

20. Digklia A, Tzika E, Voutsadakis IA. Cutaneous leukocytoclastic vasculitis associated with letrozole. J Oncol Pharm Pract. 2014;20(2):146–8.

21. Santoro S, Santini M, Pepe C, Tognetti E, Cortelazzi C, Ficarelli E, De Panfilis G. Aromatase inhibitor-induced skin adverse reactions: exemestane-related cutaneous vasculitis. J Eur Acad Dermatol Venereol. 2011;25(5):596–8.

22. Crockett JS, Burkemper NM. Grover disease (transient acantholytic dermatosis) induced by anastrozole. Cutis. 2011;88(4):175–7.

23. Chia WK, Lim YL, Greaves MW, Ang P. Toxic epidermal necrolysis in patient with breast cancer receiving letrozole. Lancet Oncol. 2006;7(2):184–5.

24. Wiechert A, Tüting T, Bieber T, Haidl G, Wenzel J. Subacute cutaneous lupus erythematosus in a leuprorelin-treated patient with prostate carcinoma. Br J Dermatol. 2008;159(1):231–3.

25. Fritzler MJ. Drugs recently associated with lupus syndromes. Lupus. 1994;3(6):455–9.

26. Erfan G, Rifaioglu EN, Kulac M, Kalayci S, Kaya S, Oznur M. Precipitation and exacerbation of psoriasiform eruption due to leuprolide acetate. J Dermatol. 2013;40(1):54–5.

27. Sakamoto R, Higashi Y, Mera K, Kanekura T, Kanzaki T. Granulomas induced by subcutaneous injection of leuprorelin acetate. J Dermatol. 2006;33(1):43–5.

28. Kawai M, Ikoma N, Yamada A, Ota T, Manabe Y, Kato M, Mabuchi T, Ozawa A, Higure T, Terachi T. A case of foreign body granuloma induced by subcutaneous injection of leuprorelin acetate—clinical analysis for 335 cases in our hospital. Tokai J Exp Clin Med. 2014;39(3):106–10.

29. Watanabe T, Yamada N, Yoshida Y, Yamamoto O. A morphological study of granulomas induced by subcutaneous injection of leuprorelin acetate. J Cutan Pathol. 2009;36(12):1299–302.

30. Fujisaki A, Kondo Y, Goto K, Morita T. Life-threatening anaphylaxis to leuprorelin acetate depot: case report and review of the literature. Int J Urol. 2012;19(1):81–4.

31. Gnanaraj J, Saif MW. Hypersensitivity vasculitis associated with leuprolide (Lupron). Cutan Ocular Toxicol. 2010;29(3):224–7.

32. Yu SS, Connolly MK, Berger TG, McCalmont TH. Dermatitis herpetiformis associated with administration of a gonadotropin-releasing hormone analog. J Am Acad Dermatol. 2006;54(2 Suppl):S58–9.

33. Grimwood RE, Guevara A. Leuprolide acetate-induced dermatitis herpetiformis. Cutis. 2005;75(1):49–52.

34. de Salins CA, Kupfer-Bessaguet I, Fleuret C, Staroz F, Plantin P. [Fixed drug eruption induced by leuprorelin]. Annales de dermatologie et de venereologie 2015;142(12):780–781.

35. Turk BG, Dereli T, Dereli D, Akalin T. Leuprolide acetate-induced leukocytoclastic vasculitis. Acta Obstet Gynecol Scandinavica. 2007;86(7):892–3.

36. Higashiyama A, Yokoyama T, Omoto Y, Habe K, Yamanaka K, Mizutani H. Flutamide-induced photoleukomelanoderma. J Dermatol. 2016;43(9):1105–6.

37. Kaur C, Thami GP. Flutamide-induced photosensitivity: is it a forme fruste of lupus. Br J Dermatol. 2003;148(3):603–4.
38. Sasada K, Sakabe J, Tamura A, Kasuya A, Shimauchi T, Ito T, Hirakawa S, Tokura Y. Photosensitive drug eruption induced by bicalutamide within the UVB action spectrum. Eur J Dermatol. 2012;22(3):402–3.
39. Swoboda A, Kasche A, Baumstark J, Worret WI, Ring J, Eberlein-König B. Vitiliginous lesions after photosensitive dermatitis due to flutamide. J Eur Acad Dermatol Venereol. 2007;21(5):681–2.
40. Reid MB, Glode LM. Flutamide induced lupus. J Urol. 1998;159(6):2098.
41. Alberto C, Konstantinou MP, Martinage C, Casassa E, Tournier E, Bagheri H, Sibaud V, Mourey L, Mazereeuw-Hautier J, Meyer N, Paul C, Bulai LC. Enzalutamide induced acute generalized exanthematous pustulosis. J Dermatol Case Rep. 2016;10(2):35–8.
42. Loprinzi CL, Kugler JW, Sloan JA, Mailliard JA, Krook JE, Wilwerding MB, Rowland KM, Camoriano JK, Novotny PJ, Christensen BJ. Randomized comparison of megestrol acetate versus dexamethasone versus fluoxymesterone for the treatment of cancer anorexia/cachexia. J Clin Oncol. 1999;17(10):3299–306.
43. Salvatore MA, Lynch PJ. Erythema nodosum, estrogens, and pregnancy. Arch Dermatol. 1980;116(5):557–8.
44. Alloo A, DeSimone JA, Kupper TS. Mycosis fungoides presenting at the site of a transdermal estradiol patch. J Am Acad Dermatol. 2012;67(5):e207–8.
45. Gregoriou S, Chiolou Z, Rigopoulos D. Beau's lines after octreotide therapy. Clin Exp Dermatol. 2009;34(8):e1020–1.
46. Rideout DJ, Graham MM. Buttock granulomas: a consequence of intramuscular injection of Sandostatin detected by In-111 octreoscan. Clin Nucl Med. 2001;26(7):650.
47. Dadzie DD, Lee EJ, Monteleone CA, Schneider SH. Desensitization treatment for hypersensitivity reaction to octreotide in an acromegalic patient. Pituitary. 2012;15(Suppl 1):S68–71.
48. Michel JL. [Eruptive naevi associated with octreotide treatment]. Annales de dermatologie et de venereologie 2011;138(10):677–680.
49. Alvarez-Escola C, Cárdenas-Salas JJ, Pelegrina B, Sanz-Valtierra A, Lecumberri B. Severe scalp hair loss in a female patient with acromegaly treated with lanreotide autogel after unsuccessful surgery. Clin Case Rep. 2015;3(11):945–8.
50. Jönsson A, Manhem P. Octreotide and loss of scalp hair. Ann Intern Med. 1991;115(11):913.
51. Nakauchi Y, Kumon Y, Yamasaki H, Tahara K, Kurisaka M, Hashimoto K. Scalp hair loss caused by octreotide in a patient with acromegaly: a case report. Endocrine J. 1995;42(3):385–9.
52. Bauzá A, Del Pozo LJ, Escalas J, Mestre F. Radiation recall dermatitis in a patient affected with pheochromocytoma after treatment with lanreotide. Br J Dermatol. 2007;157(5):1061–3.
53. Pochi PE. Acne in premature ovarian failure. Reestablishment of cyclic flare-ups with medroxyprogesterone acetate therapy. Arch Dermatol. 1974;109(4):556–7.
54. Kuno Y, Tsuji T. Acute generalized exanthematous pustulosis upon ingestion of a progesterone preparation. Acta Dermato Venereol. 1998;78(5):383.
55. Wojnarowska F, Greaves MW, Peachey RD, Drury PL, Besser GM. Progesterone-induced erythema multiforme. J Royal Soc Med. 1985;78(5):407–8.
56. Apalla Z, Tzellos T, Lallas A, Sotiriou E, Lefaki I. Possible zoledronic acid-induced dermatomyositis. Clin Exp Dermatol. 2012;37(3):309–11.
57. Tong PL, Yu LL, Chan JJ. Drug-induced dermatomyositis after zoledronic acid. Australas J Dermatol. 2012;53(4):e73–5.
58. Andreadis D, Mauroudis S, Poulopoulos A, Markopoulos A, Epivatianos A. Lip ulceration associated with intravenous administration of zoledronic acid: report of a case. Head Neck Pathol. 2012;6(2):275–8.
59. Kitagawa KH, Grassi M. Zoledronic acid-induced cutaneous B-cell pseudolymphoma. J Am Acad Dermatol. 2011;65(6):1238–40.
60. Wang J, Xu B, Yuan P, Ma F, Li Q, Zhang P, Cai R, Fan Y, Luo Y, Li Q. Phase II trial of goserelin and exemestane combination therapy in premenopausal women with locally advanced or metastatic breast cancer. Medicine. 2015;94(26):e1006.

61. Raj SG, Karadsheh AJ, Guillot RJ, Raj MH, Kumar P. Case report: systemic hypersensitivity reaction to goserelin acetate. Am J Med Sci. 1996;312(4):187–90.
62. Guillot B, Blazquez L, Bessis D, Dereure O, Guilhou JJ. A prospective study of cutaneous adverse events induced by low-dose alpha-interferon treatment for malignant melanoma. Dermatology. 2004;208(1):49–54.
63. Stafford-Fox V, Guindon KM. Cutaneous reactions associated with alpha interferon therapy. Clin J Oncol Nurs. 2000;4(4):164–8.
64. Quesada JR, Talpaz M, Rios A, Kurzrock R, Gutterman JU. Clinical toxicity of interferons in cancer patients: a review. J Clin Oncol. 1986;4(2):234–43.
65. Wolfe JT, Singh A, Lessin SR, Jaworsky C, Rook AH. De novo development of psoriatic plaques in patients receiving interferon alfa for treatment of erythrodermic cutaneous T-cell lymphoma. J Am Acad Dermatol. 1995;32(5 Pt 2):887–93.
66. Mendieta KL, Irfan M, Fernandez Faith E. Interferon-alpha induced psoriasis in a teenager. Pediatr Dermatol. 2018;35(2):e136–7.
67. Thomas R, Stea B. Radiation recall dermatitis from high-dose interferon alfa-2b. J Clin Oncol. 2002;20(1):355–7.
68. Krischer J, Pechère M, Salomon D, Harms M, Chavaz P, Saurat JH. Interferon alfa-2b-induced Meyerson's nevi in a patient with dysplastic nevus syndrome. J Am Acad Dermatol. 1999;40(1):105–6.
69. Vallés L, González M, Polo I, Enguita AB, Vanaclocha F, Ortiz-Romero PL. Lipoatrophy associated with interferon alfa adjuvant therapy for melanoma. Arch Dermatol. 2009;145(1):98–9.
70. Ruiz-Genao DP, García-F-Villalta MJ, Hernández-Núñez A, Ríos-Buceta L, Fernández-Herrera J, García-Díez A. Livedo reticularis associated with interferon alpha therapy in two melanoma patients. J Eur Acad Dermatol Venereol. 2005;19(2):252–4.
71. Pouthier D, Theissen F, Humbel RL. Lupus syndrome, hypothyroidism and bullous skin lesions after interferon alfa therapy for hepatitis C in a haemodialysis patient. Nephrol Dialysis Transpl. 2002;17(1):174.
72. Kocyigit P, Akay BN, Karaosmanoglu N. Linear IgA bullous dermatosis induced by interferon-alpha 2a. Clin Exp Dermatol. 2009;34(5):e123–4.
73. Onishi S, Nagashima T, Kimura H, Matsuyama Y, Yoshio T, Minota S. Systemic lupus erythematosus and Sjögren's syndrome induced in a case by interferon-alpha used for the treatment of hepatitis C. Lupus. 2010;19(6):753–5.
74. Gheorghe L, Cotruta B, Trifu V, Cotruta C, Becheanu G, Gheorghe C. Drug-induced Sweet's syndrome secondary to hepatitis C antiviral therapy. Int J Dermatol. 2008;47(9):957–9.
75. Sonntag M, Hodzic-Avdagic N, Bruch-Gerharz D, Neumann NJ. [Embolia cutis medicamentosa after subcutaneous injection of pegylated interferon-alpha]. Der Hautarzt; Zeitschrift fur Dermatologie, Venerologie, und verwandte Gebiete 2005;56(10):968–969.
76. Kluger N, Moguelet P, Chaslin-Ferbus D, Khosrotherani K, Aractingi S. Generalized interstitial granuloma annulare induced by pegylated interferon-alpha. Dermatology. 2006;213(3):248–9.
77. Song JS, Sohn JH, Jeong JY, Min JH, Choi WS, Kim OZ, Pyo JY. Repeated panniculitis induced by pegylated interferon alpha 2a in a patient with chronic hepatitis C. Korean J Gastroenterol. 2016;67(5):272–6.
78. North J, Mully T. Alpha-interferon induced sarcoidosis mimicking metastatic melanoma. J Cutan Pathol. 2011;38(7):585–9.
79. Mir-Bonafé JM, Blanco-Barrios S, Romo-Melgar A, Santos-Briz A, Fernández-López E. Photoletter to the editor: localized pyoderma gangrenosum after interferon-alpha2b injections. J Dermatol Case Rep. 2012;6(3):98–9.
80. Dietrich LL, Bridges AJ, Albertini MR. Dermatomyositis after interferon alpha treatment. Med Oncol. 2000;17(1):64–9.
81. Giuliani M, Mastroianni A, Di Carlo A, Donati P, Miceli M, Monini P, Rezza G. Onset of non-AIDS Kaposi sarcoma during therapy with interferon alfa-2a in an 82-year-old man with concomitant cutaneous T-cell lymphoma. Arch Dermatol. 2002;138(4):535–7.

82. Staunton MR, Scully MC, Le Boit PE, Aronson FR. Life-threatening bullous skin eruptions during interleukin-2 therapy. J Natl Cancer Inst. 1991;83(1):56–7.
83. Fellner MJ. Drug-induced bullous pemphigoid. Clin Dermatol. 1993;11(4):515–20.
84. Tranvan A, Pezen DS, Medenica M, Michelson GC, Vogelzang N, Soltani KM. Interleukin-2 associated linear IgA bullous dermatosis. J Am Acad Dermatol. 1996;35(5 Pt 2):865–7.
85. Prussick R, Plott RT, Stanley JR. Recurrence of pemphigus vulgaris associated with interleukin 2 therapy. Arch Dermatol. 1994;130(7):890–3.
86. Lee RE, Gaspari AA, Lotze MT, Chang AE, Rosenberg SA. Interleukin 2 and psoriasis. Arch Dermatol. 1988;124(12):1811–5.
87. Gaspari AA, Lotze MT, Rosenberg SA, Stern JB, Katz SI. Dermatologic changes associated with interleukin 2 administration. JAMA. 1987;258(12):1624–9.
88. Rondina A, Watson AC. Bullous Sweet's syndrome and pseudolymphoma precipitated by IL-2 therapy. Cutis. 2010;85(4):206–13.
89. Gunawardane ND, Vaghani SP, Kuzel TM, Cotliar JA. Acute generalized exanthematous pustulosis in a patient receiving high-dose recombinant interleukin-2. J Am Acad Dermatol. 2013;69(4):e183–4.
90. Abraham D, McGrath KG. Hypersensitivity to aldesleukin (interleukin-2 and proleukin) presenting as facial angioedema and erythema. Allergy Asthma Proc. 2003;24(4):291–4.
91. Chodorowska G, Czelej D, Niewiedzioł M. Interleukin-2 and its soluble receptor in selected drug-induced cutaneous reactions. Annales Universitatis Mariae Curie-Sklodowska. Sectio D: Medicina. 2003;58(2):7–13.
92. Weinstein A, Bujak D, Mittelman A, Davidian M. Erythema nodosum in a patient with renal cell carcinoma treated with interleukin 2 and lymphokine-activated killer cells. JAMA. 1987;258(21):3120–1.
93. Krigel RL, Padavic-Shaller KA, Rudolph AR, Poiesz BJ, Comis RL. Exacerbation of epidemic Kaposi's sarcoma with a combination of interleukin-2 and beta-interferon: results of a phase 2 study. J Biol Response Modifiers. 1989;8(4):359–65.
94. Blanche P, Gombert B, Rollot F, Salmon D, Sicard D. Sarcoidosis in a patient with acquired immunodeficiency syndrome treated with interleukin-2. Clin Infect Dis. 2000;31(6):1493–4.
95. Puett DW, Fuchs HA. Rapid exacerbation of scleroderma in a patient treated with interleukin 2 and lymphokine activated killer cells for renal cell carcinoma. J Rheumatol. 1994;21(4):752–3.
96. Faggiano A, Lo Calzo F, Pizza G, Modica R, Colao A. The safety of available treatments options for neuroendocrine tumors. Expert Opin Drug Saf. 2017;16(10):1149–61.
97. Jadvar H, Challa S, Quinn DI, Conti PS. One-year postapproval clinical experience with radium-223 dichloride in patients with metastatic castrate-resistant prostate cancer. Cancer Biotherapy Radiopharm. 2015;30(5):195–9.
98. Kluetz PG, Pierce W, Maher VE, Zhang H, Tang S, Song P, Liu Q, Haber MT, Leutzinger EE, Al-Hakim A, Chen W, Palmby T, Alebachew E, Sridhara R, Ibrahim A, Justice R, Pazdur R. Radium Ra 223 dichloride injection: U.S. Food and Drug Administration drug approval summary. Clin Cancer Res. 2014;20(1):9–14.
99. McGann S, Horton ER. Radium-223 dichloride: a novel treatment option for castration-resistant prostate cancer patients with symptomatic bone metastases. Ann Pharmacother. 2015;49(4):469–76.
100. Leventhal J, Young MR. Radiation dermatitis: recognition, prevention, and management. Oncology. 2017;31(12):885–7, 894–9.
101. Rosenthal A, Israilevich R, Moy R. Management of acute radiation dermatitis: a review of the literature and proposal for treatment algorithm. J Am Acad Dermatol. 2019;81(2):558–67.
102. Spałek M. Chronic radiation-induced dermatitis: challenges and solutions. Clin Cosmet Investig Dermatol. 2016;9:473–82.
103. Ristić B. Radiation recall dermatitis. Int J Dermatol. 2004;43(9):627–31.
104. Nassereddine S, Rafei H, Elbahesh E, Tabbara I. Acute graft versus host disease: a comprehensive review. Anticancer Res. 2017;37(4):1547–55.

105. Zeiser R, Blazar BR. Acute graft-versus-host disease—biologic process, prevention, and therapy. N Engl J Med. 2017;377(22):2167–79.
106. Strong Rodrigues K, Oliveira-Ribeiro C, de Abreu Fiuza Gomes S, Knobler R. Cutaneous graft-versus-host disease: diagnosis and treatment. Am J Clin Dermatol. 2018;19(1):33–50.
107. Lee SJ, Vogelsang G, Flowers ME. Chronic graft-versus-host disease. Biol Blood Marrow Transpl. 2003;9(4):215–33.
108. Roeker LE, Fox CP, Eyre TA, Brander DM, Allan JN, Schuster SJ, Nabhan C, Hill BT, Shah NN, Lansigan F, Yazdy M, Cheson BD, Lamanna N, Singavi AK, Coombs CC, Barr PM, Skarbnik AP, Shadman M, Ujjani CS, Tuncer HH, Winter AM, Rhodes J, Dorsey C, Morse H, Kabel C, Pagel JM, Williams AM, Jacobs R, Goy A, Muralikrishnan S, Pearson L, Sitlinger A, Bailey N, Schuh A, Kirkwood AA, Mato AR. Tumor lysis, adverse events, and dose adjustments in 297 venetoclax-treated CLL patients in routine clinical practice. Clin Cancer Res. 2019;25(14):4264–70.
109. Iberri DJ, Kwong BY, Stevens LA, Coutre SE, Kim J, Sabile JM, Advani RH. Ibrutinib-associated rash: a single-centre experience of clinicopathological features and management. Br J Haematol. 2018;180(1):164–6.
110. Cortes JE, Nicolini FE, Wetzler M, Lipton JH, Akard L, Craig A, Nanda N, Benichou AC, Leonoudakis J, Khoury HJ, Hochhaus A, Baccarani M, Kantarjian HM. Subcutaneous omacetaxine mepesuccinate in patients with chronic-phase chronic myeloid leukemia previously treated with 2 or more tyrosine kinase inhibitors including imatinib. Clin Lymph Myeloma Leuk. 2013;13(5):584–91.
111. Yang X, Wan M, Yu F, Wang Z. Efficacy and safety of plerixafor for hematopoietic stem cell mobilization for autologous transplantation in patients with non-Hodgkin lymphoma and multiple myeloma: a systematic review and meta-analysis. Exp Ther Med. 2019;18(2):1141–8.
112. Locke FL, Ghobadi A, Jacobson CA, Miklos DB, Lekakis LJ, Oluwole OO, Lin Y, Braunschweig I, Hill BT, Timmerman JM, Deol A, Reagan PM, Stiff P, Flinn IW, Farooq U, Goy A, McSweeney PA, Munoz J, Siddiqi T, Chavez JC, Herrera AF, Bartlett NL, Wiezorek JS, Navale L, Xue A, Jiang Y, Bot A, Rossi JM, Kim JJ, Go WY, Neelapu SS. Long-term safety and activity of axicabtagene ciloleucel in refractory large B-cell lymphoma (ZUMA-1): a single-arm, multicentre, phase 1-2 trial. Lancet Oncol. 2019;20(1):31–42.
113. O'Connor OA, Pro B, Pinter-Brown L, Bartlett N, Popplewell L, Coiffier B, Lechowicz MJ, Savage KJ, Shustov AR, Gisselbrecht C, Jacobsen E, Zinzani PL, Furman R, Goy A, Haioun C, Crump M, Zain JM, Hsi E, Boyd A, Horwitz S. Pralatrexate in patients with relapsed or refractory peripheral T-cell lymphoma: results from the pivotal PROPEL study. J Clin Oncol. 2011;29(9):1182–9.
114. Townsley DM, Scheinberg P, Winkler T, Desmond R, Dumitriu B, Rios O, Weinstein B, Valdez J, Lotter J, Feng X, Desierto M, Leuva H, Bevans M, Wu C, Larochelle A, Calvo KR, Dunbar CE, Young NS. Eltrombopag added to standard immunosuppression for aplastic anemia. N Engl J Med. 2017;376(16):1540–50.
115. Benson L. Tumor treating fields technology: alternating electric field therapy for the treatment of solid tumors. Semin Oncol Nurs. 2018;34(2):137–50.
116. Neelapu SS. Managing the toxicities of CAR T-cell therapy. Hematol Oncol. 2019;37(Suppl 1):48–52.
117. Kantoff PW, et al. Sipuleucel-T immunotherapy for castration-resistant prostate cancer. N Engl J Med. 2010;363(5):411–22.
118. Demetri GD, von Mehren M, Jones RL, Hensley ML, Schuetze SM, Staddon A, Milhem M, Elias A, Ganjoo K, Tawbi H, Van Tine BA, Spira A, Dean A, Khokhar NZ, Park YC, Knoblauch RE, Parekh TV, Maki RG, Patel SR. Efficacy and safety of trabectedin or dacarbazine for metastatic liposarcoma or leiomyosarcoma after failure of conventional chemotherapy: results of a phase III randomized multicenter clinical trial. J Clin Oncol. 2016;34(8):786–93.
119. Raman SS, Hecht JR, Chan E. Talimogene laherparepvec: review of its mechanism of action and clinical efficacy and safety. Immunotherapy. 2019;11(8):705–23.

120. Long TH, Shinohara MM, Argenyi ZB, Thompson JA, Gardner JM. Panniculitis in a patient with pathologic complete response to talimogene laherparepvec treatment for recurrent, in-transit melanoma. J Cutan Pathol. 2018;45(11):864–8.
121. Heo YA, Syed YY, Keam SJ. Pegaspargase: a review in acute lymphoblastic leukaemia. Drugs. 2019;79(7):767–77.
122. Imhof RL, Tollefson MM. Bleomycin-induced flagellate dermatitis. Mayo Clin Proc. 2019;94(2):371–2.
123. Sviggum HP, Davis MDP, Rajkumar SV, Dispenzieri A. Dermatologic adverse effects of lenalidomide therapy for amyloidosis and multiple myeloma. Arch Dermatol. 2006;142(10):1298–302.
124. Shanbhag A, Pritchard ER, Chatterjee K, Hammond DA. Highly probable drug reaction with eosinophilia and systemic symptoms syndrome associated with lenalidomide. Hosp Pharm. 2017;52(6):408–11.
125. Llamas-Velasco M, García-Martín P, Sánchez-Pérez J, Fraga J, García-Diez A. Sweet's syndrome with subcutaneous involvement associated with pegfilgrastim treatment: first reported case. J Cutan Pathol. 2013;40(1):46–9.
126. Zackheim HS. Topical carmustine (BCNU) in the treatment of mycosis fungoides. Dermatol Ther. 2003;16(4):299–302.
127. Liner K, Brown C, McGirt LY. Clinical potential of mechlorethamine gel for the topical treatment of mycosis fungoides-type cutaneous T-cell lymphoma: a review on current efficacy and safety data. Drug Design Dev Ther. 2018;12:241–54.
128. Prince GT, Cameron MC, Fathi R, Alkousakis T. Topical 5-fluorouracil in dermatologic disease. Int J Dermatol. 2018;57(10):1259–64.
129. Cheer SM, Foster RH. Alitretinoin. Am J Clin Dermatol. 2000;1(5):307–14; discussion 315–6.
130. Lowe MN, Plosker GL. Bexarotene. Am J Clin Dermatol. 2000;1(4):245–50; discussion 251–2.
131. Cantisani C, Lazic T, Richetta AG, Clerico R, Mattozzi C, Calvieri S. Imiquimod 5% cream use in dermatology, side effects and recent patents. Recent Patent Inflam Allergy Drug Discov. 2012;6(1):65–9.

Complementary and Alternative Medicine and Dermatooncology

Faraaz Zafar and Peter Lio

Abbreviations

AK   Actinic keratosis
BCC   Basal cell carcinoma
BEC   Curaderm
CAM   Complementary and alternative medicine
CDK   Cyclin-dependent kinase
CoQ10 Coenzyme Q10
DNA   Deoxyribonucleic acid
ER   Estrogen receptor
ESRD End stage renal disease
FDA   Food and Drug Administration
HFS   Hand-foot syndrome
MMS   Mohs micrographic surgery
NAC   N-Acetyl cysteine
NF-κB Nuclear Factor kappa-light-chain-enhancer of activated B cells
PDT   Photodynamic therapy
RNA   Ribonucleic acid
SCC   Squamous cell carcinoma
TEWL Transepidermal water loss
UV   Ultraviolet

F. Zafar (✉)
Department of Dermatology, University of Iowa Health Care, Iowa City, IA, USA

P. Lio
Medical Dermatology Associates of Chicago, Chicago, IL, USA

© Springer Nature Switzerland AG 2021
V. Liu (ed.), *Dermato-Oncology Study Guide*,
https://doi.org/10.1007/978-3-030-53437-0_12

Learning Objectives
1. To recognize the increasing role of complementary and alternative medicine (CAM) in society and its potential influence on dermatologic treatment and potential.
2. To highlight notable CAM with potential benefits and harms when used in the treatment and prevention of dermatologic malignancy and other dermatologic disease.
3. To highlight notable CAM with potential dermatologic adverse effects.
4. To provide healthcare providers information on these CAM in order to better educate and advise their patients in making evidence-based decisions.

Introduction

The use of complementary and alternative medicine (CAM) in general is increasing in the United States [1]. In 2007, it was estimated that 38% of adults in the United States used CAM. Several studies have looked at trends of CAM use, with many of these noting that women (up to 56% of users), individuals in higher socioeconomic brackets, and those with chronic disease were more likely to try and utilize CAM [2]. It has been noted that around 1.2% of dermatology visits involved discussion or use of CAM, similar to visits for internal medicine and family medicine [3]. In these cases, dermatitis of unspecified causes appears to be the most common diagnosis that CAM is used for [3]. Studied reasons for using CAM includes positive impressions of providers thought to have formal training in CAM [4], distrust in or lack of improvement with allopathic medications [5], and increased availability on the internet and reduced costs [6].

Many conventional allopathic therapies have undesirable side effects that are identified and documented in the literature, subsequently providing patients with greater awareness of these adverse events. Government and regulatory body documentation and warnings may also serve, perhaps unintentionally, to deter patients from pursuing such treatments. Conversely, CAMs by their very definition, are not as well studied or represented in available literature. Subsequently, many novel CAMs may be generally viewed as safe, though adverse effects and long-term safety studies are lacking. This, compounded by the fact that they are not bound by regulatory agencies to the same degree as conventional drugs, may allow them to be perceived as safer by the general public [7]. This disconnect also extends to the patient-provider relationship, and poor explanations of allopathic treatments by providers may cause confusion and aversion to utilize such options [8].

It must be noted however, that reasons for using CAM are numerous and widespread, and that this is a complex issue with no single common motivation. In the field of dermatology and in medicine as a whole, we do not have all the answers when it comes to understanding and treating disease: the term "idiopathic" still exists widely in medicine. There are innumerable alternative treatments in the world proposed both as preventative and/or treatment measures. Some have been studied and may show benefits to improving dermatologic patient health and quality of life, others may have no effect whatsoever, and some may in fact be harmful. Indeed, some may claim to treat the "root cause" of the dermatologic disease, while others may simply mask symptoms. In this chapter, we intend to highlight the treatments that have some data that show that they may be useful in an integrative approach. First however, we will cast light on well-known and often-used CAM that should be avoided. It must be stressed that this chapter is not exhaustive, but rather a sampling given the huge number of alternative treatments purported to treat and improve skin disease outcomes. Moreover, it should be recognized that our understanding of these agents is fluid and ever evolving, with potential benefits and harms continuously discovered with accumulating experience.

Agents with Evidence to Avoid

Escharotics

Background
- Escharotic agents are corrosive and caustic treatments that have been purported by CAM practitioners as well as celebrities and certain members of the general public to specifically target and destroy cancer cells [9].
- They are so called "escharotics" because of their effect in producing black, dry crusts (eschars) over the wounds that they create when applied.
- The original concept of escharotics goes back hundreds of years with the observation that certain extracts and supplements derived from nature may be used to destroy or "heal" skin lesions [10].
- The two most used ingredients are zinc chloride and bloodroot (*Sanguinaria canadensis*) [11].
  - Zinc chloride was previously used to fix tissue during the early days of Mohs micrographic surgery (MMS) by Frederick Mohs himself. In the 1930s he would use this "Mohs paste" prior to excising cancer tissue. While this is no longer used in MMS, Harry Hoxsey—an alternative cancer practitioner—developed an herbal paste he believed was effective in treating not only external dermatologic malignancy, but internal disease as well [12].

Mechanisms and Effects
- Widely known as "Black Salve", users and practitioners claim cures for squamous and basal cell carcinomas, as well as melanomas when escharotics are applied.
- Many practitioners claim that escharotics exclusively target cancer cells while leaving healthy cells harmed.
- Users may initially note that skin cancer lesions fall off given the corrosive nature of escharotics, but several studies have linked them with poor efficacy including incomplete removal of cancer, non-selective tissue damage and destruction of healthy tissue, and possible malignancy associated with the escharotic treatment itself [11, 13].

- Other issues with escharotics include patients using them on skin lesions that are not even confirmed as malignant.
  - Other recognized risks of escharotic use include chronic wounds, infection, scarring, and contamination risks of the medication due to lack of regulation in production [14].

Conclusions

- Current dermatologic and overall medical consensus on the use of escharotics are that while perceived benefits may exist, they are less effective than allopathic skin cancer treatments and have increased risk of adverse events and other negative outcomes.
- Further studies are needed to characterize the effects and long-term safety of escharotics. For now, bloodroot has been listed as a "fake cancer cure" by the U.S. Food and Drug Administration (FDA) and the public have been advised to avoid this class of treatments.

CAM with *Possible* Skin Cancer Treatment Benefit

Gossypin

Background

- Gossypin is a pentahydroxy flavone, a class of flavonoids (class of plant and fungal metabolites) found mainly in spices, red-purple fruits and vegetables, and cotton and hibiscus plants [15].
- In the past, extracts from the hibiscus plant *Hibiscus vitifolius* have been used to treat diabetes, jaundice, and inflammatory disease [16].

Mechanisms and Effects

- Some studies have shown that Gossypin inhibits the G_1-S phase during DNA replication in human melanoma cells by directing interacting with the BRAFV600E and CDK4 kinases in vitro. This thereby inhibits proliferation of the melanoma cells, reducing tumor volume and increasing survival in mice during studies [16].
  - The BRAFV600E mutation exists in up to 70% of melanomas, and so may be effective in specifically targeting these cells [16].
  - Human melanoma xenograft testing with the BRAFV600E also showed reduced tumor burden through inducing apoptosis in these cells [17].
- Another study has also found that gossypin may inhibit NF-κB, a protein complex that influences DNA transcription, proliferation of cytokines, and cell survival [18].
  - Dysregulation of NF-κB may indeed result in progression of cancer, and so this represents another mechanism through which gossypin may be used to influence melanoma.

Conclusions

- While early data is promising, actual human clinical trials are not currently ongoing, and so with most current data using in vitro human tissue and laboratory mice, there is no significant data yet available to recommend using this as an effective treatment.
- Reported safety issues with gossypin include skin irritation and corrosion, respiratory irritation, and eye irritation [19].

BEC (Curaderm)

Background

- BEC is a cream formulation also known as Curaderm. It contains high concentrations of a mixture of solasodine glycosides [20].
- Natural sources include eggplant and *Solanum sodomaeum* (Devil's Apple).

Mechanisms and Effects

- Studies have shown that it is effective as an anti-neoplastic agent, inducing apoptosis in cancer cells by arresting the cell cycle in the G_2-M phase [21].
- Its effectiveness as a cream has been noted, with studies its effect in treating lesions of actinic keratoses, as well as squamous and basal cell carcinomas [22] [23].
- When compared to a placebo group, one such study found that patients receiving BEC cream under occlusion had a 66% cure rate when treating BCC compared to 25% in the placebo group [22].
- Newer studies have also shown its possible effectiveness in treating melanoma, causing cellular necrosis in certain melanoma cell lines while having minimum effect on normal, healthy skin cells [24].
- Major adverse effects on the liver, kidneys or bone marrow have not been noted [25].

Conclusions

- Given this information, therapy with BEC may have a role in treating both localized and metastatic disease given its propensity for selectively targeting neoplastic cutaneous cancers, particularly when compared to existing treatments with more widespread systemic adverse effects [26].
- However, while this may have merit as part of an integrative approach in treating cutaneous malignancy, it must be stressed that there is no significant data suggesting its use as monotherapy.
- Currently, it remains in the experimental stage and is not FDA-approved.
    - Despite this, BEC has been available for several years, and this has given rise to creams of BEC that are marketed as alternatives to allopathic treatments such as chemotherapy, radiation, and surgical procedures.

– Often referred to as the "eggplant cancer cure", BEC has been marketed on the internet as a cure for more than cutaneous malignancy, with other claims including lung and colon cancer.

Ingenol Mebutate

Background
- Ingenol Mebutate is derived from the *Euphorbia peplus* plant (commonly known as milkweed).

Mechanisms and Effects
- Ingenol Mebutate is toxic to rapidly dividing cells, giving credence to its use against skin neoplasms [27].
  – It has been shown to produce clinical clearance in actinic keratoses that was significant when compared to placebo [27].
- A 12 month follow up study subsequently noted that topical treatment for 2–3 consecutive days on actinic keratoses produced sustained clearance at 12 months with no major reported safety concerns [28].
- Clearance does appear to be dose-dependent [29].
  – The most common side effects noted were dry skin as well as dose-related erythema [27, 29].

Conclusions
While now FDA-approved in the treatment of actinic keratoses, it has also been reported to have an effect in clearing basal cell carcinomas [29], squamous cell carcinomas, and other dermatologic conditions such as anogenital warts and actinic cheilitis [30].

Taxus brevifolia

Background
- The Pacific yew (also known as the Western yew) is a conifer tree native to the Pacific Northwest of the United States and Canada.
- It is known as the source of the chemotherapy drug paclitaxel, which already has known use in treating breast, ovarian, and lung cancer.

Mechanisms and Effects
- The mechanism by which it functions is by inhibiting microtubule assembly during the cell division process and blocking chromosomes from separating, thereby inducing apoptosis [31].
  – However, this is not solely specific for neoplastic cells and is therefore associated with many side effects including neuropathy, hand-foot syndrome, dermatitis, female infertility, and many others [32].

- While systemic paclitaxel is associated with a variety of adverse effects, topical forms also exist that minimize such issues [33].
- In vitro data currently exists supporting the use of paclitaxel-loading topicals that have strong permeation and increased antiproliferative activity in squamous cells carcinomas [34], as well as in some basal cell carcinomas [35].

Conclusions
- Once again however, such studies are more often seen in vivo and are not yet approved for current skin cancer treatment regimens.

Hypericum perforatum

Background
- Also known as Saint. John's wort, *Hypericum perforatum* is a well-known plant and medicinal herb in the world of medicine.
- Hypericin is an anthraquinone found in *H. perforatum* which along with hyperforin, is one of the main active ingredients.
- Thought to have medicinal properties since Antiquity, it has been extensively studied as a treatment for depression in humans, though its place when compared to antidepressants is controversial [36].

Mechanisms and Effects
- Topical extract of hypericin extracted from the plant has been found to induce photosensitivity in tumor types, including both pigmented and non-pigmented human melanoma cells [37], resulting in melanoma cell death when compared with photodynamic therapy (PDT).
  - The understood mechanism is via interaction with the endoplasmic reticulum of malignant melanoma cells [37].
- In addition to melanoma, non-melanoma skin cancers have also been studied with regard to hypericin.
  - One such study looked at 34 patients who had either actinic keratoses, basal cell carcinoma, and squamous cell carcinoma in situ.
  - Patients had hypericin extract placed under occlusion and subsequent PDT weekly for 6 weeks.
  - This produced complete clinical resolution in 50% of actinic keratoses, 28% of patients with superficial basal cell carcinomas, and 40% with in situ squamous cell carcinoma [38].
- Prior studies have also noted selective targeting of this treatment for neoplastic cells only [39], though a mouse model has conversely shown poor efficacy and selectivity in mouse skin [40].
- Side effects have been studied in the topical form, including increased sunburn due to photosensitivity, skin irritation, and in high doses systemic side effects including agitation, hallucination, tachycardia, and increased reflexes, to name a few [41].

Conclusions
- Overall, there is no clear consensus on *Hypericum perforatum*.
- At the time of writing, no current FDA-approved drug containing hypericin extract exists, and further studies are required to characterize the efficacy and safety profile of this source.

Curcumin

Background
- Curcumin is a chemical produced by *Curcuma longa* plants and is a principal chemical in turmeric.
- Often sold as an herbal supplement, it has numerous purported benefits including coronary artery disease, Alzheimer's disease, and cancer prevention.
- However, no human clinical trials have been performed [42].

Mechanisms and Effects
- Despite the lack of human trials, in vitro studies have noted inhibitory effects on DNA and RNA synthesis in the HeLa cell line [43].
  - Certain mouse models have noted inhibition of skin tumor carcinogenesis in mice [44–47].
- Certain adverse events have also been reported, including drug interactions, headache, diarrhea, rash, and yellow stool [48].

Conclusions
- Like most CAM in this chapter, it must be emphasized that this agent has not currently been well tested in human clinical trials.

Coenzyme Q10

Background
- Known as part of the electron transport chain in mitochondria that is involved in oxidative phosphorylation.
- Coenzyme Q10 (CoQ10) is sold as a dietary supplement.
- Currently however, it is not approved by the FDA for the treatment of any medical condition.

Mechanisms and Effects
- Given its important function within cellular regulation and function, it has been extensively studied, with low levels of CoQ10 being noted in patients with various cancers, including melanoma [49].

- Rusciani et al. also found in a prospective study that low plasma CoQ10 levels were an independent prognostic factor that could be used to estimate the risk for melanoma progression and metastases [49].
- A subsequent study looking at 32 patients with stage 1 and II melanoma receiving a combination of interferon alpha and CoQ10 over 8 years were found to have statistically significant decreases in rates of recurrence and had negligible adverse effects [50].

Conclusions
- Indeed, there does appear to be compelling evidence for further studies investigating the effects of CoQ10 and its effects on melanoma and other non-melanoma skin cancers, but currently progress appear to have stagnated.
- CoQ10 is currently sold over the counter as a supplement already, but long-term benefits and safety studies remain to be performed.

CAM that *May* Offer Skin Cancer Preventative Benefits

Coffee Plant Extract

Background
- Coffee is currently the world's most popular beverage, with millions of cups consumed daily.
- In addition to non-dermatologic health benefits, studies have also looked at the effect of coffee consumption on skin malignancy.
  - Current data is sparse and inconsistent.

Mechanisms and Effects
- One such study from 1986 looking at over 16,000 people found a significant decrease in nonmelanoma skin cancer prevalence in those who drink coffee when compared to those that do not drink coffee [51].
- A similar study looking at 93,000 women in 2007 also found this apparent protective effect, though it was mostly in women drinking six or more cups of coffee per day [52].
- Conversely, other studies have not found this association [53], and others yet still have found specific decreases in basal cell carcinoma that were again dose dependent [54].
- Such studies have detected these differences after controlling for confounding factors, but the underlying mechanism (if any) remains to be seen.

Conclusions
- Multiple studied non-dermatologic benefits are associated with coffee consumption, but its influence upon preventing skin malignancy is currently unknown.

Camellia sinensis

Background
- Like coffee, tea has long been a staple in human consumption as a daily drink, as well with its associated purported health benefits.
- Commonly used in teas are the leaves of *Camellia sinensis*, a species of shrub found in east Asia.

Mechanisms and Effects
- In dermatology there are inconsistent findings with respect to preventative benefits in skin cancer.
- One particular study looking at 1400 patients with basal cell and squamous cell carcinomas found that after controlling and matching, regular consumption of tea was associated with significantly lower risk of SCC, but weakly associated with reduced BCC risk [54].
- However, another matched study has found that there is no reduction in SCC risk with regular consumption of tea [55].
- Studies have noted increased DNA repair following exposure to UV light in patients that regularly consume green tea, through drinking or topical application [56, 57], though again the extent to which this truly prevents skin cancer is unknown, if at all.

Conclusions
- There is no current consensus on the effect of green tea consumption on skin cancer prevention, other than that further research and investigation is required.

Beta Carotene

Background
- Beta carotene is a red-orange pigment that is found in a variety of vegetables and fruits, and often associated with carrots.

Mechanisms and Effects
- While dietary antioxidants including beta carotene are associated with decreased free radical-mediated DNA damage and tumor development after exposure to UV radiation, the effect of Vitamin A and beta carotene on skin cancer prevention appears to be anecdotal [58].

Conclusions
- Long term studies over a period of numerous years have not found significant decreases in skin cancer development [59, 60].

Genistein

Background
- Similar to Gossypin, genistein is another flavone that is often found in soy, oregano, and sage.
- Genistein has often been marketed as a supplement in treating menopausal symptoms as well as reducing signs of skin aging.

Mechanisms and Effects
- Recognized as an antioxidant, it functions by inhibiting protein tyrosine kinases and has been seen to inhibit skin carcinogenesis and cutaneous aging in mice exposed to UV light, as well as photodamage in humans [61].
- Another study in mice have also shown that genistein significantly inhibited tumor multiplicity by 60–75%, and also significantly reduced inflammatory responses in skin [62].
  - It is postulated that this was likely via blockage of DNA adduct formation [63].
  - However, the decrease in tumor incidence overall was not as strong [62].
- Adverse effects have not been well studied, though some studies have suggested that it can compete with estrogen in binding to estrogen receptors, increasing the growth rate of some ER-positive breast cancers, and so should be avoided in these populations [64, 65].
- Its anti-inflammatory and antioxidant properties have been suggested in a meta-analysis to reduce side effects in cancer patients receiving chemotherapeutic treatment [66].

Conclusions
- Further investigation is still warranted and no concrete recommendations yet exist regarding its effect in preventing skin cancer.
- Given its possible association with binding at estrogen growth receptors, it should be avoided in patients with ER-positive breast cancers [62, 63]

Proanthocyanidin

Background
- Proanthocyanidin is a polyphenol derived from grape seed and many other plants.
- It is known from studies to have anti-inflammatory, anti-allergic, and anti-arthritic effects [67–69].
- Current marketing is aimed at using Proanthocyanidin for improved cardiovascular health and reducing complications in patients with inflammatory disease, including rheumatoid arthropathies, irritable bowel disease, systemic lupus erythematosus, and Behçet's disease [70].

Mechanisms and Effects

- Studies have also shown reduced carcinogenesis in mice with UV-induced skin damage [67].
- It is believed to work via inhibition of inflammation, rapid repair of DNA dimers, and stimulation of the immune system to target mutated or otherwise abnormal cells [68–70].
- Such properties may be useful in reducing incidence of skin cancer in human subjects but currently data is limited.
- It has been noted to be generally well tolerated with minimal known side effects extending beyond gastrointestinal upset and other mild symptoms [71].

Conclusions

- Indeed, there are studied benefits, but for the future large-scale clinical trials must first take place to truly assess the efficacy and safety of "grape seed extract".

Lycopene

Background

- Like beta carotene, lycopene is also a pigment found in vegetables and fruits, imparting them with red color (particularly tomatoes).
- While it does not have the same properties as vitamin A, lycopene has been regularly studied for its anticarcinogenic and antioxidant effects with regard to breast, prostate, lung, and colon cancers [72].

Mechanisms and Effects

- Similar to other antioxidants, it has been noted that lycopene may prevent DNA damage initially, and also reverse UV-induced DNA damage when applied topically [73].
- A randomized-controlled trial by Rizwan et al. in 2011 found that tomato paste rich in lycopene ingested in 20 women found reduced UV radiation-induced matrix metalloproteinases (involved in photoaging and photocarcinogenesis) [74].
- A similar study by Cooperstone et al. also noted that tomato consumption in hairless mice resulted in significantly lower tumor number when compared to controls [75].
- Major adverse effects noted in literature are gastrointestinal upset [76].
  - One such study looking at lycopene in patients with prostate cancer had a subject die of cancer-related hemorrhage, though the association with lycopene treatment is unknown [76].

Conclusions
- Despite the aforementioned studies, there are no definitive reports assessing lycopene's effects in skin cancer prevention in humans; as such, no current recommendations currently exist.

Silymarin

Background
- A flavonoid derived from the *Silybum maranum* milk thistle plant.
- The primary component of silymarin is silybin.

Mechanisms and Effects
- Silybin has been shown to have anticarcinogenic properties in prostate, cervical, and breast cancer due to its intrinsic antioxidant activity [77].
- Mouse studies have found that topical silymarin inhibited UVB-induced non-melanoma skin cancer when compared to controls [78].
- At dose dependent levels, tumor incidence, multiplicity, and volume were significantly reduced by up to 97% [79].

Conclusions
- No human clinical trials have occurred, but suggestions have been made for its use in sunscreens given its effect in mice [80].

Plant Oils

Background
- Throughout history plant oils have been used and marketed for a variety of health benefits and treatments.
- In recent years, their role in skin disease and balancing cutaneous homeostasis are being increasingly studied. Many have been investigated including olive, sunflower seed, peanut, coconut, soybean, peanut, sesame, avocado, and almond oil, to name a few.

Mechanisms and Effects
- The overall notion is that plant oils contain fatty acids and other compounds that promote antioxidative activities, upregulation of anti-microbial properties, promote wound healing, and have anticarcinogenic properties [81].
  - The latter appears to function via reduced skin inflammation and inhibition of oxidative stress pathways, with studies demonstrating this in mice [82].
- Most reactions to plant oils are contact-based allergic reactions.

Conclusions

- Plant oils have been used since known history. Both anecdotal and evidence-based information exists, but further studies are needed to characterize associations and effects.

Niacinamide

Background

- Niacinamide (also known as nicotinamide) is one of the two major forms of niacin (Vitamin B3) and found in many over the counter supplements.
- As a supplement, it is often used in treating B3 deficiency, also known as pellagra.
- While marketed in niacin deficiency and acne treatment, studies have also demonstrated a protective effect with regards to non-melanoma skin cancers.

Mechanisms and Effects

- One study looking at a total of 114,000 patients taking supplemental and dietary niacin intake for 2–4 years showed that total niacin intake was inversely associated with SCC risk, though weakly positively associated with risk of BCC and melanoma [83].
  - Additionally, a 2015 Phase 3 randomized trial of patients using oral niacinamide found that the rate of new nonmelanoma skin cancers was lowered by 23% when compared to placebo groups [84].
- Overall, evidence is conflicting. A 2018 Bayesian analysis of reported findings suggested that there is insufficient evidence to suggest its efficacy in skin cancer chemoprevention [85].
  - In terms of mechanism, a 2019 study suggested that niacinamide may function by preventing UV radiation from reducing levels of ATP and inhibiting glycolysis, thereby preventing UV-radiation induced energy crisis and therefore enhancing DNA repair.
  - This in turn reduces UV-induced suppression of immunity and may also reduce transepidermal water loss. The overall effect believed to reduce the incidence of non-melanoma skin cancers [86].
- Given that it is readily available as an oral supplement, patients should be educated on not just possible benefits, but reported adverse effects as well.
  - Studies have shown that while nicotinic acid (the other major form of B3) has many side effects including nausea, diarrhea, flushing, elevated liver enzymes, creatinine kinase, and blood glucose, as well as exacerbations of peptic ulcers and insulin resistance, these are also reported in nicotinamide, though in much lower frequency and severity [87].
  - Studies have also mentioned that it is safer in diabetic patients [88]. High dose nicotinamide has also been studied in patients at high risk of diabetes [89] and those with end-stage renal disease, particularly those requiring hemodialysis [90] [91], showing reassuring safety profiles.

Conclusions

- Given that evidence is currently conflicting and remains relatively sparse, no current recommendations exist.
- Regardless, patients should be counseled on the possible adverse effects associated with niacinamide and that while some studies show benefit, there is no clear consensus as of the current time regarding its efficacy in preventing non-melanoma skin cancer.

Polypodium leucotomos

Background

- *Polypodium leucotomos* is a fern plant native to the tropical areas of the Americas. Extracts from this plant have been used as tonics for multiple diseases, including asthma, heart disease, diabetes, and diabetes.
- *P. leucotomos* also has been studied for a variety of dermatologic conditions, including melasma, vitiligo, atopic dermatitis, photoaging, skin cancer, and psoriasis [92] [93].

Mechanisms and Effects

- Data indicate that extracts from this plant display multiple mechanisms that may be protective against skin cancers.
- Those cited include its ability to reduce UV-induced cell damage, reducing oxidative stress and DNA damage, blocking UV radiation-induced immune suppression, and by inhibiting the release of inflammatory cytokines [94].
- Clinical studies have also been performed, showing that there was a significant decrease in erythema at 24 h following sun exposure, and fewer measured sunburn cell numbers, pyrimidine dimers, and proliferating epidermal cells compared to untreated patients [95].
- Other studies have also noted the decrease in erythema and pyrimidine dimers [96, 97].
  - One study found that oral extract of 240 mg taken twice daily for 60 days was safe and effective for reducing the effects of UV radiation-induced damage [98].
- With respect to safety, few side effects are reported, though these include gastrointestinal complaints and pruritus in up to 2% of patients [99], while 90-day studies have not found any major adverse events [100].

Conclusions

- Based on such studies, it does appear that there is recognized photoprotective benefit when taking supplements derived from *P. leucotomas*, though further clinical trials and long term studies are awaited.
- While further investigation is still warranted, *P. leucotomas* supplements are already sold over the counter, and given their excellent safety profile and promising photoprotective features, providers may suggest their use.
- Definitive overall consensus in the dermatologic community, however, requires further research.

Cutaneous Oncology Issues

Hand-Foot Syndrome and Henna

Background
- Chemotherapy-induced acral erythema, also known as hand-foot syndrome (HFS), is a presentation of redness, swelling, numbness, and desquamation of the palms and soles (and sometimes the knees and elbows) of patients receiving chemotherapy, and occasionally in patients with sickle cell disease.
- A well-recognized complication of certain chemotherapeutic agents, this development may result in patients stopping therapy or requiring dose reduction, thereby reducing overall effectiveness and worsening patient outcomes.
  - However, several studies exist showing the effect of henna in reducing these untoward effects.
- Henna is an extracted product from the leaves of the *Lawsonia inermis* tree found in North Africa and Southeast Asia, and its use has been documented in Asian and African cultures for at least 3600 years.
- It is most commonly used as a natural dye for cosmetics, although medicinal properties have been noted and it has previously been used to treat jaundice, leprosy, smallpox, and other dermatologic issues such as burns and thrush [101].

Mechanisms and Effects
- The major dye product within henna is lawsone, which has been found to have anti-inflammatory, analgesic, and antipyretic effects in rats [102].
- In a case series of 6 patients with grade 3 HFS (the highest grade—ulcerative dermatitis with pain interfering with function), and 3 patients with grade 2 HFS (skin changes with pain not interfering with function) after receiving capecitabine for chemotherapy, henna dye was applied to the palms and feet and wrapped with a cloth for 5–6 h [103].
- There results:
  - Complete response was seen in 4 of the 6 with HFS, with the other 2 improving to HFS grade 1.
  - All patients with grade 2 HFS had complete responses.
  - No patients had to reduce their dose of capecitabine chemotherapy.
  - Other similar cases have also been reported [104].

Conclusions
- Such results are promising with regards to Henna and improvement of HFS.
- Further research and trials should be emphasized as necessary before recommendations can be made.

Pruritus

Multiple malignancies (non-dermatologic included) characteristically present with itching, while chemotherapeutic treatments may also induce such symptoms. Various anti-itch treatments do exist and are relatively well documented,but are often found to be lacking. The intensity of pruritus drives many patients to seek relief, both conventional and complementary/alternative in nature. Familiarity with both conventional and CAM agents can allow the physician to weigh the risks and benefits of the therapeutic options in order to optimize management of these challenging symptoms. An overview of common anti-pruritics are reviewed here.

Topical Hydrogels and Bleach

Background
- Hydrogels are gels that are composed of oxychlorine compounds, specifically hypochlorous acid and sodium hypochlorite.
- As over-the-counter preparations or as prescription devices, they have been compared with antihistamines and steroids.

Mechanisms and Effects
- Studies performed have shown that they may be 6–8 times more effective in relieving itch than creams containing hydrocortisone and diphenhydramine [105, 106].
- In patients with atopic dermatitis, pruritus has also been found to significantly decrease within 2–3 days with long-term remission [107].
- Major adverse effects noted included only mild post-application dryness that resolved with moisturizers [107].
- Hypochlorite (bleach) within some hydrogels functions as an anti-inflammatory by indirectly inhibiting the activation of NF-κB (alterations which result in the pathogenesis of several diseases in humans), and mediating itch [108].
- However, newer studies have also suggested that bleach baths may mediate improvement in skin barrier function by reducing transepidermal water loss (TEWL), in addition to reducing itch intensity [109, 110].

Conclusions
- In patients with pruritus of any kind, hydrogels present a promising treatment option when conventional therapies have not shown acceptable improvement in symptoms.

Systemic Antipruritic and Anti-pain Medications

- Outside of non-sedating antihistamines, both gabapentin and naltrexone have been studied for itch.
- An analysis of several randomized controlled trials of adults undergoing palliative care found that for uremic itch, gabapentin, nalfurafin, and naltrexone were moderately effective in treating pruritus, while in patients with cholestasis naltrexone, flumecinol, and rifampicin were found to treat associated pruritus [111].
- Common side effects noted were dizziness (naltrexone), nausea (naltrexone), insomnia (nalfurafin), and nasopharyngitis (nalfurafin) were noted [111].

Several conventional systemic medications have also been used for pruritus and pain reduction:

Mirtazapine

Background
- Mirtazapine is a drug with antihistamine, alpha2 antagonistic, and antiserotonergic activity often used as a sleep aid, to stimulate appetite, and as an antidepressant [112].

Mechanisms and Effects
- Studies in patients with pruritic have also found cases where patients responded to therapy with mirtazapine, likely due to antihistamine activity [113].
- Investigations have suggested that mirtazapine may indeed be used for treatment in patients with long-term pruritus [114, 115].
- In addition, certain adverse effects are associated with mirtazapine which must be taken into account, including drowsiness, weight gain, dry eyes, increased serum cholesterol, and a black box warning for suicidal ideation [116].

Conclusions
- Current data is promising when looking at the effects of mirtazapine for itch control.
- Further studies are required, and in the meantime, this is not approved by the FDA for this indication.

Doxepin

Background
- A tricyclic antidepressant (TCA) used to treat depression, anxiety, and other psychiatric disorders, doxepin has a place within the spectrum of treatment options for pruritus.

Mechanisms and Effects

- Its use in itch control is likely due to the H1 and H2 antihistamine properties of doxepin, with up to 800 times greater affinity for the histamine receptor compared to hydroxyzine [117].
- Systemic therapy appears to be more effective than topical treatment [117, 118].
- Patients with end stage renal disease (ESRD) and receiving dialysis may be placed on doxepin, with one study finding complete resolution in 14 patients receiving doxepin versus 2 on placebo [119].
- Like mirtazapine, doxepin is associated with several adverse effect warnings, particularly due to its antihistamine and anticholinergic effects, including sedation, dizziness, dry mouth, prolonged QT interval, and a black box warning for suicidal ideation [120].

Conclusions

- Providers should exercise caution and educate patients on using other methods to alleviate pruritus before trying doxepin, given its possible adverse effects, including suicidal ideation, for which a black-box warning has been issued.

Aprepitant

Background

- Aprepitant is a relatively recent antiemetic that blocks the neurokinin 1 (NK1) receptor and antagonizes the release of substance P.
- Aprepitant has recently been found to have some utility in pruritus.
- Substance P is also noted to mediate pain, depression, diarrhea, and pruritus [121].

Mechanisms and Effects

- One such study looked at patients with chronic pruritus due to a variety of diseases including chronic kidney disease, T cell lymphomas, Hodgkin's lymphoma, and other solid tumors who had failed conventional antipruritic therapy [121].
  - It was found that these patients had considerable relief of their pruritus and improvement in pruritus-associated symptoms such as insomnia, depression, and overall quality of life [121, 122].
- Commonly reported side effects included vertigo, drowsiness, diarrhea, dyspepsia, hypotension, bradycardia and leukopenia, though these appear to be only 1–2% higher than in patients receiving standard antiemetic therapy [121] [123].
- However, it must be noted that aprepitant is currently very expensive and not approved for treating itch at this time; thus, access can be very challenging for some patients.
  - In addition, aprepitant inhibits CYP3A4, and therefore increases plasma levels of drugs normally metabolized via this pathway [121].

Conclusions

- While FDA-approved for the prevention of chemotherapy-induced and post-operative nausea and vomiting, aprepitant has also been noted to decrease pruritus in a variety of patients.
- Further data is required before a consensus is achieved for dermatologic indications.

Butorphanol

Background

- Butorphanol is a synthetic partial agonist-antagonist of the kappa opioid and mu opioid receptors respectively.
- It is used in the management of migraines and for the management of pain during labor.
- In patients with intractable pruritus, butorphanol may also be used to alleviate these symptoms when other conventional therapies have failed.

Mechanisms and Effects

- Recent studies have pointed out its success in chronic [124], and morphine induced itch, the latter of which was studied in primates [125].
- It has also been shown that butorphanol exhibits itch suppression, reduced intensity of cowhage itch, and does not affect heat or pain sensitivity [126].
  - In this investigation, the mesolimbic circuit of the midbrain were activated to inhibit symptoms of itching when patients received butorphanol, indicating that this circuit may serve as a potential target.
- As noted with other opioid analgesics, butorphanol may display CNS effects such as sedation, confusion, dizziness, nausea, and vomiting, as well as dependency and respiratory depression.

Conclusions

- Extreme caution is recommended when using butorphanol given its classification as an opioid and lack of FDA approval for pruritus.

Acupressure

Background

- Acupressure is an alternative medicine therapy originating in Asia that is similar to acupuncture in their shared focus on improving fatigue, pain, and pruritus.
- It is performed by applying physical pressure to acupuncture points with the purported goal of clearing blockages at "meridians" that allow life energy to flow through the body [127].
- Like many CAM treatments and medications, it has been refuted as quackery, but several studies exist that have investigated its effect on muscle and joint pain and pruritus.

Mechanisms and Effects

- One pilot trial of 15 subjects with atopic dermatitis found that after 4 weeks these patients had improvement in pruritus and lichenification [128], although this may be limited by small sample size.
- In patients with ESRD on hemodialysis treatment, acupressure has also been noted to reduce pruritus significantly more when compared to control groups not receiving acupressure [129, 130].
- Minimal adverse effects are noted.

Conclusions

- Given that there are minimal adverse effects and some albeit limited literature for its efficacy, acupressure may serve as a useful adjuvant therapy for patients with low risk of harm.

Acupuncture

Background

- More well-known than acupressure, acupuncture has existed for thousands of years.
- It is a practice where thin needles are inserted into the body with claimed benefits numbering in the hundreds.
- With regard to pruritus and pain, several studies have been performed, with a range of results.

Mechanisms and Effects

- Some studies (including those with proper "sham" acupuncture controls) have shown that patients with uremic pruritus had significant lower pruritus scores when compared to controls [131, 132].
- Larger scale meta-analysis has also noted that pruritus may be alleviated with acupuncture [133].
- Conversely, while there are certainly studies supporting acupuncture in pruritus and pain, there is also evidence to the contrary demonstrating no effect of any symptom improvement with acupuncture [88].
- Furthermore, several adverse effects have also been associated with acupuncture, including infection, nerve injury, hemorrhage, skin necrosis, and pneumothorax [134].

Conclusions

- Acupuncture holds promise for pain and itch in select situations.
- While relatively safe, there remain potential adverse effects which should be considered and discussed.

Quercetin

Background

- Like gossypin and genistein, quercetin is also a plant-derived flavonoid with anti-inflammatory and antioxidant properties.

Mechanisms and Effects

- Some small studies have recently shown that prurigo nodularis [135] patients and those with generalized itching [136] may benefit from improved quality of life and reduced pruritus with quercetin, particularly in mouse studies [136].
- Small clinical trials have been performed that have shown no evidence of benefit for cancer or various other diseases [137], and the FDA has issued multiple warning letters to manufacturers against making unauthorized, unsubstantiated claims.
- If used, the general recommended dose is 400 mg oral dose twice to thrice per day, and effects are believed to be due to reduced release of histamine [138].
- No current studies have demonstrated major adverse effects, but investigations have not looked at supplementation during pregnancy and lactation and so providers should use caution.

Conclusions

- No current consensus exists despite limited promising data.
- While it appears relatively safe, data does not currently exist in pregnant and lactating patients.

N-Acetyl Cysteine (NAC)

Background

- N-acetyl cysteine (NAC) is known as a treatment for acetaminophen overdose, to loosen mucus in patients with cystic fibrosis and for hemorrhagic cystitis induced by cyclophosphamide, to name a few.
- As a dietary supplement, it may be purchased over-the-counter and is relatively inexpensive.

Mechanisms and Effects

- One such study has also noted that NAC may be useful in the treatment of trichotillomania.
  - In this study of 55 patients, those assigned to receive NAC had significantly greater reduction in hair-pulling symptoms when compared to controls [139].
- Other studies have also noted that patients with excoriation disorder and other pruritic conditions had reduced skin-picking compulsions [140].
  - It is believed that this may be due to its ability to replenish glutathione and subsequently in the nucleus accumbens of the hypothalamus, thereby reducing skin-picking behaviors [140].
- Furthermore, NAC has been found to also work as an antioxidant, inhibiting proliferation of fibroblasts and keratinocytes [141].
- Commonly reported adverse effects include urticaria, rash, and itching in IV formulations, with up to 18% of patients being at risk of developing possible anaphylaxis-like reactions [141].
  - Hence, they should also be avoided in patients with tendency to develop fluid overload, including cardiomyopathy or congestive heart failure [142].
  - In addition, oral NAC may cause vomiting in up to one third of patients and should be avoided in patients with gastrointestinal ulcers or varices [142].

Conclusions

- Studies exist showing benefit in pruritus and trichotillomania morbidity reduction.
- Further research is warranted before the dermatology community and regulatory bodies can come to a decision on its use.
- Providers should warn patients that while serious side effects of oral supplementation with NAC are rare, they must still be recognized.

Impatiens balsamina

Background

- *Impatiens balsamina*, also known as spotted snapweed, touch-me-not, and garden balsam, is a species of flowering plant native to India.
- Like many other plants in Asia, it has long been used in folk remedies for itching, in addition to constipation, gastritis, snakebites and warts, among others.

Mechanisms and Effects

- Similar to henna, *I. balsamina* also contains lawsone.
- One study in mice with dermatitis and severe scratching behavior found that those administered with extract from the petals of *I. balsamina* had decreased scratching behavior when compared to controls [143].
- Few studies regarding *I. balsamina* and dermatologic conditions exist, but from what is known there appears to be some benefit through reduction of pruritus.
- *I. balsamina* may be toxic in large doses given its high mineral content, and so should be avoided in people with a tendency to arthritis, gout, kidney stones, and hyperacidity.
  - Patients should also be aware that *I. balsamina* may display 5 alpha-reductase inhibition, hence reducing testosterone levels [144].

Conclusions

- Limited data exists for *I. balsamina* in improving symptoms of pruritus.
- However, many more studies and investigations are required before characterizing and reaching a consensus on its use.
- Patients with tendency to arthritis, gout, kidney stones, and hyperacidity should avoid its use.

Conclusions

Numerous CAM treatments are available on the market and are widely used, but the majority of them are unregulated and unapproved for dermatologic use. While some of these agents show no clinically proven benefit and in fact may produce harm, others have demonstrated promising results in clinical studies. These promising supplements and treatments may in the future be approved and serve a beneficial role in treating dermaoncologic conditions. Given the widespread and growing use of CAMs by individuals with cutaneous or systemic malignancy, it behooves the clinician to stay abreast of the risks and benefits of these therapies in order to appropriately develop partnerships with patients in order to optimally guide care.

Review Questions and Answers

1. Which of the following CAMs has the highest risk and adverse effect profile when used in an attempt to treat malignancy?
   (a) Curcumin
   (b) Ingenol Mebutate
   (c) Black Salve
   (d) Gossypin

   Answer: (c) Black Salve

2. Which of the following CAMs has FDA-approval in the treatment of actinic keratoses?
   (a) Black Salve
   (b) BEC (Curaderm)
   (c) Ingenol Mebutate
   (d) Taxus brevifolia (Paclitaxel)

   Answer: (c) Ingenol Mebutate

3. Which is the mutation most commonly seen in malignant melanoma, and may be a target for gossypin?
   (a) KIT
   (b) BRAF V600E
   (c) GNA11
   (d) MEK1

   Answer: (b) BRAF V600E

4. Which of the following CAMs for pruritus should be avoided in patients with history of gout or impaired renal function?
   (a) Impatiens balsamina
   (b) Mirtazapine
   (c) Doxepin
   (d) NAC

   Answer: (a) Impatiens balsamina

5. Which CAM, often used as an alternative for the management of depression has also been shown in some studies to produce resolution of actinic keratoses (AK)?
   (a) Gossypin
   (b) St. John's Wort (Hypericum perforatum)
   (c) Ingenol Mebutate
   (d) BEC (Curaderm)

   Answer: (b) St. John's Wort (Hypericum perforatum)

References

1. What is CAM? National Center for Complementary and Integrative Health, U.S. National Institutes of Health (NIH). 2019. http://nccam.nih.gov/health/whatiscam.
2. Straus SE. Herbal medicines—what's in the bottle? N Engl J Med. 2002;347(25):1997–8.
3. Landis ET, Davis SA, Feldman SR, Taylor S. Complementary and alternative medicine use in dermatology in the United States. J Altern Complement Med. 2014;20(5):392–8.
4. Kristoffersen AE, Stub T, Musial F, Fønnebø V, Lillenes O, Norheim AJ. Prevalence and reasons for intentional use of complementary and alternative medicine as an adjunct to future visits to a medical doctor for chronic disease. BMC Complement Altern Med. 2018;18(1):109.
5. Mitha S, Nagarajan V, Babar MG, Siddiqui MJ, Jamshed SQ. Reasons of using complementary and alternative medicines (CAM) among elderly Malaysians of Kuala Lumpur and Selangor states: an exploratory study. J Young Pharm. 2013;5(2):50–3.
6. Ventola CL. Current issues regarding complementary and alternative medicine (CAM) in the United States: part 1: the widespread use of CAM and the need for better-informed health care professionals to provide patient counseling. P T. 2010;35(8):461–8.
7. Ventola CL. Current issues regarding complementary and alternative medicine (CAM) in the United States: part 2: regulatory and safety concerns and proposed governmental policy changes with respect to dietary supplements. P T. 2010;35(9):514–22.
8. Ha JF, Longnecker N. Doctor-patient communication: a review. Ochsner J. 2010;10(1):38–43.
9. Jellinek N, Maloney ME. Escharotic and other botanical agents for the treatment of skin cancer: a review. J Am Acad Dermatol. 2005;53(3):487–95.
10. McDaniel S, Goldman GD. Consequences of using escharotic agents as primary treatment for nonmelanoma skin cancer. Arch Dermatol. 2002;138(12):1593–6.
11. Lim A. Black salve treatment of skin cancer: a review. J Dermatolog Treat. 2018;29(4):388–92.
12. Elston DM. Escharotic agents, Fred Mohs, and Harry Hoxsey. J Am Acad Dermatol. 2005;53(3):523–5.
13. Ong NC, Sham E, Adams BM. Use of unlicensed black salve for cutaneous malignancy. Med J Aust. 2014;200(6):314.
14. Cienki JJ, Zaret L. An internet misadventure: bloodroot salve toxicity. J Altern Complement Med. 2010;16(10):1125–7.
15. Bhaskaran S, Dileep KV, Deepa SS, et al. Gossypin as a novel selective dual inhibitor of V-RAF murine sarcoma viral oncogene homolog B1 and cyclin-dependent kinase 4 for melanoma. Mol Cancer Ther. 2013;12(4):361–72.
16. Wang L, Wang X, Chen H, et al. Gossypin inhibits gastric cancer growth by direct targeting of AURKA and RSK2. Phytother Res. 2019;33(3):640–50.
17. Kunnumakkara AB, Nair AS, Ahn KS, et al. Gossypin, a pentahydroxy glucosyl flavone, inhibits the transforming growth factor beta-activated kinase-1-mediated NF-kappaB activation pathway, leading to potentiation of apoptosis, suppression of invasion, and abrogation of osteoclastogenesis. Blood. 2007;109(12):5112–21.
18. Fernandez SP, Nguyen M, Yow TT, et al. The flavonoid glycosides, myricitrin, gossypin and naringin exert anxiolytic action in mice. Neurochem Res. 2009;34(10):1867–75.
19. PubChem, Gossypin Compound Summary. 2019. https://pubchem.ncbi.nlm.nih.gov/compound/Gossypin.
20. Cham BE, Daunter B, Evans RA. Topical treatment of malignant and premalignant skin lesions by very low concentrations of a standard mixture (BEC) of solasodine glycosides. Cancer Lett. 1991;59(3):183–92.
21. Cui CZ, Wen XS, Cui M, Gao J, Sun B, Lou HX. Synthesis of solasodine glycoside derivatives and evaluation of their cytotoxic effects on human cancer cells. Drug Discov Ther. 2012;6(1):9–17.
22. Francis DB, Hart LV, Wilson PR, Beardmore GL. Curaderm—or is it? Med J Aust. 1989;151(9):541–2.

23. Punjabi S, Cook LJ, Kersey P, Marks R, Cerio R. Solasodine glycoalkaloids: a novel topical therapy for basal cell carcinoma. A double-blind, randomized, placebo-controlled, parallel group, multicenter study. Int J Dermatol. 2008;47(1):78–82.
24. Al Sinani SS, Eltayeb EA, Coomber BL, Adham SA. Solamargine triggers cellular necrosis selectively in different types of human melanoma cancer cells through extrinsic lysosomal mitochondrial death pathway. Cancer Cell Int. 2016;16:11.
25. Beardmore G, Hart V, Wilson P, Francis D. Curaderm: preliminary findings. Med J Aust. 1989;150(1):46.
26. Yu S, Sheu HM, Lee CH. Extract (SR-T100) induces melanoma cell apoptosis and inhibits established lung metastasis. Oncotarget. 2017;8(61):103509–17.
27. Siller G, Gebauer K, Welburn P, Katsamas J, Ogbourne SM. PEP005 (ingenol mebutate) gel, a novel agent for the treatment of actinic keratosis: results of a randomized, double-blind, vehicle-controlled, multicentre, phase IIa study. Australas J Dermatol. 2009;50(1):16–22.
28. Lebwohl M, Shumack S, Stein Gold L, Melgaard A, Larsson T, Tyring SK. Long-term follow-up study of ingenol mebutate gel for the treatment of actinic keratoses. JAMA Dermatol. 2013;149(6):666–70.
29. Siller G, Rosen R, Freeman M, Welburn P, Katsamas J, Ogbourne SM. PEP005 (ingenol mebutate) gel for the topical treatment of superficial basal cell carcinoma: results of a randomized phase IIa trial. Australas J Dermatol. 2010;51(2):99–105.
30. Del Rosso JQ. Ingenol mebutate topical gel a status report on clinical use beyond actinic keratosis. J Clin Aesthet Dermatol. 2016;9(11 Suppl 1):S3–S11.
31. Bharadwaj R, Yu H. The spindle checkpoint, aneuploidy, and cancer. Oncogene. 2004;23(11):2016–27.
32. Marupudi NI, Han JE, Li KW, Renard VM, Tyler BM, Brem H. Paclitaxel: a review of adverse toxicities and novel delivery strategies. Expert Opin Drug Saf. 2007;6(5):609–21.
33. Bharadwaj R, Das PJ, Pal P, Mazumder B. Topical delivery of paclitaxel for treatment of skin cancer. Drug Dev Ind Pharm. 2016;42(9):1482–94.
34. Paolino D, Celia C, Trapasso E, Cilurzo F, Fresta M. Paclitaxel-loaded ethosomes®: potential treatment of squamous cell carcinoma, a malignant transformation of actinic keratoses. Eur J Pharm Biopharm. 2012;81(1):102–12.
35. Barceló R, Viteri A, Muñoz A, Gil-negrete A, Rubio I, López-Vivanco G. Paclitaxel for progressive basal cell carcinoma. J Am Acad Dermatol. 2006;54(2 Suppl):S50–2.
36. Linde K, Kriston L, Rücker G, et al. Efficacy and acceptability of pharmacological treatments for depressive disorders in primary care: systematic review and network meta-analysis. Ann Fam Med. 2015;13(1):69–79.
37. Kleemann B, Loos B, Scriba TJ, Lang D, Davids LM. St John's Wort (Hypericum perforatum L.) photomedicine: hypericin-photodynamic therapy induces metastatic melanoma cell death. PLoS One. 2014;9(7):e103762.
38. Kacerovská D, Pizinger K, Majer F, Smíd F. Photodynamic therapy of nonmelanoma skin cancer with topical hypericum perforatum extract—a pilot study. Photochem Photobiol. 2008;84(3):779–85.
39. Alecu M, Ursaciuc C, Hălălău F, et al. Photodynamic treatment of basal cell carcinoma and squamous cell carcinoma with hypericin. Anticancer Res. 1998;18(6B):4651–4.
40. Boiy A, Roelandts R, De Witte PA. Photodynamic therapy using topically applied hypericin: comparative effect with methyl-aminolevulinic acid on UV induced skin tumours. J Photochem Photobiol B Biol. 2011;102(2):123–31.
41. Knüppel L, Linde K. Adverse effects of St. John's Wort: a systematic review. J Clin Psychiatry. 2004;65(11):1470–9.
42. Nelson KM, Dahlin JL, Bisson J, Graham J, Pauli GF, Walters MA. The essential medicinal chemistry of Curcumin. J Med Chem. 2017;60(5):1620–37.
43. Huang MT, Ma W, Yen P, et al. Inhibitory effects of topical application of low doses of curcumin on 12-O-tetradecanoylphorbol-13-acetate-induced tumor promotion and oxidized DNA bases in mouse epidermis. Carcinogenesis. 1997;18(1):83–8.

44. Conney AH, Lysz T, Ferraro T, et al. Inhibitory effect of curcumin and some related dietary compounds on tumor promotion and arachidonic acid metabolism in mouse skin. Adv Enzym Regul. 1991;31:385–96.
45. Sonavane K, Phillips J, Ekshyyan O, et al. Topical curcumin-based cream is equivalent to dietary curcumin in a skin cancer model. J Skin Cancer. 2012;2012:147863.
46. Limtrakul P, Lipigorngoson S, Namwong O, Apisariyakul A, Dunn FW. Inhibitory effect of dietary curcumin on skin carcinogenesis in mice. Cancer Lett. 1997;116(2):197–203.
47. Nakamura Y, Ohto Y, Murakami A, Osawa T, Ohigashi H. Inhibitory effects of curcumin and tetrahydrocurcuminoids on the tumor promoter-induced reactive oxygen species generation in leukocytes in vitro and in vivo. Jpn J Cancer Res. 1998;89(4):361–70.
48. Lao CD, Ruffin MT, Normolle D, et al. Dose escalation of a curcuminoid formulation. BMC Complement Altern Med. 2006;6:10.
49. Rusciani L, Proietti I, Rusciani A, et al. Low plasma coenzyme Q10 levels as an independent prognostic factor for melanoma progression. J Am Acad Dermatol. 2006;54(2):234–41.
50. Rusciani L, Proietti I, Paradisi A, et al. Recombinant interferon alpha-2b and coenzyme Q10 as a postsurgical adjuvant therapy for melanoma: a 3-year trial with recombinant interferon-alpha and 5-year follow-up. Melanoma Res. 2007;17(3):177–83.
51. Jacobsen BK, Bjelke E, Kvåle G, Heuch I. Coffee drinking, mortality, and cancer incidence: results of a Norwegian prospective study. J Natl Cancer Inst. 1986;76(5):823–31.
52. Abel EL, Hendrix SO, McNeeley SG, et al. Daily coffee consumption and prevalence of nonmelanoma skin cancer in Caucasian women. Eur J Cancer Prev. 2007;16(5):446–52.
53. Corona R, Dogliotti E, D'Errico M, et al. Risk factors for basal cell carcinoma in a Mediterranean population: role of recreational sun exposure early in life. Arch Dermatol. 2001;137(9):1162–8.
54. Rees JR, Stukel TA, Perry AE, Zens MS, Spencer SK, Karagas MR. Tea consumption and basal cell and squamous cell skin cancer: results of a case-control study. J Am Acad Dermatol. 2007;56(5):781–5.
55. Asgari MM, White E, Warton EM, Hararah MK, Friedman GD, Chren MM. Association of tea consumption and cutaneous squamous cell carcinoma. Nutr Cancer. 2011;63(2):314–8.
56. Katiyar S, Elmets CA, Katiyar SK. Green tea and skin cancer: photoimmunology, angiogenesis and DNA repair. J Nutr Biochem. 2007;18(5):287–96.
57. Katiyar SK. Green tea prevents non-melanoma skin cancer by enhancing DNA repair. Arch Biochem Biophys. 2011;508(2):152–8.
58. Katta R, Brown DN. Diet and skin cancer: the potential role of dietary antioxidants in non-melanoma skin cancer prevention. J Skin Cancer. 2015;2015:893149.
59. Greenberg ER, Baron JA, Stukel TA, et al. A clinical trial of beta carotene to prevent basal-cell and squamous-cell cancers of the skin. The Skin Cancer Prevention Study Group. N Engl J Med. 1990;323(12):789–95.
60. Frieling UM, Schaumberg DA, Kupper TS, Muntwyler J, Hennekens CH. A randomized, 12-year primary-prevention trial of beta carotene supplementation for nonmelanoma skin cancer in the physician's health study. Arch Dermatol. 2000;136(2):179–84.
61. Wei H, Saladi R, Lu Y, et al. Isoflavone genistein: photoprotection and clinical implications in dermatology. J Nutr. 2003;133(11 Suppl 1):3811S–9S.
62. Wei H, Bowen R, Zhang X, Lebwohl M. Isoflavone genistein inhibits the initiation and promotion of two-stage skin carcinogenesis in mice. Carcinogenesis. 1998;19(8):1509–14.
63. Irrera N, Pizzino G, D'Anna R, et al. Dietary management of skin health: the role of genistein. Nutrients. 2017;9(6)
64. Yang X, Yang S, Mckimmey C, et al. Genistein induces enhanced growth promotion in ER-positive/erbB-2-overexpressing breast cancers by ER-erbB-2 cross talk and p27/kip1 downregulation. Carcinogenesis. 2010;31(4):695–702.
65. Ju YH, Allred KF, Allred CD, Helferich WG. Genistein stimulates growth of human breast cancer cells in a novel, postmenopausal animal model, with low plasma estradiol concentrations. Carcinogenesis. 2006;27(6):1292–9.

66. Sahin I, Bilir B, Ali S, Sahin K, Kucuk O. Soy isoflavones in integrative oncology: increased efficacy and decreased toxicity of cancer therapy. Integr Cancer Ther. 2019;18:1534735419835310.
67. Zhao J, Wang J, Chen Y, Agarwal R. Anti-tumor-promoting activity of a polyphenolic fraction isolated from grape seeds in the mouse skin two-stage initiation-promotion protocol and identification of procyanidin B5-3′-gallate as the most effective antioxidant constituent. Carcinogenesis. 1999;20(9):1737–45.
68. Katiyar SK. Dietary proanthocyanidins inhibit UV radiation-induced skin tumor development through functional activation of the immune system. Mol Nutr Food Res. 2016;60(6):1374–82.
69. Mittal A, Elmets CA, Katiyar SK. Dietary feeding of proanthocyanidins from grape seeds prevents photocarcinogenesis in SKH-1 hairless mice: relationship to decreased fat and lipid peroxidation. Carcinogenesis. 2003;24(8):1379–88.
70. Yang L, Xian D, Xiong X, Lai R, Song J, Zhong J. Proanthocyanidins against oxidative stress: from molecular mechanisms to clinical applications. Biomed Res Int. 2018;2018:8584136.
71. Sano A. Safety assessment of 4-week oral intake of proanthocyanidin-rich grape seed extract in healthy subjects. Food Chem Toxicol. 2017;108(Pt B):519–23.
72. Millsop JW, Sivamani RK, Fazel N. Botanical agents for the treatment of nonmelanoma skin cancer. Dermatol Res Pract. 2013;2013:837152.
73. Fazekas Z, Gao D, Saladi RN, Lu Y, Lebwohl M, Wei H. Protective effects of lycopene against ultraviolet B-induced photodamage. Nutr Cancer. 2003;47(2):181–7.
74. Rizwan M, Rodriguez-Blanco I, Harbottle A, Birch-Machin MA, Watson RE, Rhodes LE. Tomato paste rich in lycopene protects against cutaneous photodamage in humans in vivo: a randomized controlled trial. Br J Dermatol. 2011;164(1):154–62.
75. Cooperstone JL, Tober KL, Riedl KM, et al. Tomatoes protect against development of UV-induced keratinocyte carcinoma via metabolomic alterations. Sci Rep. 2017;7(1):5106.
76. Jatoi A, Burch P, Hillman D, et al. A tomato-based, lycopene-containing intervention for androgen-independent prostate cancer: results of a phase II study from the North Central Cancer Treatment Group. Urology. 2007;69(2):289–94.
77. Wright TI, Spencer JM, Flowers FP. Chemoprevention of nonmelanoma skin cancer. J Am Acad Dermatol. 2006;54(6):933–46.
78. Katiyar SK, Korman NJ, Mukhtar H, Agarwal R. Protective effects of silymarin against photocarcinogenesis in a mouse skin model. J Natl Cancer Inst. 1997;89(8):556–66.
79. Lahiri-Chatterjee M, Katiyar SK, Mohan RR, Agarwal R. A flavonoid antioxidant, silymarin, affords exceptionally high protection against tumor promotion in the SENCAR mouse skin tumorigenesis model. Cancer Res. 1999;59(3):622–32.
80. Katiyar SK. Silymarin and skin cancer prevention: anti-inflammatory, antioxidant and immunomodulatory effects (review). Int J Oncol. 2005;26(1):169–76.
81. Lin TK, Zhong L, Santiago JL. Anti-inflammatory and skin barrier repair effects of topical application of some plant oils. Int J Mol Sci. 2017;19(1)
82. Hakkim FL, Bakshi HA, Khan S, et al. Frankincense essential oil suppresses melanoma cancer through down regulation of Bcl-2/Bax cascade signaling and ameliorates heptotoxicity via phase I and II drug metabolizing enzymes. Oncotarget. 2019;10(37):3472–90.
83. Park SM, Li T, Wu S, et al. Niacin intake and risk of skin cancer in US women and men. Int J Cancer. 2017;140(9):2023–31.
84. Chen AC, Martin AJ, Choy B, et al. A phase 3 randomized trial of nicotinamide for skin-cancer chemoprevention. N Engl J Med. 2015;373(17):1618–26.
85. Gilmore SJ. Nicotinamide and skin cancer chemoprevention: the jury is still out. Australas J Dermatol. 2018;59(1):6–9.
86. Snaidr VA, Damian DL, Halliday GM. Nicotinamide for photoprotection and skin cancer chemoprevention: a review of efficacy and safety. Exp Dermatol. 2019;28(Suppl 1):15–22.
87. Gale EA, Bingley PJ, Emmett CL, Collier T. European nicotinamide diabetes intervention trial (ENDIT): a randomised controlled trial of intervention before the onset of type 1 diabetes. Lancet. 2004;363(9413):925–31.

88. Ernst E. Acupuncture: what does the most reliable evidence tell us? J Pain Symptom Manag. 2009;37(4):709–14.
89. Knip M, Douek IF, Moore WP, et al. Safety of high-dose nicotinamide: a review. Diabetologia. 2000;43(11):1337–45.
90. Takahashi Y, Tanaka A, Nakamura T, et al. Nicotinamide suppresses hyperphosphatemia in hemodialysis patients. Kidney Int. 2004;65(3):1099–104.
91. Young DO, Cheng SC, Delmez JA, Coyne DW. The effect of oral niacinamide on plasma phosphorus levels in peritoneal dialysis patients. Perit Dial Int. 2009;29(5):562–7.
92. Nestor M, Bucay V, Callender V, Cohen JL, Sadick N, Waldorf H. Polypodium leucotomos as an adjunct treatment of pigmentary disorders. J Clin Aesthet Dermatol. 2014;7(3):13–7.
93. Choudhry SZ, Bhatia N, Ceilley R, et al. Role of oral Polypodium leucotomos extract in dermatologic diseases: a review of the literature. J Drugs Dermatol. 2014;13(2):148–53.
94. Berman B, Ellis C, Elmets C. Polypodium leucotomos—an overview of basic investigative findings. J Drugs Dermatol. 2016;15(2):224–8.
95. Middelkamp-Hup MA, Pathak MA, Parrado C, et al. Oral Polypodium leucotomos extract decreases ultraviolet-induced damage of human skin. J Am Acad Dermatol. 2004;51(6):910–8.
96. Kohli I, Shafi R, Isedeh P, et al. The impact of oral Polypodium leucotomos extract on ultraviolet B response: a human clinical study. J Am Acad Dermatol. 2017;77(1):33–41.
97. Del Rosso JQ. Use of Polypodium leucotomas extract in clinical practice: a primer for the clinician. J Clin Aesthet Dermatol. 2016;9(5):37–42.
98. Nestor MS, Berman B, Swenson N. Safety and efficacy of Oral Polypodium leucotomos extract in healthy adult subjects. J Clin Aesthet Dermatol. 2015;8(2):19–23.
99. Winkelmann RR, Del Rosso J, Rigel DS. Polypodium leucotomos extract: a status report on clinical efficacy and safety. J Drugs Dermatol. 2015;14(3):254–61.
100. Goh CL, Chuah SY, Tien S, Thng G, Vitale MA, Delgado-Rubin A. Double-blind, placebo-controlled trial to evaluate the effectiveness of Polypodium Leucotomos extract in the treatment of melasma in Asian skin: a pilot study. J Clin Aesthet Dermatol. 2018;11(3):14–9.
101. Kazandjieva J, Grozdev I, Tsankov N. Temporary henna tattoos. Clin Dermatol. 2007;25(4):383–7.
102. Ali BH, Bashir AK, Tanira MO. Anti-inflammatory, antipyretic, and analgesic effects of Lawsonia inermis L. (henna) in rats. Pharmacology. 1995;51(6):356–63.
103. Yucel I, Guzin G. Topical henna for capecitabine induced hand-foot syndrome. Invest New Drugs. 2008;26(2):189–92.
104. Ilyas S, Wasif K, Saif MW. Topical henna ameliorated capecitabine-induced hand-foot syndrome. Cutan Ocul Toxicol. 2014;33(3):253–5.
105. Papoiu AD, Chaudhry H, Hayes EC, Chan YH, Herbst KD. TriCalm(®) hydrogel is significantly superior to 2% diphenhydramine and 1% hydrocortisone in reducing the peak intensity, duration, and overall magnitude of cowhage-induced itch. Clin Cosmet Investig Dermatol. 2015;8:223–9.
106. Kircik L. The effect of desonide hydrogel on pruritus associated with atopic dermatitis. J Drugs Dermatol. 2014;13(6):725–8.
107. Draelos ZD. Antipruritic hydrogel for the treatment of atopic dermatitis: an open-label pilot study. Cutis. 2012;90(2):97–102.
108. Leung TH, Zhang LF, Wang J, Ning S, Knox SJ, Kim SK. Topical hypochlorite ameliorates NF-κB-mediated skin diseases in mice. J Clin Invest. 2013;123(12):5361–70.
109. Perez-Nazario N, Yoshida T, Fridy S, De Benedetto A, Beck LA. Bleach baths significantly reduce itch and severity of atopic dermatitis with no significant change in S aureus colonization and only modest effects on skin barrier function. J Invest Dermatol. 2015;135:S37.
110. Hon KL, Tsang YC, Lee VW, et al. Efficacy of sodium hypochlorite (bleach) baths to reduce Staphylococcus aureus colonization in childhood onset moderate-to-severe eczema: a randomized, placebo-controlled cross-over trial. J Dermatolog Treat. 2016;27(2):156–62.
111. Siemens W, Xander C, Meerpohl JJ, Antes G, Becker G. Drug treatments for pruritus in adult palliative care. Dtsch Arztebl Int. 2014;111(50):863–70.

112. Anttila SA, Leinonen EV. A review of the pharmacological and clinical profile of mirtazapine. CNS Drug Rev. 2001;7(3):249–64.
113. Davis MP, Frandsen JL, Walsh D, Andresen S, Taylor S. Mirtazapine for pruritus. J Pain Symptom Manag. 2003;25(3):288–91.
114. Hundley JL, Yosipovitch G. Mirtazapine for reducing nocturnal itch in patients with chronic pruritus: a pilot study. J Am Acad Dermatol. 2004;50(6):889–91.
115. Khanna R, Boozalis E, Belzberg M, Zampella JG, Kwatra SG. Mirtazapine for the treatment of chronic pruritus. Medicines (Basel). 2019;6(3).
116. Jilani TN, Gibbons JR, Faizy RM, Saadabadi A. Mirtazapine. StatPearls; 2020.
117. Smith PF, Corelli RL. Doxepin in the management of pruritus associated with allergic cutaneous reactions. Ann Pharmacother. 1997;31(5):633–5.
118. Steinhoff M, Cevikbas F, Ikoma A, Berger TG. Pruritus: management algorithms and experimental therapies. Semin Cutan Med Surg. 2011;30(2):127–37.
119. Pour-Reza-Gholi F, Nasrollahi A, Firouzan A, Nasli Esfahani E, Farrokhi F. Low-dose doxepin for treatment of pruritus in patients on hemodialysis. Iran J Kidney Dis. 2007;1(1):34–7.
120. Almasi A, Meza CE. Doxepin. StatPearls; 2019.
121. He A, Alhariri JM, Sweren RJ, Kwatra MM, Kwatra SG. Aprepitant for the treatment of chronic refractory pruritus. Biomed Res Int. 2017;2017:4790810.
122. Huh JW, Jeong YI, Choi KH, Park HJ, Jue MS. Treatment for refractory pruritus using oral aprepitant. Ann Dermatol. 2016;28(1):124–5.
123. Hauser JM, Azzam JS, Kasi A. Antiemetic medications. StatPearls; 2019.
124. Dawn AG, Yosipovitch G. Butorphanol for treatment of intractable pruritus. J Am Acad Dermatol. 2006;54(3):527–31.
125. Lee H, Naughton NN, Woods JH, Ko MC. Effects of butorphanol on morphine-induced itch and analgesia in primates. Anesthesiology. 2007;107(3):478–85.
126. Papoiu ADP, Kraft RA, Coghill RC, Yospovitch G. Butorphanol suppression of histamine itch is mediated by nucleus accumbens and septal nuclei. A pharmacological fMRI study. J Invest Dermatol. 2015;135(2):560–8.
127. Adams A, Eschman J, Ge W. Acupressure for chronic low back pain: a single system study. J Phys Ther Sci. 2017;29(8):1416–20.
128. Lee KC, Keyes A, Hensley JR, et al. Effectiveness of acupressure on pruritus and lichenification associated with atopic dermatitis: a pilot trial. Acupunct Med. 2012;30(1):8–11.
129. Kiliç Akça N, Taşçi S, Karataş N. Effect of acupressure on patients in Turkey receiving hemodialysis treatment for uremic pruritus. Altern Ther Health Med. 2013;19(5):12–8.
130. Yan CN, Yao WG, Bao YJ, et al. Effect of auricular acupressure on uremic pruritus in patients receiving hemodialysis treatment: a randomized controlled trial. Evid Based Complement Alternat Med. 2015;2015:593196.
131. Che-yi C, Wen CY, Min-tsung K, Chiu-ching H. Acupuncture in haemodialysis patients at the Quchi (LI11) acupoint for refractory uraemic pruritus. Nephrol Dial Transplant. 2005;20(9):1912–5.
132. Badiee Aval S, Ravanshad Y, Azarfar A, Mehrad-Majd H, Torabi S, Ravanshad S. A systematic review and meta-analysis of using acupuncture and acupressure for uremic pruritus. Iran J Kidney Dis. 2018;12(2):78–83.
133. Yu C, Zhang P, Lv ZT, et al. Efficacy of acupuncture in itch: a systematic review and meta-analysis of clinical randomized controlled trials. Evid Based Complement Alternat Med. 2015;2015:208690.
134. Ernst E, Lee MS, Choi TY. Acupuncture: does it alleviate pain and are there serious risks? A review of reviews. Pain. 2011;152(4):755–64.
135. Pennesi CM, Neely J, Marks AG, Basak SA. Use of isoquercetin in the treatment of Prurigo nodularis. J Drugs Dermatol. 2017;16(11):1156–8.
136. Oku H, Ueda Y, Ishiguro K. Antipruritic effects of the fruits of Chaenomeles sinensis. Biol Pharm Bull. 2003;26(7):1031–4.
137. EFSA. Scientific opinion on the substantiation of health claims related to quercetin and protection of DNA, proteins and lipids from oxidative damage (ID 1647), "cardiovascular

system" (ID 1844), "mental state and performance" (ID 1845), and "liver, kidneys" (ID 1846) pursuant to article 13(1) of regulation (EC) no 1924/2006. EFSA J. 2011;9(4):1–15.
138. Maramaldi G, Togni S, Pagin I, et al. Soothing and anti-itch effect of quercetin phytosome in human subjects: a single-blind study. Clin Cosmet Investig Dermatol. 2016;9:55–62.
139. Grant JE, Odlaug BL, Kim SW. N-acetylcysteine, a glutamate modulator, in the treatment of trichotillomania: a double-blind, placebo-controlled study. Arch Gen Psychiatry. 2009;66(7):756–63.
140. Grant JE, Chamberlain SR, Redden SA, Leppink EW, Odlaug BL, Kim SW. N-acetylcysteine in the treatment of excoriation disorder: a randomized clinical trial. JAMA Psychiatry. 2016;73(5):490–6.
141. Adil M, Amin SS, Mohtashim M. N-acetylcysteine in dermatology. Indian J Dermatol Venereol Leprol. 2018;84(6):652–9.
142. Ershad M, Vearrier D. N acetylcysteine. StatPearls; 2019.
143. Oku H, Ishiguro K. Antipruritic and antidermatitic effect of extract and compounds of Impatiens balsamina L. in atopic dermatitis model NC mice. Phytother Res. 2001;15(6):506–10.
144. Ishiguro K, Oku H, Kato T. Testosterone 5alpha-reductase inhibitor bisnaphthoquinone derivative from Impatiens balsamina. Phytother Res. 2000;14(1):54–6.